GLOBAL STUDIES

CHINA

ELEVENTH EDITION

Dr. Suzanne Ogden
Northeastern University

OTHER BOOKS IN THE GLOBAL STUDIES SERIES

- Africa
- Europe
- India and South Asia
- Japan and the Pacific Rim
- Latin America
- The Middle East
- Russia, the Eurasian Republics, and
  Central/Eastern Europe

McGraw-Hill/Dushkin Company
2460 Kerper Boulevard, Dubuque, Iowa 52001
Visit us on the Internet—http://www.dushkin.com

Staff

Larry Loeppke	*Managing Editor*
Jill Peter	*Senior Developmental Editor*
Lori Church	*Permissions Coordinator*
Maggie Lytle	*Cover*
Tara McDermott	*Designer*
Kari Voss	*Typesetting Supervisor/Co-designer*
Jean Smith	*Typesetter*
Sandy Wille	*Typesetter*
Karen Spring	*Typesetter*

Sources for Statistical Reports

U.S. State Department *Background Notes* (2003)

C.I.A. *World Factbook* (2002)

World Bank *World Development Reports* (2002/2003)

UN *Population and Vital Statistics Reports* (2002/2003)

World Statistics in Brief (2002)

The Statesman's Yearbook (2003)

Population Reference Bureau *World Population Data Sheet* (2002)

The World Almanac (2003)

The Economist Intelligence Unit (2003)

Copyright

Cataloging in Publication Data
Main entry under title: Global Studies: China. 11th ed.
     1. Africa—History—1976–. 2. Taiwan—History—1945–.
I. Title:China. II. Ogden, Suzanne, *comp.*
ISBN 0–07–319872–2   ISSN 1050–2025

Eleventh Edition

Printed in the United States of America   1234567890BAHBAH54   Printed on Recycled Paper

CHINA

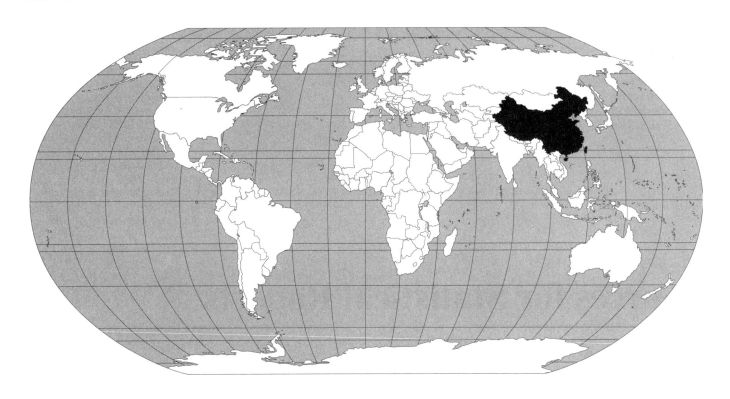

AUTHOR/EDITOR

Dr. Suzanne Ogden

Dr. Suzanne Ogden is a professor in the Political Science Department at Northeastern University, and a research associate at the Fairbank Center for East Asian Studies, Harvard University. She writes primarily about development, democracy, and political culture in China. Her most recent manuscript, *Inklings of Democracy in China,* was published by Harvard University Press in 2002. She is now working on the impact of stakeholder participation in the management of China's river basins on environmental sustainability.

Contents

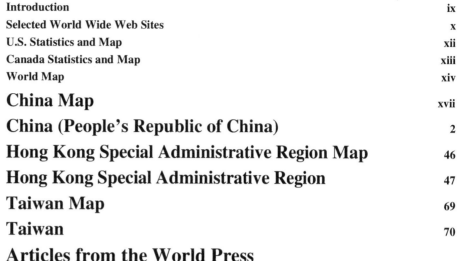

Articles from the World Press

China Articles

Hong Kong Articles

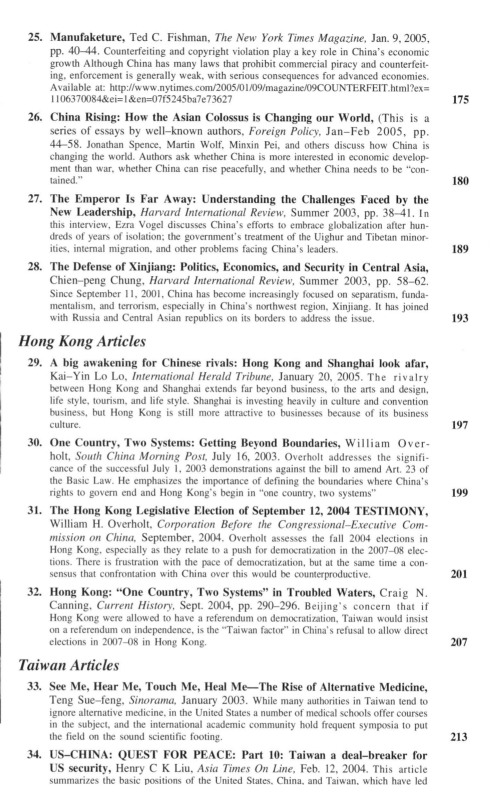

Taiwan Articles

Using *Global Studies: China*

THE GLOBAL STUDIES SERIES

The Global Studies series was created to help readers acquire a basic knowledge and understanding of the regions and countries in the world. Each volume provides a foundation of information—geographic, cultural, economic, political, historical, artistic, and religious—that will allow readers to better assess the current and future problems within these countries and regions and to comprehend how events there might affect their own well-being. In short, these volumes present the background information necessary to respond to the realities of our global age.

Each of the volumes in the Global Studies series is crafted under the careful direction of an author/editor—an expert in the area under study. The author/editors teach and conduct research and have traveled extensively through the regions about which they are writing.

MAJOR FEATURES OF THE GLOBAL STUDIES SERIES

The Global Studies volumes are organized to provide concise information on the regions and countries within those areas under study. The major sections and features of the books are described here.

Country Reports

Concise reports are written for each of the countries within the region under study. These reports are the heart of each Global Studies volume. *Global Studies: China, Eleventh Edition,* contains 3 country reports: People's Republic of China, Hong Kong, and Taiwan.

The country reports are composed of five standard elements. Each report contains a detailed map that visually positions the country among its neighboring states; a summary of statistical information; a current essay providing important historical, geographical, political, cultural, and economic information; a historical timeline, offering a convenient visual survey of a few key historical events; and four "graphic indicators," with summary statements about the country in terms of development, freedom, health/welfare, and achievements.

A Note on the Statistical Reports

The statistical information provided for each country has been drawn from a wide range of sources. (The most frequently referenced are listed on page ii.) Every effort has been made to provide the most current and accurate information available. However, sometimes the information cited by these sources differs to some extent; and, all too often, the most current information available for some countries is somewhat dated. Aside from these occasional difficulties, the statistical summary of each country is generally quite complete and up to date. Care should be taken, however, in using these statistics (or, for that matter, any published statistics) in making hard comparisons among countries. We have also provided comparable statistics for the United States and Canada, which can be found on pages 8 and 9.

World Press Articles

Within each Global Studies volume is reprinted a number of articles carefully selected by our editorial staff and the author/editor from a broad range of international periodicals and newspapers. The articles have been chosen for currency, interest, and their differing perspectives on the subject countries. There are 19 articles in *Global Studies: China, Eleventh Edition.*

The articles section is preceded by an annotated table of contents. This resource offers a brief summary of each article.

WWW Sites

An extensive annotated list of selected World Wide Web sites can be found on the facing page (7) in this edition of *Global Studies: China.* In addition, the URL addresses for country-specific Web sites are provided on the statistics page of most countries. All of the Web site addresses were correct and operational at press time. Instructors and students alike are urged to refer to those sites often to enhance their understanding of the region and to keep up with current events.

Glossary, Bibliography, Index

At the back of each Global Studies volume, readers will find a glossary of terms and abbreviations, which provides a quick reference to the specialized vocabulary of the area under study and to the standard abbreviations used throughout the volume.

Following the glossary is a bibliography that lists general works, national histories, and current-events publications and periodicals that provide regular coverage on China.

The index at the end of the volume is an accurate reference to the contents of the volume. Readers seeking specific information and citations should consult this standard index.

Currency and Usefulness

Global Studies: China, like the other Global Studies volumes, is intended to provide the most current and useful information available necessary to understand the events that are shaping the cultures of the region today.

This volume is revised on a regular basis. The statistics are updated, regional essays and country reports revised, and world press articles replaced. In order to accomplish this task, we turn to our author/editor, our advisory boards, and—hopefully—to you, the users of this volume. Your comments are more than welcome. If you have an idea that you think will make the next edition more useful, an article or bit of information that will make it more current, or a general comment on its organization, content, or features that you would like to share with us, please send it in for serious consideration.

Selected World Wide Web Sites for *Global Studies: China*

(Some Web sites continually change their structure and content, so the information listed here may not always be available. Check our Web site at: http://www.dushkin.com/online/ —Ed.)

GENERAL SITES

Access Asia
http://www.accessasia.org
Asia Intelligence Home Page
http://www.asiaint.com/
Asia News
http://www.asianews.it/view.php?l=en&art=537
Asia Source (Asia Society)
http://www.asiasource.org
Asia Times
http://www.atimes.com/
Asia Resources on the World Wide Web
http://www.aasianst.org/wwwchina.htm
Beijing International
http://www.ebeijing.gov.cn/Tour/default.htm
BBC News, China
http://news.bbc.co.uk/1/hi/in_depth/asia_pacific/2004/china/default.stm
Brookings Institution, Center for Northeast Asian Studies
http://www.brookings.edu/fp/cnaps/center_hp.htm

PEOPLE'S REPUBLIC OF CHINA

Carnegie Endowment for International Peace: China Program
http://www.ceip.org/files/events/events.asp?pr=16&EventID=674
China Business Information Center
http://www.cbiz.cn/
The China Daily
http://www.ceip.org/files/events/events.asp?pr=16&EventID=674
China Development Brief: Index of International NGOS in China
http://www.chinadevelopmentbrief.com/dingo/index.asp
China Digital News
http://www.ceip.org/files/events/events.asp?pr=16&EventID=674
China Elections and Governance
http://chinaelections.org/en/default.asp
The China Journal
http://rspas.anu.edu.au/ccc/journal.htm
China Law and Governance Review China's Ministry of Foreign Affairs
http://www.fmprc.gov.cn/eng/wjb/zzjg/ldmzs/gjlb/3488/t17367.htm
China Study Group
http://www.chinastudygroup.org/
CSIS International Security Program
http://www.chinatopnews.com/MainNews/English/
China Related Web Sites
http://orpheus.ucsd.edu/chinesehistory/othersites.html
China's Official Gateway to News & Information
http://www.china.org.cn
China Law Home Page
http://www.qis.net/chinalaw/
Chinese Law in English
http://www.qis.net/chinalaw/lawtran1.htm
Chinese Military Power Research Sites
http://www.comw.org/cmp/links.html

Cold War International History Project
http://wwics.si.edu/index.cfm?topic_id=1409&fuseaction=topics.home
Congressional-Executive Commission on China
http://www.cecc.gov/
East Turkestan Information Center
http://www.uygur.org/enorg/h_rights/human_r.htm
Foreign Policy in Focus
http://www.fpif.org/index.html
Foreign Policy In Focus Policy Brief Missile Defense & China
http://www.fpif.org/briefs/vol6/v6n03taiwan.html
http://www.uschinaedu.org-Program.asp
Human Rights in China
http://iso.hrichina.org/public/index
Inside China Today—Groups Urge EU To Censure China At UN Over Rights Learn Chinese with Homestay in China
http://www.lotusstudy.com/
Mainland Affairs Council Malaysia News Center–China News
http://news.newmalaysia.com/world/china/
Modern East-West Encounters
http://www.thescotties.pwp.blueyonder.co.uk/ew-asiapacific.htm
National Committee on U.S. China Relations
http://www.ncuscr.org/
Needham Research Institute, Cambridge, England
http://www.nri.org.uk/
New Malaysia
http://news.newmalaysia.com/world/china/
People's Daily Online
http://english.peopledaily.com.cn/
SCMP.com - Asia's leading English news channel
http://www.scmp.com/
Sinologisches Seminar, Heidelberg University
http://www.sino.uni-heidelberg.de/
SinoWisdom
http://www.sinowisdom.com/main.htm
Status of Population and Family Planning Programme in China by Province Tiananmen Square, 1989, The Declassified Story: A National Security Archive Briefing Book
http://www.gwu.edu/~nsarchiv/NSAEBB/NSAEBB16/documents/index.html
The Chairman Smiles - Chinese Posters 1966-1976
http://www.iisg.nl/exhibitions/chairman/chnintro2.html
The Chinese Military Power Page–The Commonwealth Institute
http://www.comw.org/cmp/
U.S. China Education Programs
http://www.fpif.org/briefs/vol6/v6n03taiwan.html
U.S. INTERNATIONAL TRADE COMMISSION
http://www.usitc.gov/
UCSD Modern Chinese History Site
http://www.usitc.gov/
United Nations: China's Millennium Goals, Progress
http://www.unchina.org/MDGConf/html/reporten.pdf
United Nations Human Development Reports, China
http://hdr.undp.org/
US-China Education and Culture Exchange Center
www.uschinaedu.edu
World Link Education's China Programs
http://www.worldlinkedu.com/?source=overture&OVRAW=List%20of%20Chinese%20language%20television%20channels&OVKEY=chinese%20language&OVMTC=advanced

GLOBAL STUDIES

Xinhua Net
http://news.xinhuanet.com/english/
Yahoo! News and Media Newspapers By Region Countries China
http://dir.yahoo.com/News_and_Media/Newspapers/By_Region/Countries/China/

HONG KONG

Chinese University of Hong Kong
http://www.usc.cuhk.edu.hk/uscen.asp
Civic Exchange, Christine Loh's Newsletter
http://www.civic-exchange.org/n_home.htm
Hong Kong Special Administrative Region Government Information
http://www.info.gov.hk/eindex.htm
Hong Kong Transition Project, 1982-2007
http://www.hkbu.edu.hk/~hktp/

South China Morning Post
http://www.scmp.com/

TAIWAN

Mainland Affairs Council
http://www.mac.gov.tw/
Taiwan Economic and Cultural Representative Office in the U.S.
http://www.tecro.org/
Taiwan Headlines
http://english.www.gov.tw/index.jsp?id=10
Taipei Times
http://www.taipeitimes.com/News/
Taiwan News
http://www.etaiwannews.com/Taiwan/

See individual country report pages for additional Web sites.

The United States (United States of America)

GEOGRAPHY

Area in Square Miles (Kilometers):
3,717,792 (9,629,091) (about 1/2 the size of Russia)

Capital (Population): Washington, DC (3,997,000)

Environmental Concerns: air and water pollution; limited freshwater resources, desertification; loss of habitat; waste disposal; acid rain

Geographical Features: vast central plain, mountains in the west, hills and low mountains in the east; rugged mountains and broad river valleys in Alaska; volcanic topography in Hawaii

Climate: mostly temperate, but ranging from tropical to arctic

PEOPLE
Population

Total: 293,027,571

Annual Growth Rate: 0.92%

Rural/Urban Population Ratio: 24/76

Major Languages: predominantly English; a sizable Spanish-speaking minority; many others

Ethnic Makeup: 77% white; 13% black; 4% Asian; 6% Amerindian and others

Religions: 56% Protestant; 28% Roman Catholic; 2% Jewish; 4% others; 10% none or unaffiliated

Health

Life Expectancy at Birth: 74 years (male); 80 years (female)

Infant Mortality: 6.63/1,000 live births

Physicians Available: 1/365 people

HIV/AIDS Rate in Adults: 0.6%

Education

Adult Literacy Rate: 97% (official)

Compulsory (Ages): 7–16; free

COMMUNICATION

Telephones: 181,599,900 main lines

Daily Newspaper Circulation: 238/1,000 people

Televisions: 776/1,000 people

Internet Users: 159 million (2002)

TRANSPORTATION

Highways in Miles (Kilometers): 3,906,960 (6,261,154)

Railroads in Miles (Kilometers): 149,161 (240,000)

Usable Airfields: 14,807

Motor Vehicles in Use: 206,000,000

GOVERNMENT

Type: federal republic

Independence Date: July 4, 1776

Head of State/Government: President George W. Bush is both head of state and head of government

Political Parties: Democratic Party; Republican Party; others of relatively minor political significance

Suffrage: universal at 18

MILITARY

Military Expenditures (% of GDP): 3.3%

Current Disputes: various boundary and territorial disputes; "war on terrorism"

ECONOMY

Per Capita Income/GDP: $37,800/$10.082 trillion

GDP Growth Rate: 3.1%

Inflation Rate: 2.3%

Unemployment Rate: 6%

Population Below Poverty Line: 12%

Natural Resources: many minerals and metals; petroleum; natural gas; timber; arable land

Agriculture: food grains; feed crops; fruits and vegetables; oil-bearing crops; livestock; dairy products

Industry: diversified in both capital and consumer-goods industries

Exports: $714.5 billion (primary partners Canada, Mexico, Japan)

Imports: $1.26 trillion (primary partners Canada, China, Mexico)

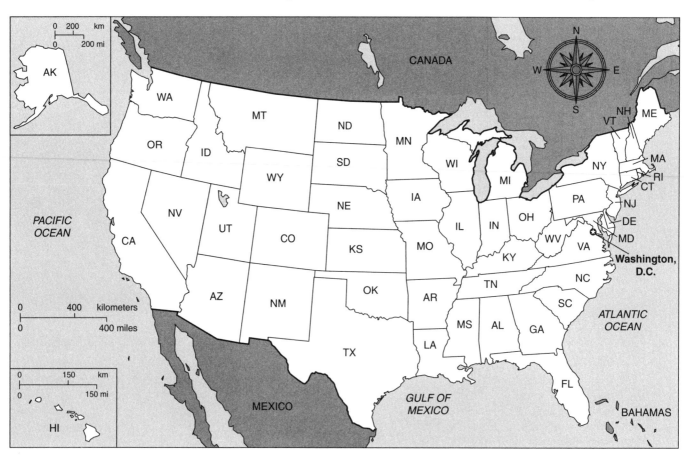

Canada

GEOGRAPHY

Area in Square Miles (Kilometers):
3,850,790 (9,976,140) (slightly larger than the United States)

Capital (Population): Ottawa (1,094,000)

Environmental Concerns: air and water pollution; acid rain; industrial damage to agriculture and forest productivity

Geographical Features: permafrost in the north; mountains in the west; central plains; lowlands in the southeast

Climate: varies from temperate to arctic

PEOPLE

Population

Total: 32,507,874

Annual Growth Rate: 0.92%

Rural/Urban Population Ratio: 23/77

Major Languages: both English and French are official

Ethnic Makeup: 28% British Isles origin; 23% French origin; 15% other European; 6% others; 2% indigenous; 26% mixed

Religions: 46% Roman Catholic; 36% Protestant; 18% others

Health

Life Expectancy at Birth: 76 years (male); 83 years (female)

Infant Mortality: 4.95/1,000 live births

Physicians Available: 1/534 people

HIV/AIDS Rate in Adults: 0.3%

Education

Adult Literacy Rate: 97%

Compulsory (Ages): primary school

COMMUNICATION

Telephones: 19,950,900 main lines

Daily Newspaper Circulation: 215/1,000 people

Televisions: 647/1,000 people

Internet Users: 16,840,000 (2002)

TRANSPORTATION

Highways in Miles (Kilometers): 559,240 (902,000)

Railroads in Miles (Kilometers): 22,320 (36,000)

Usable Airfields: 1,419

Motor Vehicles in Use: 16,800,000

GOVERNMENT

Type: confederation with parliamentary democracy

Independence Date: July 1, 1867

Head of State/Government: Queen Elizabeth II; Prime Minister Paul Martin

Political Parties: Progressive Conservative Party; Liberal Party; New Democratic Party; Bloc Québécois; Canadian Alliance

Suffrage: universal at 18

MILITARY

Military Expenditures (% of GDP): 1.1%

Current Disputes: maritime boundary disputes with the United States

ECONOMY

Currency ($U.S. equivalent): 1.39 Canadian dollars = $1

Per Capita Income/GDP: $29,800/$875 billion

GDP Growth Rate: 1.7%

Inflation Rate: 3%

Unemployment Rate: 7%

Labor Force by Occupation: 74% services; 15% manufacturing; 6% agriculture and others

Natural Resources: petroleum; natural gas; fish; minerals; cement; forestry products; wildlife; hydropower

Agriculture: grains; livestock; dairy products; potatoes; hogs; poultry and eggs; tobacco; fruits and vegetables

Industry: oil production and refining; natural-gas development; fish products; wood and paper products; chemicals; transportation equipment

Exports: $273.8 billion (primary partners United States, Japan, United Kingdom)

Imports: $238.3 billion (primary partners United States, European Union, Japan)

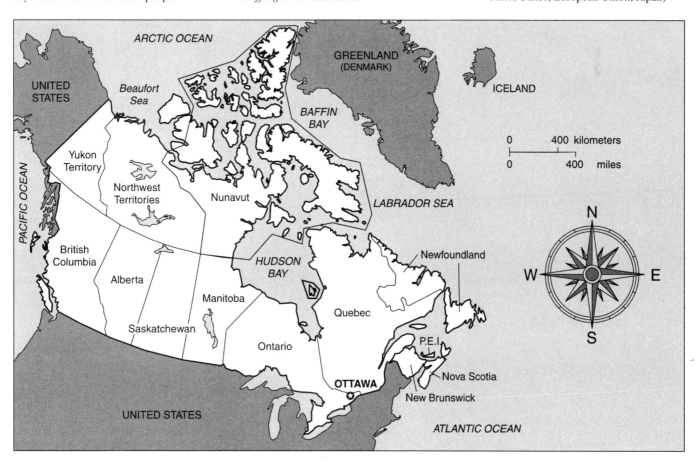

This map is provided to give you a graphic picture of where the countries of the world are located, the relationship they have with their region and neighbors, and their positions relative to major trade and power blocs. We have focused on certain areas to illustrate these crowded regions more clearly. China is shaded for emphasis.

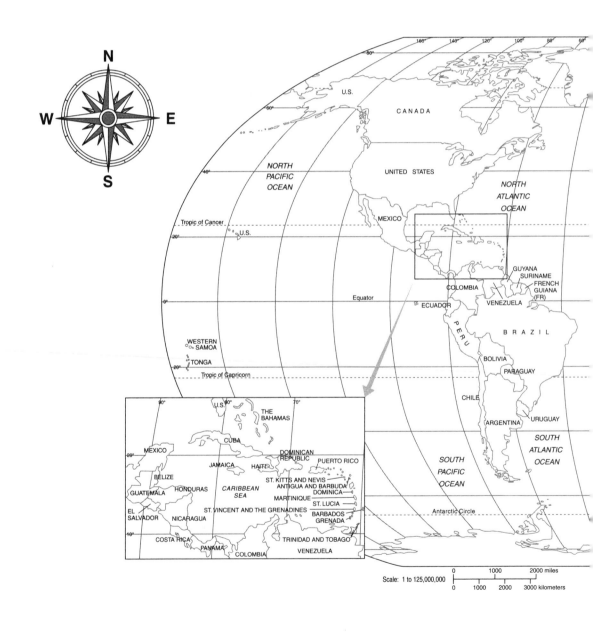

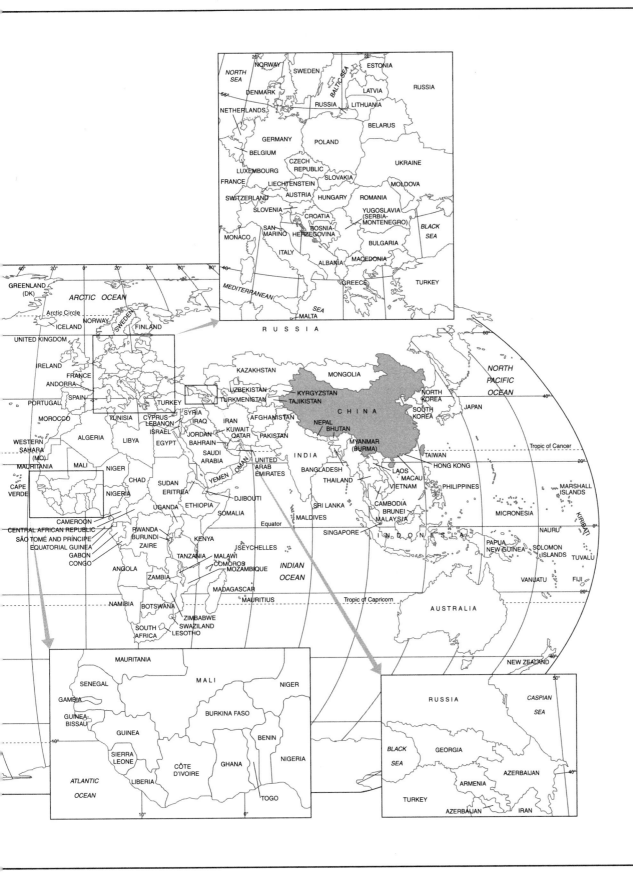

China Map

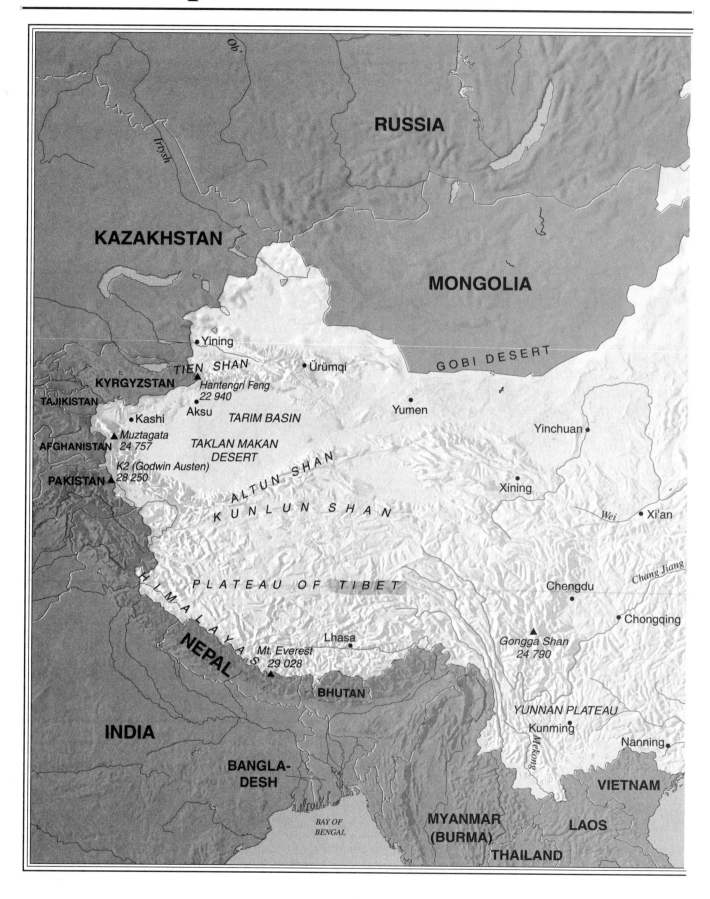

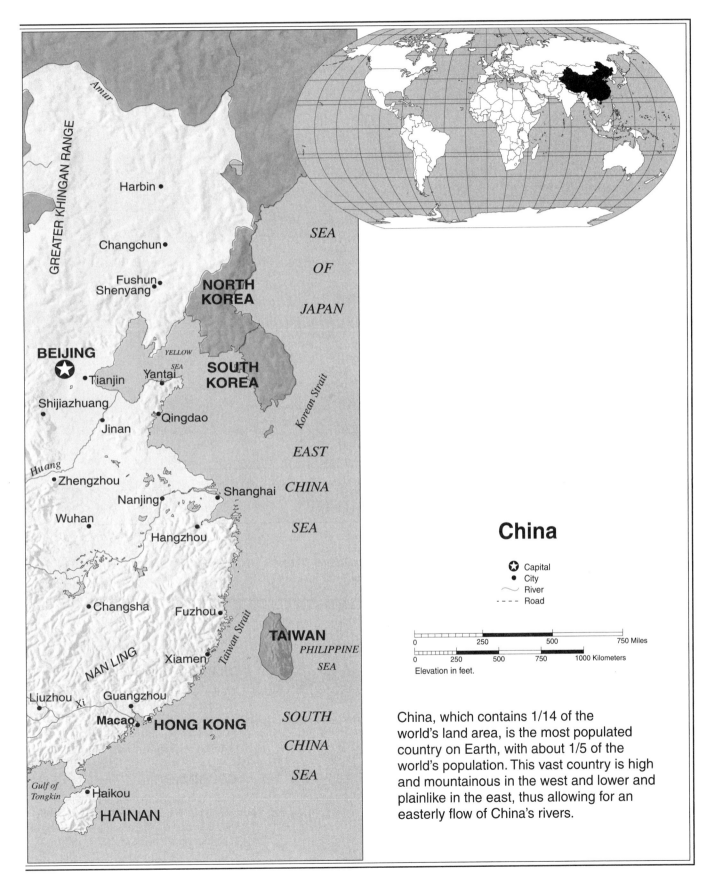

China

⭐ Capital
● City
〰 River
- - - - Road

Elevation in feet.

China, which contains 1/14 of the world's land area, is the most populated country on Earth, with about 1/5 of the world's population. This vast country is high and mountainous in the west and lower and plainlike in the east, thus allowing for an easterly flow of China's rivers.

China (People's Republic of China)

People's Republic of China Statistics

GEOGRAPHY

Area in Square Miles (Kilometers):
3,705,386 (9,596,960) (about the same size as the United States)

Capital (Population): Beijing 6,619,000 (city proper) 9,376,200 (metro area)

Environmental Concerns: air and water pollution; water shortages; desertification; trade in endangered species; acid rain; loss of agricultural land; deforestation

Geographical Features: mostly mountains, high plateaus, and deserts in the west; plains, deltas, and hills in the east

Climate: extremely diverse, from tropical to subarctic

PEOPLE

Population

Total: 1,298,847,624
Annual Growth Rate: 0.93%

Rural/Urban Population Ratio: 38.6/61.4
Major Languages: Standard Chinese (Putonghua) or Mandarin; Yue (Cantonese); Wu (Shanghainese); Minbei (Fuzhou); Minuan (Hokkien-Taiwanese); Xiang; Gan; Hahka

Ethnic Makeup: 91.9% Han; 8.1% minority groups

Religions: Taoism, Buddhism, Islam, Christianity, ancestor worship, animism, and numerous sects

Health

Life Expectancy at Birth: 70.4 years (male); 73.72 years (female)

Infant Mortality: 25.28/1,000 live births

Physicians Available: 1/628 people; 5,535,682 health care personnel

HIV/AIDS Rate in Adults: less than 0.1%

Education

Adult Literacy Rate: 86%
Compulsory Years of Schooling: 9

COMMUNICATION

Telephones: 263,000,000 main lines
Daily Newspaper Circulation: 42.3/1,000 people
Televisions: 400 million
Internet Users: 79,500,000

TRANSPORTATION

Highways in Kilometers: 1,402,698
Railroads in Kilometers: 70,058
Usable Airfields: 507
Motor Vehicles in Use: more than 11 million privately owned cars

GOVERNMENT

Type: one-party Communist state
Independence Date: Declared a Republic on January 1, 1912; People's Republic of China established October 1, 1949
Head of State/Government: President Hu Jintao; Premier Wen Jiabao

Political Parties: Chinese Communist Party; eight registered small parties controlled by the CCP

Suffrage: universal at 18 in village and urban district elections

MILITARY

Military Expenditures (% of GDP): 3.5-5.0% (FY2003)

Current Disputes: minor border disputes with a few countries, and potentially serious disputes over Spratly and Paracel Islands with several countries

ECONOMY

Currency ($ U.S. equivalent): 8.28 yuan = $1

Per Capita Income/GDP: $4,300/$5.56 trillion

GDP Growth Rate: 9.1%

Inflation Rate: 1.2%

Unemployment Rate: about 10% in urban areas. 10% in rural areas

Labor Force by Occupation: 50% agriculture; 22% services; 28% industry

Natural Resources: coal; petroleum; iron ore; tin; tungsten; antimony; lead; zinc; vanadium; magnetite; uranium; hydropower

Agriculture: food grains; cotton; oilseed; pork; fish; tea; potatoes; peanuts

Industry: iron and steel; coal; machinery; light industry; textiles and apparel; food processing; consumer durables and electronics; telecommunications; armaments

Exports: $436.1 billion (primary partners United States, Hong Kong, South Korea, Japan, Germany)

Imports: $397.4 billion (primary partners Japan, Taiwan, South Korea, Germany)

SUGGESTED WEB SITES

http://www.odci.gov/cia/
publications/factbook/geos/
ch.html

http://www.exxun.com

http://www.china.org

http://www.infoplease.com

http://www.state.gov

People's Republic of China Country Report

Chinese civilization originated in the Neolithic Period, which began around 5000 B.C., but scholars know little about it until the Shang Dynasty, which dates from about 2000 B.C. By that time, the Chinese had already developed the technology and art of bronze casting to a high standard; and they had a sophisticated system of writing with ideographs, in which words are portrayed by picturelike characters. From the fifth to the third centuries B.C., the level of literature and the arts was comparable to that of Greece in the Classical Period, which flourished at the same time. Stunning breakthroughs occurred in science, a civil service evolved, and the philosopher Confucius developed a highly sophisticated system of ethics for government and moral codes for society. Confucian values were dominant until the collapse of the Chinese imperial system in 1911, but even today they influence Chinese thought and behavior in China, and in Chinese communities throughout the world.

From several hundred years B.C. until the 15th century, China was the world's leader in technology, had the largest economy, and enjoyed the highest GDP per capita income in the world. By 1500, however, the government had closed China's doors to broad international trade, and "Europe's" GDP per capita surpassed China's. Still, it remained the world's largest economy, accounting for some 30 percent of the world's GDP in 1820. Over the next 130 years, war, revolution, and invasions ate away at China's productive capabilities. By the time the Chinese Communist Party came to power in October 1949, China's share of the world's GDP had dropped to 5 percent.[1]

The historical baggage that China carried into the period of the People's Republic of China in 1949 was, then, substantial. China had fallen from being one of the world great empires—not just from an economic perspective, but also from technological and scientific perspectives—before 1500, to one of the poorest countries in the world. When the Chinese Communists interpreted that history through the lens of Marxism and Leninism, they saw feudalism, capitalism, and imperialism as the cause of China's problem. They saw China as a victim both of exploitation within their own society, and from abroad.

The Chinese Empire

By 221 B.C., the many feudal states ruled by independent princes had been conquered by Qin Shi Huang Di, the first ruler of a unified Chinese Empire. He established a system of governmental institutions and a concept of empire that continued in China until A.D. 1911. Although China was unified from the Qin Dynasty onward, it was far less concrete than the term *empire* might indicate. China's borders really reached only as far as its cultural influence did. Thus China contracted and expanded according to whether or not other groups of people accepted the Chinese ruler and culture as their own.

Those peoples outside "China" who refused to acknowledge the Chinese ruler as the "Son of Heaven" or pay tribute to him were called "barbarians." In part, the Great Wall, which stretches more than 2,000 miles across north China and was built in stages between the third century B.C. and the seventeenth century A.D., was constructed in order to keep marauding "barbarians" out of China.

Nevertheless, they frequently invaded China and occasionally even succeeded in subduing the Chinese—as in the Yuan (Mongol) Dynasty (1279–1368) and, later, the Qing (Manchu) Dynasty (1644–1911).

However, the customs and institutions of the invaders eventually yielded to the powerful cultural influence of the Chinese. Indeed, in the case of the Manchus, who seized control of the Chinese Empire in 1644 and ruled until 1911, their success in holding onto the throne for so long may be due in part to their willingness to assimilate Chinese ways and to rule through existing Chinese institutions, such as the Confucian-ordered bureaucracy. By the time of their overthrow, the Manchu rulers were hardly distinguishable from the ethnic Han Chinese in their customs, habits, and beliefs. When considering today's policies toward the numerous minorities who inhabit such a large expanse of the People's Republic of China, it should be remembered that the central Chinese government's ability to absorb minorities was key to its success in maintaining a unified entity called China (*Zhongguo*—"the Central Kingdom") for more than 2,000 years.

The Imperial Bureaucracy

A distinguishing feature of the political system of imperial China was the civil-service examinations through which government officials were chosen. These examinations tested knowledge of the moral principles embodied in the classical Confucian texts. Although the exams were, in theory, open to all males in the Chinese Empire, the lengthy and rigorous preparation required meant that, in practice, the

(New York Public Library)

CONFUCIUS: CHINA'S FIRST "TEACHER"

Confucius (551–479 B.C.), whose efforts to teach the rulers of China how to govern well were spurned, spent most of his life teaching his own disciples. Yet 300 years later, Confucianism, as taught by descendants of his disciples, was adopted as the official state philosophy. The basic principles of Confucianism include hierarchical principles of obedience and loyalty to one's superiors, respect for one's elders, and filial piety; principles and practices for maintaining social order and harmony; and the responsibility of rulers to exercise their power benevolently.

sons of the wealthy and powerful with access to a good education had an enormous advantage. Only a small percentage of those who began the process actually passed the examinations and received an appointment in the imperial bureaucracy. Some of those who were successful resided in the capital to advise the emperor, while others were sent as the emperor's agents to govern throughout the far-flung realm.

The Decline of the Manchus

The vitality of Chinese institutions and their ability to respond creatively to new problems came to an end during the Man-

chu Dynasty (1644–1911). A stagnant agricultural system incapable of supporting the burgeoning population and the increasing exploitation of the peasantry who comprised the vast majority of Chinese people led to massive internal rebellions and the rise of warlords. As the imperial bureaucracy became increasingly corrupt and incompetent, the Manchu Dynasty gradually lost the ability to govern effectively.

China's decline in the nineteenth century was exacerbated by a social-class structure that rewarded those who could pass the archaic, morality-based civil service examinations rather than those who had expertise in science and technology and could thereby contribute to China's development. An inward-looking culture contributed to the malaise by preventing the Chinese from understanding the dynamism of the Industrial Revolution then occurring in the West. Gradually, the barriers erected by the Manchu rulers to prevent Western culture and technology from polluting the ancient beauty of Chinese civilization crumbled, but too late to strengthen China to resist the West's military onslaughts.

The Opium War (1839–1842)

In the early nineteenth century, the British traded with China, but it was primarily a one-way trade. The British nearly drained their coffers buying Chinese silk, tea, and porcelain; China's self-satisfied rulers found little of interest to purchase from the rapidly industrializing British. The British were also frustrated by China's refusal to recognize the British Empire as an equal of the Chinese Empire, and to open up ports to trade with them along China's extensive coastline and rivers.

Opium, produced in the British Empire's colony India, proved to be the one product that the Chinese were willing to purchase, and it reversed the trade balance in favor of the British. Eventually they used the Chinese attack on British ships carrying opium as an excuse for declaring war on the decrepit Chinese Empire. The Opium War ended with defeat for the Chinese and the signing of the Treaty of Nanjing. This treaty ceded the island of Hong Kong to the British as a colony and allowed them to establish trading posts.

Subsequent wars with the British and other European powers brought further concessions—the most important of which was the Chinese granting of additional "treaty ports" to Europeans. The Chinese had hoped that they could contain and control Europeans within these ports. Although that was true to a degree, this penetration of China led to the spread of Western values that challenged the stag-

nant, and by then morally decayed, Chinese Empire. As the West and, by the late nineteenth century, Japan, nibbled away at China, the Manchu rulers made a last-ditch effort at reform to strengthen and enrich China. But it was too late, and the combination of internal decay, provincialism, revolution, and foreign imperialism finally toppled the Manchu Dynasty. Thus ended more than 2,000 years of imperial rule in China.

REPUBLICAN CHINA

The 1911 Revolution, which led to the collapse of Manchu rule and derived its greatest inspiration from Chinese nationalist Sun Yat-sen, resulted in the establishment of the Republic of China (R.O.C.). It was, however, a "republic" only in name, for China was unable to successfully transfer Western forms of democratic governance to China. This was in no small part because of China's inability to remain united and to maintain law and order. China had been briefly united under the control of the dominant warlord of the time, Yuan Shikai; but with his death in 1916, China was again torn apart by the resurgence of contending warlords, internal political decay, and further Japanese territorial expansion in China. Efforts at reform failed in the context of China's weakness and internal division.

Chinese intellectuals searched for new ideas from abroad to strengthen their nation during the vibrant May Fourth period and New Culture Movement (spanning the period from roughly 1917 through the mid-1920s). In the process, the Chinese invited influential foreigners such as the English mathematician, philosopher, and socialist Bertrand Russell, and the American philosopher and educator John Dewey, to lecture in China. Thousands of Chinese traveled, worked, and studied abroad. It was during this period that ideas such as liberal democracy, syndicalism, guild socialism, and communism were put forth as possible solutions to China's many problems.

The Founding of the Chinese Communist Party

It was during that period, in 1921, that a small Marxist study group in Shanghai founded the Chinese Communist Party (CCP). The Moscow-based Comintern (Communist International) advised this highly intellectual but politically and militarily powerless group to join with the more militarily powerful Kuomintang (KMT or Nationalist Party, led first by Sun Yat-sen and, after his death in 1925, by Chiang Kai-shek) until it could gain strength and break away to establish themselves as an independent party. Thus it was with the support of the Communists in a "united front" with the

OPIUM AS A PRETEXT FOR WAR

Although the opium poppy is native to China, large amounts of opium were shipped to China by Great Britain's East India Company, from the British colony of India. Eventually India exported so much opium to China that 5 to 10 percent of its revenues derived from its sale.

By the late 1700s, the Chinese government had officially prohibited first the smoking and selling of opium, and later its importation or domestic production. But because the sale of opium was so profitable—and also because so many Chinese officials were addicted to it—the Chinese officials themselves illegally engaged in the opium trade. As the number of addicts grew and the Chinese government became more corrupted by its own unacknowledged participation in opium smuggling, so grew the interest of enterprising Englishmen in smuggling it into China for financial gain.

The British government was primarily interested in establishing an equal diplomatic and trade relationship with the Chinese to supplant the existing one, in which the Chinese court demanded that the English kowtow to the Chinese emperor. Britain was interested in expanded trade with China. But it also wanted to secure legal jurisdiction over its nationals residing in China to protect them against Chinese practices of torture of those suspected of having committed a crime.

China's efforts to curb the smuggling of opium and the Chinese refusal to recognize the British as equals reached a climax in 1839, when the Chinese destroyed thousands of chests of opium smuggled in on a British ship. This served as an ideal pretext for the British to attack China with their sophisticated gunboats (pictured below destroying a junk in Canton's (Guangzhou) harbor). Ultimately their superior firepower gave victory to the British.

Thus the so-called Opium War (1839–1842) ended with defeat for the Chinese and the signing of the Treaty of Nanjing, which ceded the island of Hong Kong to the British and allowed them to establish trading posts on the Chinese mainland.

(New York Public Library)

Nationalists that Chiang Kai-shek conquered the warlords and reunified China under one central government. Chiang felt threatened by the Communists' ambitions to gain political power, however, so in 1927 he executed all but the few Communists who managed to escape.

The Long March

China actually had few native capitalists, and hence only a small urban proletariat in China, so the aim of the CCP to carry out a Marxist revolution by uniting urban workers to "overthrow the capitalist class" made little sense. Nevertheless, the remaining members of the CCP continued to take their advice from Moscow; and they tried to organize an orthodox Marxist, urban-based movement of industrial workers twice more. Because the cities were completely controlled by the KMT, the CCP found it difficult to organize the workers, and ultimately the KMT's police and military forces decimated the ranks of the Communists. It is a

testimony to the appeal of communism in that era that the CCP managed to recover its strength each time. Indeed, the growing power of the CCP was such that Chiang Kai-shek considered the CCP, rather than the invading Japanese, to be the main threat to his complete control of China.

Eventually the Chinese Communist leaders agreed that an urban strategy could not succeed. Lacking adequate military power to confront the Nationalists, however, they retreated. In what became known as the Long March (1934–1935), they traveled more than 6,000 miles from the southeast, through the rugged interior and onto the windswept, desolate plains of Yan'an in northern China.

It was during this retreat, in which as many as 100,000 followers perished, that Mao Zedong staged his contest for power within the CCP. With his victory, the CCP reoriented itself toward a rural strategy and attempted to capture the loyalty of China's

peasants, then comprising some 85 percent of China's total population. Mao saw the downtrodden peasantry as the major source of support for the revolutionary overthrow of Chiang Kai-shek's government. Suffering from an oppressive and exploitative system of landlord control, disillusioned with the government's unwillingness to carry out land reform, and desirous of owning their own land, the peasantry looked to the CCP for leadership. Slowly the CCP started to gain control over China's vast countryside.

United Against the Japanese

In 1931, Japan invaded China and occupied Manchuria, the three northeastern provinces. In 1937, Japan attacked again, advancing southward to occupy China's heartland. Although the CCP and KMT were determined to destroy each other, Japan's threat to spread its control over the rest of China caused them to agree to a second "united front," this time for the pur-

MAO ZEDONG: CHINA'S REVOLUTIONARY LEADER

(New York Public Library)

Mao Zedong (1893–1976) came from a moderately well-to-do peasant family and, as a result, received a very good education, as compared to the vast majority of the Chinese. Mao was one of the founders of the Chinese Communist Party in 1921, but his views on the need to switch from an orthodox Marxist strategy, which called for the party to seek roots among the urban working class, to a rural strategy centered on the exploited peasantry were spurned by the leadership of the CCP and its sponsors in Moscow.

Later, it became evident that the CCP could not flourish in the Nationalist-controlled cities, as time and again the KMT quashed the idealistic but militarily weak CCP. Mao appeared to be right: "Political power grows out of the barrel of a gun."

The Communists' retreat to Yan'an in central China at the end of the Long March was not only for the purpose of survival but also for regrouping and forming a stronger Red Army. There the followers of the Chinese Communist Party were taught Mao's ideas about guerrilla warfare, the importance of winning the support of the people, principles of party leadership, and socialist values. Mao consolidated his control over the leadership of the CCP during the Yan'an period and led it to victory over the Nationalists in 1949.

From that time onward, Mao became a symbol of the new Chinese government, of national unity, and of the strength of China against foreign humiliation. In later years, although his real power was eclipsed, the party maintained the public illusion that Mao was the undisputed leader of China.

In his declining years, Mao waged a struggle, in the form of the "Cultural Revolution," against those who followed policies antagonistic to his own—a struggle that brought the country to the brink of civil war and turned the Chinese against one another. The symbol of Mao as China's "great leader" and "great teacher" was used by those who hoped to seize power after him: first the minister of defense, Lin Biao, and then the "Gang of Four," which included Mao's wife.

Mao's death in 1976 ended the control of policy by the Gang of Four. Within a few years, questions were being raised about the legacy that Mao had left China. By the 1980s, it was broadly accepted throughout China that Mao had been responsible for a full 20 years of misguided policies. Since the Tiananmen Square protests of 1989, however, there has been a resurgence of nostalgia for Mao. This nostalgia is captured in such aspects of popular culture as a tape of songs about Mao entitled "The Red Sun"—an all-time best-selling tape in China, at 5 million copies—that captures the Mao cult and Mao mania of the Cultural Revolution; and in a small portrait of Mao that virtually all car owners and taxi drivers hang over their rear-view mirrors for "good luck." Many Chinese long for the "good old days" of Mao's rule, when crime and corruption were at far lower levels than today and when there was a sense of collective commitment to China's future. But they do not long for a return to the mass terror of the Cultural Revolution, for which Mao also bears responsibility.

pose of halting the Japanese advance. Both the KMT and the CCP had ulterior motives, but according to most accounts, the Communists contributed more to the national wartime efforts. The Communists organized guerrilla efforts to peck away at the fringes of Japanese-controlled areas while Chiang Kai-shek, head of the KMT, retreated to the wartime capital of Chongqing (Chungking). His elite corps of troops and officers kept the best of the newly arriving American supplies for themselves, leaving the rank-and-file Chinese to fight against the Japanese in cloth boots and with inferior equipment. It was not the Nationalist Army but, rather, largely the unstinting efforts and sacrifices of the Chinese people and the American victory over Japan that brought World War II to an end in 1945. With the demobilization of the Japanese, however, Chiang Kai-shek was free once again to focus on defeating the Communists.

The Communists Oust the KMT

It seemed as if the Communists' Red Army had actually been strengthened through its hard fighting during World War II, turning itself into a formidable force. Meanwhile, the relatively soft life of the KMT military elite during the war did not leave it well prepared for civil war against the Red Army. Chiang Kai-shek relied on his old strategy of capturing China's cities, but the Communists, who had gained control over the countryside by winning the support of the vast peasantry, surrounded the cities. Like besieged fortresses, the cities eventually fell to Communist control. By October 1949, the CCP could claim control over all of China, except for the island of Taiwan. It was there that the Nationalists' political, economic, and military elites, with American support, had fled.

Scholars still dispute why the Red Army ultimately defeated the Nationalist Army. They cite as probable reasons the CCP's promises to undertake land reform; the Communists' more respectful treatment of the peasantry as they marched through the countryside (in comparison to that of the KMT soldiers); the CCP's more successful appeal to the Chinese sense of nationalism; and Chiang Kaishek's unwillingness to undertake reforms that would benefit the peasantry, advance economic development, and control corruption. Even had the KMT made greater efforts to reform, however, any wartime government confronted with the demoralization of the population ravaged by war, inflation, economic destruction, and the humiliation of a foreign occupation would have found it difficult to maintain the loyal support of its people. Even the middle class eventually deserted the KMT. Many of those industrial and commercial capitalists who had supported the Nationalists now joined with the CCP to rebuild China. Others, however, stayed behind only because they were unable to flee to Hong Kong or Taiwan.

One thing is clear: The Chinese Communists did not gain victory because of support from the Soviet Union; for the Soviets, who were anxious to be on the winning side in China, chose to give aid to the KMT until it was clear that the Communists would win. Furthermore, the Communists' victory was due not to superior weapons but, rather, to a superior strategy, support from the Chinese people, and (as Mao Zedong believed) a superior political "consciousness." It was because of the Communist victory over a technologically superior army that Mao thereafter believed in the superiority of "man over weapons" and that the support of the people was essential to an army's victory. The relationship of the soldiers to the people is, Mao said, like the relationship of fish to water—without the water, the fish will die.

THE PEOPLE'S REPUBLIC OF CHINA

The Red Army's final victory came rapidly—far faster than anticipated. Suddenly China's large cities fell to the Communists, who now found themselves in charge of a nation of more than 600 million people. They had to make critical decisions about how to unify and rebuild the country. They were obligated, of course, to fulfill their promise to redistribute land to the poor and landless peasantry in return for the peasants' support of the Communists during the Civil

War. The CCP leaders were, however, largely recruited from among the peasantry; and like revolutionary fighters everywhere, knew how to make a revolution but had little experience with governance. So, rejected by the Western democratic/capitalist countries because of their embrace of communism, and desperate for aid and advice, the Communists turned to the Soviet Union for direction and support. They did this in spite of the Soviet leader Joseph Stalin's fickle support of the Chinese Communists throughout the 1930s and '40s.

The Soviet Model

In the early years of CCP rule, China's leaders "leaned to one side" and followed the Soviet model of development in education, the legal system, the economic system, and elsewhere. The Soviet economic model favored capital-intensive industrialization, but all the Soviet "aid" had to be repaid. Furthermore, following the Soviet model required a reliance on Soviet experts and well-educated Chinese, whom the Communists were not sure they could trust. Without Soviet support in the beginning, however, it is questionable whether the CCP would have been as successful as it was in developing China in the 1950s.

The Maoist Model

China soon grew exasperated with the limitations of Soviet aid and the inapplicability of the Soviet model to Chinese circumstances. China's preeminent leader, Mao Zedong, proposed a Chinese model of development more appropriate to Chinese circumstances. What came to be known as the "Maoist model" took account of China's low level of development, poverty, and large population. Mao hoped to substitute China's enormous manpower for expensive capital equipment by organizing people into ever larger working units.

In 1958, in what became known as the "Great Leap Forward," Mao Zedong launched his model of development. It was a bold scheme to rapidly accelerate the pace of industrialization so that China could catch up with the industrialized states of the West. In the countryside, land was merged into large communes, untested and controversial planting techniques were introduced, and peasant women were engaged fully in the fields in order to increase agricultural production. The communes became the basis for industrializing the countryside through a program of peasants building their own "backyard furnaces" to smelt steel. The Maoist model assumed that those people possessing a proper revolutionary, or "red" (communist), consciousness—that is, a commitment to achieving communism— would be able to produce more than those

who were "expert" but lacked revolutionary consciousness. In the cities, efforts to increase industrial production through longer work days, and overtaxing industrial equipment, likewise led to a marked decline in production and industrial wastage.

The Maoist model of extreme "egalitarianism"—captured in the Chinese expression "all eat out of the same pot"— and "continuous revolution, was a rejection of the Soviet model of development, which Mao came to see as an effort to hold the Chinese back from more rapid industrialization. In particular, the Soviets' refusal to give the Chinese the most advanced industrial-plant equipment and machinery, or to share nuclear technology with them, made Mao suspicious of their intentions.

Sino–Soviet Relations Sour

For their part, the Soviets believed that the Maoist model was doomed to failure. The Soviet leader Nikita Khrushchev denounced the Great Leap Forward as "irrational"; but he was equally distressed at what seemed a risky scheme by Mao Zedong to bring the Soviets and Americans into direct conflict over the Nationalist-controlled Taiwan Strait. The combination of what the Soviets viewed as Mao's irrational economic policy and his risk-taking confrontation with the United States prompted the Soviets to abruptly withdraw their experts from China in 1959. They packed up their bags, along with spare parts for Soviet-supplied machinery and blueprints for unfinished factories, and returned home.

The Soviets' withdrawal, combined with the disastrous decline in production resulting from the policies of the Great Leap Forward and several years of bad weather, set China's economic development back many years. Population figures now available indicate that millions died in the years from 1959 to 1962, mostly from starvation and diseases caused by malnutrition. The catastrophic consequences of the Great Leap Forward resulted in the leadership paying no more than lip service to Mao Zedong's ideas. The Chinese people were not told that Mao Zedong bore blame for their problems, but the Maoist model was abandoned for the time being. More pragmatic leaders took over the direction of the economy, but without further support from the Soviets. Not until 1962 did the Chinese start to recover their productivity gains of the 1950s.

By 1963, the Sino–Soviet split had become public, as the two Communist powers found themselves in profound disagreement over a wide range of issues: whether socialist countries could use capitalist methods, such as free markets, to advance economic development; appropriate policies toward the United States; whether China or the Soviet Union could claim to follow Marxism-Leninism more faithfully, entitling it to lead the Communist world. The Sino–Soviet split was not healed until the late 1980s. By then, neither country was interested in claiming Communist orthodoxy.

The Cultural Revolution

In 1966, whether Mao Zedong hoped to provoke an internal party struggle and regain control over policy, or (as he alleged) to re-educate China's exploitive, corrupt, and oppressive officials in order to restore a revolutionary spirit to the Chinese people and to prevent China from abandoning socialism, Mao launched what he termed the "Great Proletarian Cultural Revolution." He called on China's youth to "challenge authority," particularly "those revisionists in authority who are taking the capitalist road." If China continued along its "revisionist" course, he said, the achievements of the Chinese revolution would be undone. China's youth were therefore urged to "make revolution."

(China Pictorial)

The radical leaders of China's Cultural Revolution, who came to be known as the Gang of Four, were brought to trial in late 1980. Here they are pictured (along with another radical, Chen Boda, fourth from the right, who was not part of the Gang) in a Beijing courtroom, listening to the judge pass sentence. The Gang of Four are the first three (from right to left) standing in the prisoner's dock: Jiang Qing, Yao Wenyuan, Wang Hongwen; and on the far left, Zhang Chunqiao.

THE GANG OF FOUR

The current leadership of the Chinese Communist Party views the Cultural Revolution of 1966–1976 as having been a period of total chaos that brought the People's Republic of China to the brink of political and economic ruin. While Mao Zedong is criticized for having begun the Cultural Revolution with his mistaken ideas about the danger of China turning "capitalist," the major blame for the turmoil of those years is placed on a group of extreme radicals labeled the "Gang of Four."

The Gang of Four consisted of Jiang Qing, Mao's wife, who began playing a key role in China's cultural affairs during the early 1960s; Zhang Chunqiao, a veteran party leader in Shanghai; Yao Wenyuan, a literary critic and ideologue; and Wang Hongwen, a factory worker catapulted into national prominence by his leadership of rebel workers during the Cultural Revolution. By the late 1960s, these four individuals were among the most powerful leaders in China. Drawn together by common political interests and a shared belief that the Communist Party should be relentless in ridding China of suspected "capitalist roaders," they worked together to keep the Cultural Revolution on a radical course. One of their targets had been Deng Xiaoping, who emerged as China's paramount leader in 1978, after the members of the Gang of Four had been arrested.

Although they had close political and personal ties to Mao and derived many of their ideas from him, Mao became quite disenchanted with them in the last few years of his life. He was particularly displeased with the unscrupulous way in which they behaved as a faction within the top levels of the party. Indeed, it was Mao who coined the name Gang of Four, as part of a written warning to the radicals to cease their conspiracies and obey established party procedures.

The Gang of Four hoped to take over supreme power in China following Mao's death, on September 9, 1976. However, their plans were upset less than a month later, when other party and army leaders had them arrested—an event that is now said to mark the formal end of the "10 bad years" Cultural Revolution. Removing the party's most influential radicals from power set the stage for the dramatic reforms that have become the hallmark of the post-Mao era in China.

In November 1980, the Gang of Four were put on trial in Beijing. They were charged with having committed serious crimes against the Chinese people and accused of having had a hand in "persecuting to death" tens of thousands of officials and intellectuals whom they perceived as their political enemies. All four were convicted and sentenced to long terms in prison.

(New York Public Library)

RED GUARDS: ROOTING OUT THOSE "ON THE CAPITALIST ROAD"

During the Cultural Revolution, Mao Zedong called upon the country's young people to "make revolution." Called "Mao's Red Guards," their ages varied, but for the most part they were teenagers.

Within each class and school, various youths would band together in a Red Guard group that would take on a revolutionary-sounding name and would then carry out the objective of challenging people in authority. But the people in authority—especially schoolteachers, school principals, bureaucrats, and local leaders of the Communist Party—initially ignored the demands of the Red Guards that they reform their "reactionary thoughts" or eliminate their "feudal" habits.

The Red Guards initially had no real weapons and could only threaten. Since they were considered just misdirected children by those under attack, their initial assaults had little effect. But soon the frustrated Red Guards took to physically beating and publicly humiliating those who stubbornly refused to obey them. Since Mao had not clearly defined precisely what should be their objectives or methods, the Red Guards were free to believe that the ends justified extreme and often violent means. Moreover, many Red Guards took the opportunity to take revenge against authorities, such as teachers who had given them bad grades. Others (like those pictured above wearing masks, to guard against the influenza virus while simultaneously concealing their identities) would harangue crowds on the benefits of Maoism and the evils of foreign influence.

The Red Guards went on rampages throughout the country, breaking into people's houses and stealing or destroying their property, harassing people in their homes in the middle of the night, stopping girls with long hair and cutting it off on the spot,

destroying the files of ministries and industrial enterprises, and clogging up the transportation system by their travels throughout the country to "make revolution." Different Red Guard factions began to fight with one another, each claiming to be the most revolutionary.

Mao eventually called on the army to support the Red Guards in their effort to challenge "those in authority taking the capitalist road." This created even more confusion, as many of the Red Guard groups actually supported the people they were supposed to be attacking. But their revolutionary-sounding names and their pretenses at being "Red" (Communist) confused the army. Moreover, the army was divided within itself and did not particularly wish to overthrow the Chinese Communist Party authorities, the main supporters of the military in their respective areas of jurisdiction.

Since the schools had been closed, the youth of China were not receiving any formal education during this period. Finally, in 1969, Mao called a halt to the excesses of the Red Guards. They were disbanded and sent home. Some were sent to work in factories or out to the countryside to labor in the fields with the peasants. But the chaos set in motion during the Cultural Revolution did not come to a halt until the arrest of the Gang of Four, some 10 years after the Cultural Revolution had begun.

During the "10 bad years," when schools were either closed or operating with a minimal program, children received virtually no formal education beyond an elementary school level. Although this meant that China's development of an educated elite in most fields came to a halt, nevertheless it resulted in well over a 90 percent basic literacy rate among the Chinese raised in that generation.

Such vague objectives invited abuse, including personal feuds and retribution for alleged past wrongs. Determining just who was "Red" and committed to the Communist revolution, and who was "reactionary" itself generated chaos, as people tried to protect themselves by attacking others—including friends and relatives.

During that period, people's cruelty to each other was immeasurable. People were psychologically, and sometimes physically, tortured until they "admitted" to their "rightist" or "reactionary" behavior. Murders, suicides, ruined careers, and broken families were the debris left behind by this effort to "re-educate" those who had strayed from the revolutionary path. It is estimated that approximately 10 percent of the population—that is, *100 million people*—became targets of the Cultural Revolution, and that tens of thousands lost their lives during these years of political violence.

The Cultural Revolution attacked Chinese traditions and cultural practices as being feudal and outmoded. It also destroyed the authority of the Chinese Communist Party, through prolonged public attacks on many of its most respected leaders. Policies changed frequently in those "10 bad years" from 1966 to 1976, as first one faction and then another gained the upper hand. Few leaders escaped unscathed. Ultimately, the Chinese Communist Party and Marxist-Leninist ideology were themselves the victims of the Cultural Revolution. By the time the smoke cleared, the legitimacy of the CCP had been destroyed, and the people could no longer accept the idea that the party leaders were infallible. Both traditional Chinese morality and Marxist-Leninist values had been thoroughly undermined.

Reforms and Liberalization

With the death of Mao Zedong and the subsequent arrest of the politically radical "Gang of Four" (which included Mao's wife) in 1976, the Cultural Revolution came to an end. Deng Xiaoping, a veteran leader of the CCP who had been purged twice during the "10 bad years," was "rehabilitated" in 1977.

By 1979, China once again set off down the road of construction and put an end to the radical Maoist policies of "continuous revolution" and the idea that it was more important to be "red" than "expert." Saying that he did not care whether the cat was black or white, as long as it caught mice, Deng Xiaoping pursued more pragmatic, flexible policies in order to modernize China. He deserves credit for opening up China to the outside world and to reforms that led to the liberalization of both the economic and the political

spheres. When he died in 1997, Deng left behind a country that, despite some setbacks and reversals, had already traveled a significant distance down the road to liberalization and modernization.

In spite of the fact that Deng Xiaoping followed policies that were more pragmatic than revolutionary, and more "expert" than "red," and in spite of Mao Zedong's clear responsibility for precipitating policies that were devastating to the Chinese people, Mao has never been defrocked in China; for to do so would raise serious questions about the legitimacy of the CCP. China's leaders have admitted that, beginning with the Anti-Rightest Campaign of 1957 and the Great Leap Forward of 1958, Mao made "serious mistakes"; but the CCP insists that these errors must be seen within the context of his many accomplishments and his commitment, even if sometimes misdirected, to Marxism-Leninism. In contrast to the Gang of Four and others who were condemned as "counter-revolutionaries," Mao has been called a "revolutionary" who made "mistakes." As recently as the Fourth Plenum of the 16th Party Congress of the CCP in 2004, Mao Thought (the Chinese adaptation of Marxism-Leninism to Chinese conditions) remained enshrined in the party's constitution as providing the foundation for continued CCP rule.

The Challenge of Reform

The erosion of traditional Chinese values, then of Marxist-Leninist values and faith in the Chinese Communist Party's leadership, and finally of Mao Thought, left China without any strong belief system. Such Western values as materialism, capitalism, individualism, and freedom swarmed into this vacuum to undermine both Communist ideology and the traditional Chinese values that had provided the glue of society. Deng Xiaoping's prognosis had proven correct: The "screen door" through which Western science and technology (and foreign investments) could flow into China was unable to keep out the annoying "insects" of Western values. The screen door had holes that were too large to prevent this invasion.

China's leadership in the reform period has not been united. The less pragmatic, more ideologically oriented "conservative" or "hard-line" leadership (who in the new context of reforms could be viewed as hard-line ideologues of a Maoist vintage) challenged the introduction of liberalizing economic reforms precisely because they threatened to undo China's earlier socialist achievements and erode Chinese culture. To combat the negative side effects of introducing free-market values and institutions, China's leadership launched a number of "mass

campaigns": the campaign in the 1980s against "spiritual pollution"—the undermining of Chinese values;[2] a repressive campaign following the brutal crackdown against those challenging the leadership in Tiananmen Square in 1989; on-going campaigns against corruption; and campaigns to "strike hard" against crime and to "get civilized."[3]

Since 1979, in spite of setbacks, China's leadership has been able to keep the country on the path of liberalization. As a result, the economy has had an average annual growth rate of 9.5% for the last 25 years. It is now ranked as the sixth largest economy in the world. China has dramatically reformed the legal and political system as well, even though much work remains to be done. When Deng Xiaoping died in 1997, he was succeeded in a peaceful transition by Jiang Zemin, another committed economic reformer. In turn, Jiang stepped down from his position as party leader in 2001, as president in 2002, and as the head of the Military Affairs Commission in 2004. China's leaders now operate within what is a younger and increasingly well-institutionalized and better-educated system of collective leadership. The problems that the leadership faces as a consequence of China's rapid modernization and liberalization are formidable: massive and growing unemployment; increasing crime, corruption, and social dislocation; a lack of social cohesion; and challenges to the CCP's monopoly on power. The forces of rapid growth and social and economic modernization have taken on a momentum of their own. China's increasing involvement in the international community has also put into motion seemingly uncontrollable forces, some of which are destablizing, and others that are contributing to demands for political reform. As will be noted below in the discussion of these issues, the real concern for China is not whether China will engage in further reform, but whether it can maintain stability in the context of this potentially destabilizing international and domestic environment.

The Student and Mass Movement of 1989

Symbolism is very important in Chinese culture; the death of a key leader is a particularly significant moment. In the case of Hu Yaobang, the former head of the CCP, his sudden death in April 1989 became symbolic of the death of liberalizing forces in China. The deceased leader's career and its meaning were touted as symbols of liberalization, even though his life was hardly a monument to liberal thought. More conservative leaders in the CCP had removed him from his position as the CCP's general-secretary in part because he had offended their cultural sensibilities. Apart from everything else, Hu's suggestion that the Chinese turn in their chopsticks for forks, and not eat food out of a common dish because it spread disease, were culturally offensive to them.

Hu's death provided students with a catalyst to place his values and policies in juxtaposition with those of the then increasingly conservative leadership.[4] The students' reassessment of Hu Yaobang's career, in a way that rejected the party's evaluation, was in itself a challenge to the authority of the CCP's right to rule China. The students' hunger strike in Tiananmen Square—essentially in front of party headquarters—during the visit of the Soviet Union President, Mikhail Gorbachev, to China was, even in the eyes of ordinary Chinese people, an insult to the Chinese leadership. Many Chinese later stated that the students went too far, as by humiliating the leadership, they humiliated *all* Chinese.

Part of the difficulty in reaching an agreement between the students and China's leaders was that the students' demands changed over time. At first they merely wanted a reassessment of Hu Yaobang's career. But quickly the students added new demands: dialogue between the government and the students (with the students to be treated as equals with top CCP leaders), retraction of an offensive *People's Daily* editorial, an end to official corruption, exposure of the financial and business dealings of the central leadership, a free press, the removal of the top CCP leadership, and still other actions that challenged continued CCP rule.

The students' hunger strike, which lasted for one week in May, was the final straw that brought down the wrath of the central leadership. Martial law was imposed in Beijing. When the citizens of Beijing resisted its enforcement and blocked the armies' efforts to reach Tiananmen Square to clear out the hunger-strikers, both students and CCP leaders dug in; but both were deeply divided bodies. Indeed, divisions within the student-led movement caused it to lose its direction; and divisions within the central CCP leadership incapacitated it. For two weeks, the central leadership wrangled over who was right and the best course of action. On June 4, the "hard-liners" won out, and they chose to use military power rather than a negotiated solution with the students.

Did the students make significant or well-thought-out statements about "democracy" or realistic demands on China's leaders? The short and preliminary answer is no; but then, is this really the appropriate question to ask? One could argue that what the students *said* was less important than what they *did*: They mobilized the population of China's capital and other major cit-

ies to support a profound challenge to the legitimacy of the CCP's leadership. Even if workers believed that "You can't eat democracy," and even if they participated in the demonstrations for their *own* reasons (such as gripes about inflation and corruption), they did support the students' demand that the CCP carry out further political reforms. This was because the students successfully promoted the idea that if China had had a democratic system rather than authoritarian rule, the leadership would have been more responsive to the workers' bread-and-butter issues.

Repression Within China Following the Crackdown

By August 1989, the CCP leadership had established quotas of "bad elements" for work units and identified 20 categories of people to be targeted for punishment. But people were more reluctant than in the past to follow orders to expose their friends, colleagues, and family members, not only because such verdicts had often been reversed at a later time, but also because many people questioned the CCP's version of what happened in Beijing on June 4. Although the citizenry worried about informers,[5] there seemed to be complicity from top to bottom, whether inside or outside the ranks of the CCP, in refusing to go along with efforts to ferret out demonstrators and sympathizers with the prodemocracy, anti-party movement. Party leaders below the central level appeared to believe that the central government's leadership was doomed; for this reason, they dared not carry out its orders. Inevitably, there would be a reversal of verdicts, and they did not want to be caught in it.

As party leaders in work units droned on in mandatory political study sessions about Deng Xiaoping's important writings, workers wondered how long it would be before the June 4 military crackdown would be condemned as a "counterrevolutionary crime against the people." Individuals in work units had to fill out lengthy questionnaires. A standard one had 24 questions aimed at "identifying the enemy." Among them were such questions as, "What did you think when Hu Yaobang died?" "When Zhao Ziyang went to Tiananmen Square, what did you think? Where were you?" At one university, each faculty member's questionnaire had to be verified by two people (other than one's own family) or the individual involved would not be allowed to teach.[6]

As part of the repression that followed the military crackdown in June 1989, the government carried out arrests of hundreds of those who participated in the demonstrations. During the world's absorption with

the Persian Gulf War in 1991, the government suddenly announced the trials and verdicts on some of China's best known leaders of the 1989 demonstrations. Of those who were summarily executed, available information indicates that almost all were workers trying to form labor unions. All the other known 1989 student and dissident leaders were eventually released, although some were deported to the West as a condition of their release. The government has also occasionally re-arrested 1989 protesters for other activities. In 1998, for example, some former protesters made bold attempts to establish a new party to challenge Chinese Communist Party rule. Although their efforts to register this new party were at first tolerated, several were later arrested, tried, and sentenced to prison. Finally, as discussed below, the government has attempted to ferret out and arrest activist leaders of the Falun Gong.

In spite of these important exceptions, many repressive controls were relaxed, and China's press is steadily expanding the parameters of allowable topics and opinions. Although there are occasional arrests of individuals who are blatantly challenging CCP rule, the leadership is more focused on harnessing the talents of China's best and brightest for the country's modernization than it is on controlling dissent. No longer a revolutionary party, the CCP is focused on governing and developing China.

THE PEOPLE OF CHINA

Population Control

By 2005, China's population was estimated to be about 1.3 billion. In the 1950s, Mao had encouraged population growth, as he considered a large population to be a major source of strength: Cheap human labor could take the place of expensive technology and equipment. No sustained attempts to limit Chinese population occurred until the mid-1970s. Even then, because there were no penalties for those Chinese who ignored them, population control programs were only marginally successful.

In 1979, the government launched a serious birth-control campaign, rewarding couples giving birth to only one child with work bonuses and priority in housing. The only child was later to receive preferential treatment in university admissions and job assignments (a policy eventually abandoned). Couples who had more than one child, on the other hand, were to be penalized by a 10 percent decrease in their annual wages, and their children would not be eligible for free education and health care benefits.

The one-child policy in China's major cities has been rigorously enforced, to the

The Chinese government has made great efforts to curb the country's population growth by promoting the merits of the one-child family. Today, China has an average annual population growth rate of under one percent.

point where it is almost impossible for a woman to get away with a second pregnancy. Who is allowed to have a child, as well as when she may give birth, is rigidly controlled by the woman's work unit. Furthermore, with so many state-owned enterprises now paying close to half of their entire annual wages as "bonuses," authorities have come up with additional sanctions to ensure compliance. Workers are usually organized in groups of 10 to 30 individuals. If any woman in the group gives birth to more than one child, *the entire group* will lose its annual bonus. With such overwhelming penalties for the group as a whole, pressures for a couple not to give birth to a second child are enormous.

To ensure that any unauthorized pregnancy does not occur, women who have already given birth are required to stand in front of x-ray machines (fluoroscopes) to verify that their IUDs (intrauterine birth-control devices) are still in place. Abortions can and will be performed throughout the period of a woman's unsanctioned pregnancy. (The moral issues that surround

abortions for some Christians are not concerns for the Chinese.)

The effectiveness of China's birth-control policy in the cities is not merely attributable to the surveillance by state-owned work units, neighborhood committees, and the "granny police" who watch over the families in their residential areas. Changed social attitudes also play a critical role, and urban Chinese now accept the absolute necessity of population control in their overcrowded cities. Well-to-do entrepreneurs who live in private housing, however, can avoid population-control measures because they are not part of any public housing or work unit. Nevertheless, even they are unlikely to have more than two children.

The one-child policy in China's cities has led to a generation of remarkably spoiled children. Known as "little emperors," these only children are the center of attention of six anxious adults (the parents and two sets of grandparents), who carefully scrutinize their every movement. It has led to the overuse of medical services by these parents and grandparents, who

rush their only child/grandchild to the doctor at the first signs of a sniffle or sore throat. It has also led to overfed, even obese, children. Being overweight used to be considered a hedge against bad times, and the Chinese were initially pleased that their children were becoming heavier. A common greeting showing admiration had long been, "You have become fat!" But as contemporary urban Chinese adopt many of the values associated in the developed world with becoming wealthier, they are changing their perspectives on weight. Jane Fonda–style exercise programs are now a regular part of Chinese television, and weight-loss salons and fat farms are coming into vogue for China's well-fed middle class. Still, most people view the major purpose of exercise as staying healthy and keeping China a strong nation, not looking attractive.

The one-child policy has led to other serious concerns. One is the demographic crunch presented by having too few young people to support the large number of elderly people in future years. Now that those born into one-child families are coming of marital age, China has changed its population-control policy to allow for married couples who both come from one-child families to have two children. Although female infanticide is illegal, it sometimes happens, especially in rural areas. China's orphanages absorb some of the unwanted girl babies, now much sought after in the West.

The other concern is the aborting of female fetuses. The sex of fetuses is usually known through the widespread use in China of ultrasound machines. To use these machines to reveal the sex of the child is illegal, but for a very small bribe, doctors will usually let the mother know the sex. So the government has promulgated several new laws and has investigated several thousand cases of alleged abuse of sex-identification of fetuses. It is now even considering rewriting the Criminal Law to prohibit the sex identification of the fetus, especially when it is followed by the aborting of a female fetus.[7]

The decline in the ratio of women to men has in turn led to another demographic crisis: an insufficient number of brides. Apart from societal unhappiness, this has led to a sharp increase in the kidnapping of young women as well as the practice of selling girls as brides in rural marketplaces as they came of marriageable age at the turn of the 21st century.[8]

In the vast rural areas of China, where some three quarters of the population still live, efforts to enforce the one-child policy have met with less success than in the cities, because the benefits and punishments are not as relevant for peasants. After the communes were disbanded in the early 1980s and families were given their own land to till, peasants wanted sons, to do the heavy farm labor. As a result, the government's policy in the countryside became more flexible. In some villages, if a woman gives birth to a girl and decides to have another child in hopes of having a boy, she may pay the government a substantial fee (usually an amount more than the entire annual income of the family) in order to be allowed to do so. Yet in an ironic reflection of this still very male-dominant society, today's farming, which is far more physically demanding and less lucrative than factory jobs, is increasingly left to the women, while the men go off to towns and cities to make their fortunes.

It is estimated that at least several million peasants have taken steps to ensure that their female offspring are not counted toward their one-child (and now, in some places, two-child) limit: One strategy is for a pregnant woman simply to move to another village to have her child. Since the local village leaders are not responsible for women's reproduction when they are

not from their own village, women are not harassed into getting an abortion in other villages. If the child is a boy, the mother can simply return to her native village, pay a fine, and register him; if a girl, the mother can return and not register the child. Thus a whole generation of young girls is growing up in the countryside without ever having been registered with the government. Since, except for schooling, peasants have few claims to state-supplied benefits anyway, they may consider this official nonexistence of their daughters a small price to pay for having as many children as necessary until giving birth to a boy. And if this practice is as common as some believe, it may mean that China will not face quite such a large demographic crisis in the ratio between males and females as has been projected.

Males continue to be more valued in Chinese culture because only sons are permitted to carry on traditional Chinese family rituals and ancestor worship. This is unbearably painful—in fact, unacceptable—for families without sons, who feel that their entire ancestral history, often recorded over several hundred years on village-temple tablets, is coming to an end. As a result, a few villages have changed the very foundations of ancestral worship: They now permit daughters to continue the family lineage down the female line. The government itself is encouraging this practice, and it is also changing certain other family-related policies, such as who is responsible under the law for taking care of their parents. It used to be the son, meaning that parents whose only child was a girl could not expect to be supported in their old age. Now, both sons and daughters are deemed responsible. Furthermore, it is hoped that a new system of social security and pensions for retired people will gradually lead the state and employers to absorb the responsibility for caring for the elderly.

China's strict population-control policies have been effective: Since 1977, the population has grown at an *average* annual rate of 1.1 percent, one of the lowest in the developing world. (As of 2004, the annual growth rate was 0.93 percent—slightly less than the replacement rate.) Unfortunately, even this low rate has worked out to an average annual increase of China's population of more than 15 million people. This is an ever-growing drain on the country's limited resources and poses a challenge, and perhaps a threat, to its future economic development. It has also proven damaging to the environment as the growing population demands an ever-higher standard of living. The steady rises in pollution of China's air, water, and land, and the depletion of nonrenewable resources and energy as China struggles to address these needs, are already

causing ecological crises that the government is having difficulty addressing.

Women

It is hardly surprising that overlaying (but never eradicating) China's traditional culture with a communist ideology in which men and women are supposed to be equal has generated a bundle of contradictions. Under Chinese Communist Party rule, women have long had more rights and opportunities than women in almost any other developing country, and in certain respects, their rights and opportunities have even outpaced those of women in some of the developed countries.[9] Although Chinese women have rarely broken through the "glass ceiling" to the highest levels of the workplace or the ruling elite, and have often been given "women's work," their pay scale is similar to that of men. This is in an economy where the gap between the highest and lowest paid, whether male or female, was small until the last few decades. Furthermore, an ideological morality that insisted on respect for women as equals (with both men and women being addressed as "comrades") combined with a de-emphasis on the importance of sexuality, resulted in at least a superficial respect for women that was rare before the Communist period.

The economic reforms that began in 1979 have, however, precipitated changes in the manner in which women are treated, and in how women act. While many women entrepreneurs and workers benefit as much as the men from economic reforms, there have also been certain throwbacks to earlier times that have undercut women's equality. Women are now treated much more as sex objects than they used to be; and while some women revel in their new freedom to beautify themselves, some companies will hire only women who are perceived as physically attractive, and many enterprises are now using women as "window dressing." For example, women dressed in *qipao*—the traditional, slim-fitting Chinese dress slit high on the thigh—stand outside restaurants and other establishments to entice customers. At business meetings, many women have become mere tea-pourers. In newspapers, employment ads for Chinese enterprises often state in so many words that only young, good-looking women need apply.

The emphasis on profits and efficiency since the reforms has also made state-run enterprises reluctant to hire women because of the costs in maternity benefits (including three to 12 months of maternity leave at partial or even full pay) and because mothers are still more likely than fathers to be in charge of sick children and the household. Under the socialist system, where the purpose of an enterprise was not necessarily to make profits but to fulfill such socialist objectives as the equality of women and full employment, women fared better. Economic reforms have provided enterprise managers with the excuse they need not to hire women. Whatever the real reason, they can always claim that their refusal to hire more women or to promote them is justified: Women are more costly, or less competent, or less reliable.

National Minorities

Ninety-two percent of the population is Han Chinese. Although only 8 percent is classified as "national minorities," they occupy more than 60 percent of China's geographical expanse. These minorities inhabit almost all of the border areas, including Tibet, Inner Mongolia, and Xinjiang Province. The stability and allegiance of the border areas are important for China's national security. Furthermore, China's borders with the many neighboring countries are poorly defined, and members of the same minority usually live on both sides of the borders.

ACHIEVEMENTS

Over several thousands of years, Chinese culture has produced a rich heritage in literature, painting, ceramics, music, and opera. In the early decades of communism, the arts were marshaled in the service of ideology and lost some of their dynamism. But the introduction of the market economy in the 1980s revived the arts, and today's China produces a wide variety of high-quality art and cultural productions, including film. Since 1949, literacy has increased dramatically and now stands at 73 percent—the highest in Chinese history. Perhaps China's greatest achievement under communism has been its ability to maintain unity and stability, providing the longest period of peace and prosperity to the Chinese people in hundreds of years.

To address this issue, China's central government has pursued policies designed to get the minorities on the Chinese side of the borders to identify with the Han Chinese majority. Rather than admitting to this objective of undermining distinctive national identities, the CCP leaders have phrased the policies in terms of getting rid of the minorities' "feudal" customs, such as religious practices, which are contrary to the "scientific" values of socialism. Teaching children their native language was often prohibited. At times these policies have been brutal and have caused extreme bitterness among the minorities, particularly the Tibetans and the large number of peoples who practice Islam. The extreme policies of the "10 bad years"

that encouraged the elimination of the "four olds" led to the wanton destruction of minority cultural artifacts, temples, mosques, texts, and statuary.

In the 1980s, the Deng Ziaoping leadership conceded that Beijing's harsh assimilation policies had been ill conceived, and it tried to gain the loyalty of the national minorities through more culturally sensitive policies. Minority children are now taught their own language in schools, alongside the "national" language (Mandarin). By the late 1980s, however, the loosening of controls had led to further challenges to Beijing's control, so the central government tightened up security in Xinjiang (China's far-northwestern province) and reimposed marital law in Tibet in an effort to quell protests and riots against Beijing's discriminatory policies. Martial law was lifted in Tibet in 1990, but security has remained tight ever since. The terrorist attacks on the United States on September 11, 2001, have led to even greater surveillance and controls on those minority groups that practice Islam and are believed to have ties with terrorist organizations in the Middle East.

Tibet

The Dalai Lama is the most important spiritual leader of the Tibetans, but he lives in exile in India, where he fled after a Chinese crackdown on Tibetans in 1959. He has stepped up his efforts to reach some form of accommodation with China; and at long last, in 2002, his delegation met in secrecy in Beijing to address the issue of greater autonomy for Tibet. The Dalai Lama insists that he is not pressing for independence, only for autonomy; and that as long as he is in charge, Tibetans will use only non-violent methods to gain greater autonomy. The Dalai Lama also asserts that more Tibetan control over their own affairs is necessary to protect their culture from extinction. Regardless of the Dalai Lama's intentions and promises, the people who surround him are far more militant and feel that autonomy will be only the first step toward independence from China.

The major threat to Tibetan culture at this point comes not from efforts by China's government to assimilate Tibetans into Han culture but, rather, from highly successful Chinese entrepreneurs who, under economic liberalization policies, have taken over many of the commercial and entrepreneurial activities of Tibet. Ironically, the Tibetan feelings about the Chinese mirror the feelings of the Chinese toward the West: The Tibetans want Chinese technology and commercial goods, but not the values that come with the people providing those goods and technology. And

among Tibetans, as among the Chinese, the young are more likely to want to become part of the modern world, to be modern and hip, and to leave behind traditional culture and values. Young Tibetans in Lhasa have been swept up in efforts to make money, a pleasure somewhat reduced by the fact that the increasingly large number of Chinese entrepreneurs in Lhasa usually make higher profits than they do. If Tibetan culture is to survive, Tibetans need to take on a modern identity, one that allows them to be both Tibetan and modern at the same time. Otherwise, by its sheer dynamism, Han Chinese culture may well overwhelm Tibetan culture.[10]

As suggested above, not all Tibetans accept the Dalai Lama's preferred path of nonviolence. In 1996, for the first time in decades, Beijing admitted that there were isolated bombing incidents and violent clashes between anti-Chinese Tibetans (reportedly armed) and Chinese authorities. The government, in response, sealed off most monasteries in Lhasa, the capital of Tibet. Beginning in the late 1990s, however, China decided to restore many of Tibet's monasteries—in part to placate Tibetans, in part to attract tourists. China has also made it far easier for foreigners to travel to Tibet. By 2005 it had almost completed a new highway across the length of the vast Qinghai Province plateau to connect Tibet with the rest of China. Construction has also begun on a U.S.$60 billion project to build a railroad through the Tibetan mountains to the capital, Lhasa.

As a result of an intense effort to alleviate poverty in Tibet. Tibet now receives more financial aid from the central Chinese government than any other province or autonomous region in China. Tibetans are better fed and clothed than they were; but Tibet still remains China's poorest administrative area. Generous state subsidies and a growing tourist industry seem unable to generate development. This is largely because of disastrous centrally-conceived policies, a bloated administrative structure, a large Chinese military presence to house and feed, disdain for Tibetan culture, and incompetent Han cadres who have little understanding of local issues and rarely speak Tibetan.[11] But Tibet's land-locked, remote location and its lack of arable land certainly exacerbate problems in development.

Regardless of what the Chinese do to develop Tibet, it is viewed with suspicion: New roads or railroads? The better for the Chinese to exploit Tibetan resources and even to invade. Encouraging tourism in Tibet or sending Tibetans to higher quality Chinese schools outside of Tibet? The better to destroy Tibetan culture. Allowing Chinese entrepreneurs to do business in Tibet or desperately poor youth from neigh-

(United Nations photo/John Isaac)

This family lives in a commune in Inner Mongolia. China's central government has long attempted to undermine distinctive national identities on and around its borders.

boring provinces to go to Tibet for work? The better to take away jobs from Tibetans. Projects to develop Tibet's infrastructure for development? The better to destroy its environment. In short, Tibetans tend to regard all of Beijing's policies, and Han cadres in Tibet, with suspicion.

Much of the recent anger in Tibet against China's central government arose from Beijing's decision in 1995 not to accept the Tibetan Buddhists' choice of a young boy as the reincarnation of the former Panchen Lama,[12] the second-most-important spiritual leader of the Tibetans. Instead of accepting the Tibetans' recommendation for the next Panchen Lama, chosen according to traditional Tibetan Buddhist ritual, Beijing substituted its own six-year old candidate. The Tibetans' choice, meanwhile, is living in seclusion somewhere in Beijing, under the watchful eye of the Chinese. China's concern is that any new spiritual leader could become a focus for a new

push for Tibetan independence—an eventuality it wishes to avoid.

Inner Mongolia

Inner Mongolia (an autonomous region under Beijing's control) lies on the southern side of Mongolia, which is an independent state. Beijing's concern that the Mongolians in China would want to unite with Mongolia led to a policy that diluted the Mongol population with what has grown to be an overwhelming majority of Han Chinese. According to the 2000 national census, the national minority (largely Mongol) population was only 4.93 million—a mere 20.76% of the total population of 23.76 million. Inner Mongolia's capital, Huhhot, is essentially a Han city, and assimilation of Mongols into Han culture in the capital is almost complete. Mongolians are dispersed throughout the vast countryside as shepherds, herdsmen, and farmers and retain many of their ethnic traditions and practices.

Events in (Outer) Mongolia have led China's central leadership to keep a watchful eye on Inner Mongolia,. In 1989, Mongolia's government—theoretically independent but in fact under Soviet tutelage until—decided to permit multiparty rule at the expense of the Communist Party's complete control; and in democratic elections held in 1996, the Mongolian Communist Party was ousted from power.

Beijing has grown increasingly concerned that these democratic inklings might spread to their neighboring cousins in Inner Mongolia, with a resulting challenge to one-party CCP rule. As with the Islamic minorities, China's leadership worries that the Mongols in Inner Mongolia may try to secede from China and join with the independent state of Mongolia because of their shared culture. So far, however, those Inner Mongolians who have traveled to Mongolia have been surprised by the relative lack of development there and have remained fairly quiet about seceding.

Nevertheless, privatization of the economy, combined with an insensitivity to Mongolian culture led to major demonstrations against the Han Chinese-dominated government in 2004. Contributing to these problems was the purchase by a Han Chinese of the leading Mongolian museum in the capital, Huhhot, which houses artifacts from the Altaic folklore of the Mongols, and of Ghengis Khan, the leader of the Mongols in the 12th century. The fact that it is a Han Chinese who now owns the museum, and plans to modernize it to make it more attractive to tourists, has offended Mongol sensibilities.

Muslim Minorities

In the far northwest, the predominantly Muslim population—particularly the Uyghur minority—of Xinjiang Province continues to challenge the authority of China's central leadership. The loosening of policies aimed at assimilating the minority populations into the Han (Chinese) culture has given a rebirth to Islamic culture and practices, including prayer five times a day, architecture in the Islamic style, traditional Islamic medicine, and teaching Islam in the schools. With the dissolution of the Soviet Union in 1991 into 15 independent states, the ties between the Islamic states on China's borders (Kazakhstan, Kyrgyzstan, and Tajikistan, as well as Afghanistan and Pakistan) have accelerated rapidly.

Beijing is concerned that China's Islamic minorities may find that they have more in common with these neighboring Islamic nations than with the Chinese Han majority and may attempt to secede from China. Beijing's concern was heightened in late 1992, when Turkey held a summit conference at which it announced that the next century would be the "century of Islam." Subsequent signs of a growing, worldwide Islamic movement have exacerbated Beijing's anxieties about controlling China's Islamic minorities. The 9/11 attacks by Muslim fundamentalists, followed by the U.S.–led "war on terrorism" throughout the world, have led China to intensify its efforts to root out Islamic radicals. Although some analysts argue that the war on terrorism has given Beijing an ideal pretext for cracking down on what is a legitimate desire for national independence by Muslims in Xinjiang, others accept Beijing's view that those using violence (including bombs) are "terrorists," not "freedom fighters."

Uyghurs who have engaged in violence are not motivated by religious fanaticism, but rather, a desire to achieve a concrete, pragmatic goal: Xinjiang's secession from China. Still, in the last decade, they have received funding from the Islamic world, including, it is believed, from terrorist groups located therein. It is unlikely, however, that the Uyghurs accept the tenets of Islamic fundamentalists, or that they view their struggle against Chinese rule as a struggle of good against evil. (Indeed, it has often been noted that Islam is much more moderate, tolerant, and progressive as it spreads eastward.) Evidence that the Uyghurs are engaged in an out-and-out political struggle for independence from Chinese rule would be that Uyghur violence in Xinjiang has not been in form of terrorist attacks on the local Han population but rather, on the state structure of the governing Han. So it could be argued that, except for the 1997 attacks in Beijing on innocent bystanders, and several bus bombings in Xinjiang, Uyghur violence is not "terrorism" as it is now defined.

China's nearly 9 million Hui—Han Chinese who practice Islam—are also classified as a "national minority;" but over many centuries, they have become so integrated into mainstream Chinese culture that at this point in history their only remaining distinct characteristic is their practice of Islam. They speak standard Chinese and live together with other Han. Although a large number of Hui live in one autonomous region, Ningxia, they are also spread throughout China. In general, in spite of shared Islamic beliefs, they do not identify with Uyghur nationalism, which is seen as particular to Uyghur ethnicity, and not to a broader Islamic identity.

Religion

Confucianism

Confucianism is the "religion" most closely associated with China. It is not, however, a religion in Western terms, as there is no place for gods, the afterlife, or most other beliefs associated with formal religions. But like most religions, it does have a system of ethics for human relationships; and it adds to this what most religions do not have, namely, principles for good governance that include the hierarchical ordering of relationships, with obedience and subordination of those in lower ranks to those in higher ranks.

The Chinese Communists rejected Confucianism until the 1980s, but not because they saw it as an "opiate of the masses." (That was Karl Marx's description of religion, which he viewed as a way of trapping people in a web of superstitions, robbing them of their money and causing them to passively endure their miserable lives on Earth.) Instead, they denounced Confucianism for providing the ethical rationale for a system of patriarchy that allowed officials to insist on obedience from subordinates. During the years in which "leftists" such as the Gang of Four set the agenda, moreover, the CCP rejected Confucianism for its emphasis on education as a criterion for joining the ruling elite. Instead, the CCP favored ideological commitment— "redness"—as the primary criterion for ruling. The reforms since 1979, however, have emphasized the need for an educated elite, and Confucian values of hard work and the importance of the family are frequently referred to. The revival of Confucian values has, in fact, provided an important foundation for China's renewed emphasis on nationalism as a substitute for the now nearly defunct values of Marxism-Leninism-Mao Thought.

Buddhism

Buddhism has remained important among some of the largest of the national minorities, notably the Tibetans and Mongols. The CCP's efforts to eradicate formal Buddhism have been interpreted by the minorities as national oppression by the Han Chinese. As a result, the revival of Buddhism since the 1980s has been associated with efforts by Tibetans and Mongolians to assert their national identities and to gain greater autonomy in formulating their own policies. Under the influence of the more moderate policies of the Deng Xiaoping reformist leadership, the CCP reconsidered its efforts to eliminate religion. The 1982 State Constitution permits religious freedom, whereas previously only atheism was allowed. The state has actually encouraged the restoration of Buddhist temples, in part because of Beijing's awareness of the continuing tensions caused by its efforts to deny minorities their respective religious practices, and in part because of a desire to attract both tourists and money to the minority areas.

But Buddhism is far more widespread than Tibet and Inner Mongolia. Indeed, popular Buddhism, which is full of stories and Buddhist mythology, is pervasive throughout the rural population—and even among some urban populations. And popular Buddhist beliefs even are worked into many of the sects, cults, and folk religions in China. Today, Buddhist temples are frequented by increasingly large numbers of Chinese, who go there to propitiate their ancestors and to pray for good health and more wealth.

Folk Religions

For most Chinese, folk religions are far more important than any organized religion.[13] The CCP's best efforts to eradicate folk religions and to impart in their place an educated "scientific" viewpoint have failed. Animism—the belief that nonliving things have spirits that should be respected through worship—continues to be practiced by China's vast peasantry. Ancestor worship—based on the belief that the living can communicate with the dead and that the dead spirits to whom sacrifices are ritually made have the ability to bring a better (or worse) life to the living—absorbs much of the income of China's peasants. The costs of offerings, burning paper money, and using shamans and priests to perform rituals that will heal the sick, appease the ancestors, and exorcise ghosts (who are often poorly treated ancestors returned to haunt their descendants) at times of birth, marriage, and death, can be financially burdensome. But the willingness of peasants to spend money on traditional religious folk practices is contributing to the reconstruction of practices prohibited in earlier decades of Communist rule.

Taoism, Qigong, and Falun Gong

Taoism, which requires its disciples to renounce the secular world, has had few adherents in China since the early twentieth century. But during the repression that followed the crackdown on Tiananmen Square's prodemocracy movement in 1989, many Chinese who felt unable to speak freely turned to mysticism and Taoism. *Qigong,* the ancient Taoist art of deep breathing, had by 1990 become a national pastime. Some 30 Taoist priests in China took on the role of national soothsayers, portending the future of everything from the weather to China's political leadership. What these priests said—or were believed to have said—quickly spread through a vast rumor network in the cities. Meanwhile, on Chinese Communist Party–controlled television, qigong experts swallow needles and thread, only to have the needles subsequently come out of their noses perfectly threaded. It is widely believed that, with a sufficient concentration of *qi* (vital energy or breath), a practitioner may literally knock a person to the ground.[14] The revival of Taoist mysticism and meditation, folk religion, and formal religions suggests a need to find meaning from religion to fill the moral and ideological vacuum created by the near-collapse of Communist values.

Falun Gong ("Wheel of Law"), which the government has declared a "sect"—and hence not entitled to claim a constitutional right to practice religion freely—has been charged with involvement in a range of illegal activities. Falun Gong is a complex mixture of Buddhism, Taoism, and qigong practices —the last relying on many ideas from traditional Chinese medicine. According to its adherents, the focus is on

healing and good health, but it also has a millennial component, predicting the end of the world and a bad ending for those who are not practitioners. According to the government, the sect's practices can endanger people's health and have in fact caused the deaths of hundreds. It also accuses the sect of being a front for antigovernment political activities.

HEALTH/WELFARE

The Communist government has overseen dramatic improvements in the provision of social services for the masses. Largely because of its emphasis on preventive medicine and sanitation, life expectancy has increased from 45 years in 1949 to 72 years (overall) today. The government has successfully eradicated many childhood diseases and has made great strides against other diseases, such as malaria and tuberculosis. The privatization of medicine in recent years has, however, meant that many no longer have guaranteed access to medical treatment.

In 1999, thousands of Falun Gong adherents, some from distant provinces, suddenly materialized in front of CCP headquarters, on the edge of Tiananmen Square in Beijing. Hundreds were arrested, but most were soon released and sent back to their home provinces. Others were sent to labor camps or jailed, and some died while incarcerated.[15] In many state-owned work units, officials continue to have weekly meetings to discuss the dangers of Falun Gong, to encourage followers to end their participation in Falun Gong, and to root out its leaders.

Religious practice often provides the foundation for illegal or "black" societies. Falun Gong and a number of other sects have been accused of using their organizations as fronts for drugs, smuggling, prostitution, and other illegal activities. Religious sects and black societies are widely believed to provide the basis of power for candidates for office. In the countryside, religion can become a tool of the family clans, who sometimes use it to pressure villagers to vote for their candidates.

Christianity

Christianity, which was introduced in the nineteenth and early twentieth centuries by European missionaries, has several million known adherents; and its churches, which were often used as warehouses or public offices after the Communist victory in 1949, have been reopened for religious practice. Bibles in several editions are available for purchase in many large city bookstores. A steady stream of Christian

proselytizers flow to China in search of new converts. Today's churches are attended as much by the curious as by the devout. As with eating Western food in places such as McDonald's and Kentucky Fried Chicken, attending Christian churches is a way that some Chinese feel they can participate in Western culture. Some Chinese want to become Christians because they see that in the West, Christians are rich and powerful. They believe that Christianity helps explain the wealth and power of Western capitalists, and hope that converting to Christianity will ensure them of greater wealth and power.

The government generally permits mainstream Christian churches to practice in China, but it continues to exercise one major control over Roman Catholics: Their loyalty must be declared to the state, not to the pope. The Vatican is prohibited from involvement in China's practice of Catholicism, and Beijing does not recognize the Vatican's appointment of bishops and cardinals for China as valid. Underground "house churches," primarily for smaller Christian sects, offshoots of mainstream Protestant religions, and papal Catholics, are forbidden. Nevertheless, they seem to flourish as officials busy themselves with addressing far more pressing social isues.

Since the mid-1990s, the government has tried to clamp down on nonmainstream Christian churches as well as religious sects, arresting and even jailing some of their leaders. They have justified their actions on the grounds that, as in the West, some of the churches are involved in practices that endanger their adherents; some are actually involved in seditious activities against the state; and some are set up as fronts for illegal activities, including gambling, prostitution, and drugs.

Marxism-Leninism-Mao Zedong Thought

In general, Marxists are atheists. They believe that religions hinder the development of "rational" behavior and values that are so important to modernization. Yet societies seem to need some sort of spiritual, moral, and ethical guidance. For Communist party–led states, Marxism is believed to be adequate to fill the role of moral, if not spiritual, guidance. In China, however, Marxism-Leninism was reshaped by Mao Zedong Thought to accommodate for Chinese conditions. Paramount among these conditions was that China was a predominantly peasant society, not a society in which there was a capitalist class exploiting large numbers of urban workers. Thus the package of ideology in China became known as Marxism-Leninism-Mao Zedong Thought. The Chinese leadership believed that it provided the ethical values necessary

DENG XIAOPING = TENG HSIAO-P'ING. WHAT IS PINYIN?

Chinese is the oldest of the world's active languages and is now spoken and written by more people than any other modern language. Chinese is written in the form of characters, which have evolved over several thousand years from picture symbols (like ancient Egyptian hieroglyphics) to the more abstract forms now in use. Although spoken Chinese varies greatly from dialect to dialect (for example, Mandarin, Cantonese, Shanghai-ese), the characters used to represent the language remain the same throughout China. Dialects are really just different ways of pronouncing the same characters.

There are more than 50,000 different Chinese characters. A well-educated person may be able to recognize as many as 25,000 characters, but basic literacy requires familiarity with only a few thousand.

Since Chinese is written in the form of characters rather than by a phonetic alphabet, Chinese words must be transliterated so that foreigners can pronounce them. This means that the sound of the character must be put into an alphabetic approximation.

Since English uses the Roman alphabet, Chinese characters are Romanized. (We do the same thing with other languages that are based on non-Roman alphabets, such as Russian, Greek, Hebrew, and Arabic.) Over the years, a number of methods have been developed to Romanize the Chinese language. Each method presents what the linguists who developed it believe to be the best way of approximating the sound of Chinese characters. *Pinyin* (literally, "spell

⊙	⊖	⊟	日	rì sun
☽	☽	♀	月	yuè moon
彳	勹	几	人	rén person
⽊	⽊	⽊	木	mù tree

Chinese characters are the symbols used to write Chinese. Modern Chinese characters fall into two categories: one with a phonetic component, the other without it. Most of those without a phonetic component developed from pictographs. From ancient writing on archaeological relics we can see their evolution, as in the examples shown (from left to right) above.

sounds"), the system developed in the People's Republic of China, has gradually become the most commonly accepted system of Romanizing Chinese.

However, other systems are still used in areas such as Taiwan. This can cause some confusion, since the differences between Romanization systems can be quite significant. For example, in pinyin, the name of China's dominant leader is spelled Deng Xiaoping. But the Wade-Giles system, which was until the 1980s the Romanization method most widely used by Westerners, transliterates his name as Teng Hsiao-p'ing. Same person, same characters, but a difference in how to spell his name in Roman letters.

to guide China toward communism; and it was considered an integrated, rational thought system.

Nevertheless, this core of China's Communist political ideology exhibited many of the trappings of religions. It included scriptures (the works of Marx, Lenin, and Mao, as well as party doctrine); a spiritual head (Mao); and ritual observances (particularly during the Cultural Revolution, when Chinese were forced to participate in the political equivalent of Bible study each day). Largely thanks to the shaping of this ideology by Maoism, it included moral axioms that embodied traditional Chinese— and, some would say, Confucian—values that resemble teachings in other religions. Thus the moral of Mao's story of "The Foolish Old Man Who Wanted to Remove a Mountain" is essentially identical to the Christian principle "If you have faith you can walk on water," based on a story in the New Testament. Like this teaching, the essence of Mao Zedong Thought was concerned with the importance of a correct moral (political) consciousness.

In the 1980s, the more pragmatic leadership focused on liberalizing reforms and encouraged the people to "seek truth

from facts" rather than from Marxism-Leninism-Mao Zedong Thought. As a result, the role of this political ideology declined, in spite of efforts by more conservative elements in the political leadership to keep it as a guiding moral and political force. Participants in the required weekly "political study" sessions in most urban work units abandoned any pretense of interest in politics. Instead, they focused on such issues as "how to do our work better" (that is, how to become more efficient and make a profit) that were in line with the more pragmatic approach to the workplace. Nevertheless, campaigns like the "get civilized" and anticorruption ones retain a strong moralistic tone.

Ideology has not, therefore, been entirely abandoned. In the context of modernizing the economy and raising the standard of living, the current leadership is still committed to building "socialism with Chinese characteristics." Marxist-Leninist ideology is still being reformulated in China; but it is increasingly evident that few true believers in communism remain. Rarely does a Chinese leader even mention Marxism-Leninism in a speech. Leaders instead focus on modernization and becoming more effi-

cient; they are more likely to discuss interest rates and trade balances than ideology. Fully aware that they need something to replace their own nearly defunct guiding ideological principles, however, and fearing that pure materialism and consumerism are inadequate substitutes, China's leaders seem to be relying on patriotism and nationalism as the key components of a new ideology whose primary purpose is very simple: economic modernization and support of the leadership of the Chinese Communist Party.

Undergirding China's nationalism is a fierce pride in China's history, civilization, and people. Insult, snub, slight, or challenge China, and the result is certain to be a country united behind its leadership, against the offender. To oppose the CCP or its objective of modernization is viewed as "unpatriotic." China's nationalism, on the other hand, is fired by antiforeign sentiments. These sentiments derive from the belief that foreign countries are—either militarily, economically, or through insidious cultural invasion—attempting to hurt China or to intervene in China's sovereign affairs by telling China's rulers how to govern properly. This is most notably the case whenever the Western countries condemn China for its human-rights record. U.S. support for Taiwan, and the U.S. bombing of the Chinese Embassy in Belgrade, Yugoslavia, during the Kosovo War in May 1999, fueled Chinese nationalism and injected even more tension into Sino–American relations. China's decision to join with the United States in its war on terrorism since 9/11 has, however, resulted in a toned-down nationalism and a less strident approach to international relationships. China's growing entanglement in a web of international economic and political relationships has also contributed to a softening of its nationalistic stance.

Language

By the time of the Shang Dynasty, which ruled in the second millennium B.C., the Chinese had a written language based on "characters." Over 4,000 years, these characters, or "ideographs," have evolved from being pictorial representations of objects or ideas into their present-day form. Each character usually contains a phonetic element and one (or more) of the 212 symbols called "radicals" that help categorize and organize them.[16] Before the May Fourth Movement, only a tiny elite of highly educated men could read these ideographs, which were organized in the difficult grammar of the classical style of writing, a style that in no way reflected the spoken language. All this changed with language reform in the 1920s: The classical style was abandoned, and the written language became almost identical in its structure to the spoken language.

Increasing Literacy

When the Chinese Communists came to power in 1949, they decided to facilitate the process of becoming literate by allowing only a few thousand of the more than 50,000 Chinese characters in existence to be used in printing newspapers, official documents, and educational materials. However, since a word is usually composed of a combination of two characters, these few thousand characters form the basis of a fairly rich vocabulary: Any single character may be used in numerous combinations in order to form many different words. The Chinese Communists have gone even further in facilitating literacy by simplifying thousands of characters, often reducing a character from more than 20 strokes to 10 or even fewer.

In 1979, China adopted a new system, *pinyin,* for spelling Chinese words and names. This system, which uses the Latin alphabet of 26 letters, was created largely for foreign consumption and was not widely used within China. The fact that so many characters have the same Romanization (and pronunciation), plus cultural resistance, have thus far resulted in ideographs remaining the basis for Chinese writing. There are, as an example, at least 70 different Chinese ideographs that are pronounced *zhu,* but each means something different. Usually the context is adequate to indicate which word is being used. But when it may not be clear which of many homonyms is being used, Chinese often use their fingers to draw the character in the air.

Something of a national crisis has emerged in recent years over the deleterious effect of computer use on the ability of Chinese to write Chinese characters. Computers are set up to write Chinese characters by choosing from multiple Chinese words whose sound is rendered into a Latin alphabet. As a result, computer users no longer need to remember how to write the many strokes in Chinese characters—they simply scroll down to the correct Chinese character under the sound of, say, "tang." Then they press the "enter" key. The problem is that, without regular practice writing out characters, it is easy to forget how to write them—even for Chinese people. It is a problem akin to the loss of mathematical skills due to the use of calculators and computers.

Spoken Chinese

The Chinese have shared the same written language over the last 2,000 years, regardless of which dialect of Chinese they spoke. (The same written characters were simply pronounced in different ways, depending on the dialect.) Building a sense of national unity was difficult, however, when people needed interpreters to speak with someone living even a few miles away. After the Communist victory in 1949, the government decided that all Chinese would speak the same dialect in order to facilitate national unity. A majority of the delegates to the National People's Congress voted to adopt the northern dialect, Mandarin, as the national language, and required all schools to teach in Mandarin (usually referred to as "standard Chinese").

In the countryside, however, it has been difficult to find teachers capable of speaking and teaching Mandarin; and at home, whether in the countryside or the cities, the people have continued to speak their local dialects. The liberalization policies that began in 1979 have had as their by-product a discernible trend back to speaking local dialects, even in the workplace and on the streets. Whereas a decade ago a traveler could count on the national language being spoken in China's major cities, this is no longer the case. As a unified language is an important factor in maintaining national cohesion, the re-emergence of local dialects at the expense of standard Chinese threatens China's fragile unity.

One force that is slowing this disintegration is television, for it is broadcast almost entirely in standard Chinese. As there is a growing variety of interesting programming available, it may be that most Chinese will make an effort to maintain or even acquire the ability to understand standard Chinese.

Education

The People's Republic of China has been remarkably successful in educating its people. Before 1949, less than 20 percent of the population could read and write. Today, nine years of schooling are compulsory. In the larger cities, 12 years of schooling is becoming the norm, with children attending either a vocational middle school or a college-preparatory school. Computer use is increasingly common in urban schools.

It is difficult to enforce the requirement of nine years of school in the impoverished countryside. Still, close to 90 percent of those children living in rural areas attend at least primary school. Village schools, however, often lack rudimentary equipment such as chairs and desks. Rural education also suffers from a lack of qualified teachers, as any person educated enough to teach can probably get a better-paying job in the industrial or commercial sector. But the situation is in flux, because as rural families are having fewer children than before, more can now afford school fees. And their objective is different from the past: not to prepare their children for farming, but to prepare them for working in factories and office jobs in the towns and cities. So, at the same time that the collective basis for funding schools has deteriorated, in some villages many more farmers are able and willing to pay for school tuition.

At the other end of the spectrum, many more students are now pursuing a college-oriented curriculum than will ever go on to college. From 1998 to 2004, China doubled the number of students it admitted to universities, from 2.1 to 4.2 million students.[17] Nevertheless, only about 5 percent of the senior middle school graduates will pass the university entrance examinations and be admitted. As a result, many who had prepared for a college curriculum are inappropriately educated for the workplace. The government is attempting to augment vocational training for high school students, but it is also increasing the number of slots available in colleges and universities. Private high schools and colleges are becoming increasingly popular as parents try to optimize the chances for their only

(United Nations photo)

Communes were disbanded by the early 1980s. In some areas, however, farmers have continued to work their land as a single unit in order to benefit from the economies of scale of large tracts of land.

child to climb the academic ladder in order to gain social and economic success.

Political Education

Until the reforms that began in 1979, the content of Chinese education was suffused with political values and objectives. A considerable amount of school time—as much as 100 percent during political campaigns—was devoted to political education. Often this amounted to nothing more than learning by rote the favorite axioms and policies of the leading faction in power. When that faction lost power, the students' political thought had to be reoriented in the direction of the new policies and values. History, philosophy, literature, and even foreign languages and science were laced with a political vocabulary.

The prevailing political line has affected the balance in the curriculum between political study and the learning of skills and scientific knowledge. Beginning in the 1960s, the political content of education increased dramatically until, during the Cultural Revolution, schools were shut down. When they reopened in the early 1970s, politics dominated the

curriculum. When Deng Xiaoping and the "modernizers" consolidated their power in the late 1970s, this tendency was reversed. During the 1980s, in fact, schools jettisoned the study of political theory because both administrators and teachers wanted their students to do well on college-entrance examinations, which by then focused on academic subjects. As a result, students refused to clog their schedules with the study of political theory and the CCP's history. The study of Marxism and party history was revived in the wake of the events of Tiananmen Square in 1989, with the entering classes for many universities required to spend the first year in political study and indoctrination, sometimes under military supervision; but this practice was abandoned after two years. Today, political study has again been confined to a narrow part of the curriculum, in the interest of giving students an education that will help advance China's modernization.

Study Abroad

Since 1979, when China began to promote an "open door" policy, more than 100,000

PRC students have been sent for a university education to the United States, and tens of thousands more have gone to Europe and Japan. China has sent so many students abroad in part because the quality of education had seriously deteriorated during the "10 bad years" that included the Cultural Revolution, and in part because China's limited number of universities can accommodate only a tiny percentage of all high school graduates. Although an increasingly large number of Chinese universities are able to offer graduate training, talented Chinese students still travel abroad to receive advanced degrees.

Chinese students who have returned home have not always met a happy fate. Many of those educated abroad who were in China at the time of the Communists' victory in 1949, or who returned to China thereafter, were not permitted to hold leadership positions in their fields. Ultimately they were the targets of class-struggle campaigns and purges in the 1950s, '60s, and '70s, precisely because of their Western education. For the most part, those students who returned to China in the 1980s found that they could not be promoted because of the continuation of a system of seniority.

DENG XIAOPING: TAKING A "PRACTICAL" APPROACH TO CHINA'S PROBLEMS

Deng Xiaoping, a controversial figure throughout his political career, was twice purged from power. Deng became the dominant figure in Chinese politics after he returned to power in 1978. Under his leadership, China implemented policies of economic liberalization, including the market system of supply and demand for distributing goods, services, and resources. These policies replaced many of China's centrally planned economic policies. Deng's statement that "I don't care whether the cat is black or white as long as it catches mice" illustrates his practical, nonideological approach to modernizing China. In other words, Deng did not care if he used capitalist methods, as long as they helped modernize China faster than socialist methods.

Deng's economic-liberalization policies frequently were blamed when problems such as inflation and corruption occurred in the 1980s and 1990s. Those opposing his policies used this as their rationale for retreating from economic liberalization twice during the 1980s. Since 1992, however, when Deng reasserted the need to move ahead with economic liberalization, the leadership has fairly steadily implemented Deng's policies.

Deng did not hold any official post in his final years of life; but, from 1992 until he died in February 1997, his policies were not successfully challenged. Because the liberalizing reforms he introduced have enjoyed such remarkable success, it is unlikely they will be undone any time soon, regardless of who leads China.

(Hsin-Hua News Agency)

Since 1992, however, when Deng Xiaoping announced a major shift in economic and commercial policy to support just about anything that would help China become rich and powerful, the government has offered students significant incentives to return to China, including excellent jobs, promotions, good salaries, and even the chance to start new companies. Chinese students educated abroad are also recruited for their expertise and understanding of the outside world by the rapidly multiplying number of joint ventures in China, and by universities that are establishing their own graduate programs. Today, fully one third of students educated abroad return to live in China.

The Chinese government also now sees those Chinese who do stay abroad and become citizens of other countries as forming critical links for China to the rest of the world. They have become the bridges over which contracts, loans, and trade flow to China, and are viewed as a positive asset. Finally, like immigrants elsewhere, Chinese who settle abroad tend to send remittances back to their families in China. These remittances amount to hundreds of millions of U.S. dollars in foreign currency each year and are valuable not just to the family recipients but also to the government's bank reserves.

Chinese studying abroad learn much about liberal democratic societies. Those who have returned to China bring with them the values at the heart of liberal-democratic societies. While this does not necessarily mean that they will demand the democratization of the Chinese political system, they do bring pluralistic liberal-democratic ideas to their own institutions. They have been instrumental in introducing institutions such as "think tanks," have encouraged debate within their own fields, and have been insistent that China remain open to the outside world through the Internet, travel, conferences, and communications.

THE ECONOMIC SYSTEM

A Command Economy

Until 1979, the Chinese had a centrally controlled command economy. That is, the central leadership determined the economic policies to be followed and allocated all of the country's resources—labor, capital, land, and raw materials. It also determined how much each enterprise, and even each individual, would be allocated for production and consumption. Once the Chinese Communist Party leadership determined the country's political goals and the correct ideology, the State Planning Commission and the State Economic Com-

mission would then decide how to implement these objectives through specific policies for agriculture and industry and the allocation of resources. This is in striking contrast to a capitalist laissez-faire economy, in which government control over both consumers and producers is minimal and market forces of supply and demand play the primary role in determining the production and distribution of goods.

The CCP leadership adopted the model of a centralized planned economy from the Soviet Union. Such a system was not only in accord with the Leninist model of centralized state governance; it also made sense for a government desperate to unify China after more than 100 years of internal division, instability, and economic collapse. Historically, China suffered from the ability of large regions to evade the grasp of central control over such matters as currency and taxes. The inability of the Nationalist government to gain control over the country's economy in the 1930s and early 1940s undercut its power and contributed to its failure to win control over China. Thus, the Chinese Communist Party's decision to centralize economic decision making after 1949 helped the state to function as an integrated whole.

Over time, however, China's highly centralized economy became inefficient and too inflexible to address the complexity of the country's needs. Although China possesses a large and diverse economy, with a broad range of resources, topography, and climate, its economic planners made policy as if it were a uniform, homogeneous whole. Merely increasing production was itself considered a contribution to development, regardless of whether a market for the products existed or whether the products actually helped advance modernization.

State planning agencies, without the benefit of market research or signals from the marketplace, determined whether or not a product should be manufactured, and in what quantity. For example, the central government might set a goal for a factory to manufacture 5 million springs per year—without knowing if there was even a market for them. The factory management did not care, as the state was responsible for marketing the products and paid the factory's bills. If the state had no buyer for the springs, they would pile up in warehouses; but rarely would production be cut back, much less a factory be closed, as this would create the problem of employing the workers cut from the factory's payroll. Economic inefficiencies of this sort were often justified because socialist political objectives such as full employment were being met. Even today the state worries about shutting down a state-owned factory that is losing money, because it creates unemployment. In turn, unemployment leads to popular anger and provides a volatile, unstable environment, ripe for public political protest. Quality control was similarly not as important an issue as it should have been for state-run industries in a centrally planned economy. Until market reforms began in 1979, the state itself allocated all finished products to other industries that needed them. If a state-controlled factory made defective parts, the industry using them had no recourse against the supplier, because each factory had a contract with the state, not with other factories. It was the state that would pay for additional parts to be made, so the enterprises did not bear the costs.

As a result, China's economic development under the centralized political leadership of the CCP occurred by fits and starts. Much waste resulted from planning that did not take into account market factors of supply and demand. Centrally set production quotas took the place of profit-and-loss issues in the allocation of resources. Although China's command economy was able to meet the country's most important industrial needs, problems like these took their toll over time. Enterprises had little incentive to raise productivity, quality, or efficiency when doing so

did not affect their budgets, wages, or funds for expansion.

Disastrous Agricultural Programs

By the late 1950s, central planning was causing significant damage to the agricultural sector. Regardless of geography or climate, China's economic planners repeatedly ordered the peasants to restructure their economic production units according to one centralized plan. China's peasants, who had supported the CCP in its rise to power before 1949 in order to acquire their own land, had enthusiastically embraced the CCP's fulfillment of its pledge of "land to the tillers" after the Communists took over in 1949. But in 1953, the leadership, motivated by a belief that small-scale agricultural production could not meet the production goals of socialist development, ordered all but 15 percent of the arable land to be pooled into "lower-level agricultural producer cooperatives" of between 300 and 700 workers. The remaining 15 percent of land was to be set aside as private plots for the peasants, and they could market the produce from these plots in private markets throughout the countryside. Then, in 1956, the peasants throughout the country were ordered into "higher-level agricultural producer cooperatives" of 10 times that size, and the size of the private plots allotted to them was reduced to 5 percent of the cooperatives' total land.

Many peasants felt cheated by these wholesale collectivization policies. When in 1958 the central leadership ordered them to move into communes 10 times larger still than the cooperatives they had just joined, they were irate. Mao Zedong's Great Leap Forward policy of 1958 forced all peasants in China to become members of large communes: enormous economic and administrative units consisting of between 30,000 and 70,000 people. Peasants were required to relinquish their private plots, and turn over their private utensils, as well as their household chickens, pigs, and ducks, to the commune. Resisting this mandate, many peasants killed and ate their livestock. Since private enterprise was no longer permitted, home handicraft industries ground to a halt.

CCP chairman Mao's vision for catching up with the West was to industrialize the vast countryside. Peasants were therefore ordered to build "backyard furnaces" to smelt steel. Lacking iron ore, much less any knowledge of how to make steel, and under the guidance of party cadres who themselves were ignorant of steelmaking, the peasants tore out metal radiators, pipes, and fences. Together with pots and pans, they were dumped into their furnaces. Almost none of the final smelted product was

usable. Finally, the central economic leadership ordered all peasants to eat in large, communal mess halls. This was reportedly the last straw for a people who valued family above all else. Being deprived of time alone with their families for meals, the peasants refused to cooperate further in agricultural collectivization.

When the catastrophic results of the Great Leap Forward policy poured in, the CCP retreated—but it was too late. Three subsequent years of bad weather, combined with the devastation wreaked by these policies and the Soviet withdrawal of all assistance, brought economic catastrophe. Demographic data indicate that in the "three bad years" from 1959 to 1962, more than 20 million Chinese died from starvation and malnutrition-related diseases.

By 1962, central planners had condoned peasants returning to production and accounting units that were smaller than communes. Furthermore, peasants were again allowed to farm a small percentage of the total land as private plots, to raise domestic animals for their own use, and to engage in household handicrafts. Free markets, at which the peasantry could trade goods from private production, were reopened. The commune structure was retained throughout the countryside, however; and until the CCP leadership introduced the contract responsibility system in 1979, it provided the infrastructure of rural secondary school education, hospitals, and agricultural research.

Other centrally determined policies, seemingly oblivious to reality, have compounded the P.R.C.'s difficulties in agriculture. Maoist policies carried out during both the Great Leap Forward and the Cultural Revolution included attempts to plant three crops per year in areas that for climatic reasons can only support two (as the Chinese put it, "Three times three is not as good as two times five"); and "close planting," which often more than doubled the amount of seed planted, with the result that all of it grew to less than full size or simply wilted for lack of adequate sunshine and nutrients.

A final example of centrally determined agricultural policy bringing catastrophe was the decision during the Cultural Revolution that "the whole country should grow grain." The purpose was to establish China's self-sufficiency in grain. Considering China's immense size and diverse climates, soil types, and topography, a policy ordering everyone to grow the same thing was doomed to failure. Peasants plowed under fields of cotton and cut down rubber plantations and fruit orchards, planting grain in their place. China's planners were largely CCP cadres, not economic experts, and they ignored overwhelming evidence

(United Nations photo)

COMMUNES: PEASANTS WORK OVERTIME DURING THE GREAT LEAP FORWARD

In the socialist scheme of things, communes are considered ideal forms of organization for agriculture. They are supposed to increase productivity and equality, reduce inefficiencies of small-scale individual farming, and bring modern benefits to the countryside more rapidly through rural industrialization.

These objectives are believed to be attained largely through the economies of scale of communes; that is, it is presumed that things done on a large scale are more efficient and cost-effective than when done on a small scale. Thus, using tractors, harvesters, trucks, and other agricultural machinery makes sense when large tracts of land can be planted with the same crops and plowed at one time. Similarly, small-scale industries may be based on the communal unit of 30,000 to 70,000 people, since, in such a large work unit, some people can take care of agricultural needs for the entire commune, leaving others to work in commune-based industries.

Because of its size, a commune may also support other types of organizations that smaller work units would find impossible to support, both financially and otherwise. A commune, for example, can support a hospital, a high school, an agricultural-research organization, and, if the commune is wealthy enough, even a "sports palace" and a cultural center for movies and entertainment.

During the Great Leap Forward, launched in 1958, peasants were—much against their will—forced into these larger agricultural and administrative units. They were particularly distressed that their small remaining private plots were, like the rest of their land, collectivized. Communal kitchens, run by women, were to prepare food for everyone while other workers went about their duties. Peasants were told that they had to eat in the communal mess halls rather than in the privacy of their own homes.

When the combination of bad policies and bad weather led to a severe famine, widespread peasant resistance forced the government to retreat from the Great Leap Forward policy and abandon the communes. But a modified commune system remained intact in much of China until the late 1970s, when the government ordered communes to be dissolved. A commune's collective property was then distributed to the peasants belonging to it, and a system of "contract responsibility" was launched. Today, with the exception of a few communes that refused to be dissolved, agricultural production is no longer collectivized. Individual households are again, as before 1953, engaged in small-scale agricultural production on private plots of land.

(Xinhua News Agency)

Under the economic-liberalization program, shops such as this one in Sichuan were allowed to prosper.

that grain would not grow well in all areas and that China would have to import everything that it had replaced with grain, at far greater costs than it would have paid for importing just grain. Peasant protests were futile in the face of local-level Communist Party leaders who hoped to advance their careers by implementing central policy. The policy of self-sufficiency in grain was abandoned only with the arrest of the Gang of Four in 1976.

Economic Reforms:
Decentralization and Liberalization

In 1979, the Deng Xiaoping undertook reform and liberalization of the economy, a critical component of its modernization program. The government tried to maintain centralized state control of the direction of policy and the distribution and pricing of strategic and energy resources, while decentralizing decision making down to the level of township and village enterprises

(TVEs). Decentralization was meant to facilitate more rational decision making, based on local conditions, needs, and efficiency. Although the state has retained the right to set overall economic priorities, TVEs (as well as larger state-owned enterprises) have been encouraged to respond to local market forces of supply and demand. Centrally determined quotas and pricing for many products have been phased out, and enterprises now contract with one another rather than the state. Thus the government's role as the go-between in commercial transactions—and as the central planner for everything in the economy—is gradually disappearing.

Today most collective and individual enterprises, instead of fulfilling centrally determined production quotas, meet contractual obligations that they negotiate with officials. After fulfilling their contracts and paying their taxes, enterprises may use any remaining profits to expand

their production facilities, improve equipment, and award bonuses. Effectively, they function as private enterprises. Some 70 percent of China's gross national product (GNP) is now produced by these nonstate enterprises. They compete with state-run enterprises to supply goods and services, and if they are not profitable, they go bankrupt.

On the other hand, state-run enterprises that operate at a loss may continue to be subsidized by the government out of fear of the destabilizing impact of a high level of unemployment. Subsidies to unprofitable enterprises consume a significant portion of the state's budget. China's membership in the World Trade Organization (WTO) as of 2001 will, however, force China's large state-owned enterprises to become efficient, or else perish in the face of foreign competition. In the meantime, the government continues to close down the most unprofitable state-run enterprises, and it

(United Nations photo/A. Holcombe)

With the highly centralized or command economy in place until the 1980s, China's manufacturing energy was focused on production, with little regard to need or markets. These women in Shanghai were producing mechanical toys for no defined market.

counts on the entrepreneurial sector and foreign investors to create the jobs to absorb the unemployed.

In addition, under a carefully managed scheme, the government has started to sell off some state-run industries to TVEs, the private sector, and foreign investors. Whoever buys them must in most cases guarantee some sort of livelihood, even if not full employment, to the former employees of the state-run enterprises. The new owners, however, have far more freedom to make a profit than did the enterprises when they were state-owned. The state is also slowly introducing state-run pension and unemployment funds to care for those workers who lose their jobs when state-owned enterprises are shut down.

The agricultural sector is now almost complete market-driven. Under the contract responsibility system, individual households to which the formerly collective lands and production tools were dis-

tributed are responsible for planning and carrying out production on their own land. The "10,000 yuan" household (about U.S. $1,200), once a measure of extraordinary wealth in China, has now become a realizable goal for many peasants. Free markets are booming in China, and peasants may now produce whatever they can get the best price for in the market, once they have filled their grain quotas to the state. Wealthy rural towns are springing up throughout the agriculturally rich and densely populated east coast, as well as along China's major transportation routes.

Problems Created by Economic Reforms

Disputes Over Private Property

The downside of privatization and marketization is a distinct increase in disputes among villagers. There are disputes over who gets to use the formerly collectively owned goods, such as tractors, harvesters, processing equipment, and boats

for transporting goods at harvest time. And with formerly collective fields turned into a patchwork of small private plots, villagers often protest when others cross their land with farm equipment. Theoretically, the land that is "leased" to the peasants is collectively owned by the village, but in practice the land is treated as the private property of the peasants. Those who choose to leave their land may contract it out to others, so some peasants have amassed large tracts of land suitable for large-scale farm machinery. To encourage development, the government has permitted land to be leased for as long as 30 years and for leased rights to be inherited. Furthermore, peasants have built houses on what they consider their own land, itself a problem because it is usually arable land. Nevertheless, the village councils also have some ability to reallocate land so that soldiers and others who settle in the villages can receive adequate land to farm.

CHINA'S SPECIAL ECONOMIC ZONES

In 1979, China opened four "Special Economic Zones" (SEZs) within the territory of the People's Republic of China as part of its program of far-reaching reform of the socialist economy. The SEZs were allowed a great deal of leeway in experimenting with new economic policies. For example, Western management methods, including the right to fire unsatisfactory workers (something unknown under the Soviet-style centrally planned economy), were introduced into SEZ factories. Laws and regulations on foreign investment were greatly eased in the SEZs in order to attract capital from abroad. Export-oriented industries were established with the goal of earning large amounts of foreign exchange in order to help China pay for the imported technology needed to hasten modernization. To many people, the SEZs looked like pockets of capitalism inside the socialist economy of the P.R.C.; indeed, they are often referred to as "mini-Hong Kongs."

The largest of the Special Economic Zones is Shenzhen, which is located just across the border from the Hong Kong New Territories. The transformation of Shenzhen over the last two decades from a sleepy little rural town to a large, modern urban center and one of China's major industrial cities has been phenomenal. The city now boasts broad avenues and China's tallest skyscrapers, and the standard of living is the highest in the country. But its air and water quality has steadily deteriorated.

But with growth and prosperity have come numerous problems. The pace of construction has gotten out of hand, outstripping the ability of the city to provide adequate services to the growing population. Speculation and corruption have been rampant, and crime is a more serious problem in

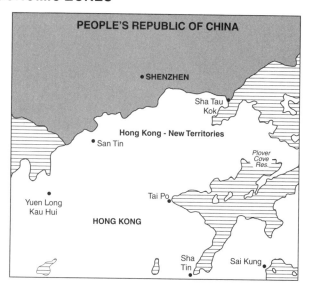

Shenzhen is the largest of China's SEZs. It is located close to the Hong Kong New Territories.

Shenzhen than it is elsewhere in China. Strict controls on immigration have been implemented to stem the flood of people who are attracted to Shenzhen in the hopes of making their fortune.

Nevertheless, the success of Shenzhen and other Special Economic Zones has led the leadership to expand the concept of SEZs throughout the country. The special privileges, such as lower taxes, that foreign businesses and joint ventures could originally enjoy only in these zones were expanded first along the coast and then to the interior, so that it too could benefit from foreign investment. Eventually, all of these special privileges will be eradicated.

Instability and Crime

With the growth of free enterprise in the rural towns since 1979, some 60 million to 100 million peasants have left the land to work for higher pay in small-scale rural industry. It is not just that the wages are higher, but also that the peasants are so heavily taxed by the government and arbitrary local taxes and fees that they cannot make a living on the land any more. And, in addition to those who leave because they seek higher wages are growing numbers of peasants whose land has been confiscated by the local government for development. Rarely compensated adequately, if at all, for their land, they are forced to leave the villages for towns and cities. Others roam the country searching for work.

For some, especially those able to find employment in the booming construction industry in many cities, this new system has meant vast personal enrichment. But tens of millions of unemployed migrants clog city streets, parks, and railroad sta-

tions. They have been joined by tens of millions of workers from bankrupt state-owned enterprises. Together, they have contributed to a vast increase in criminality and social instability. According to Chinese government reports, there were more than 58,000 "mass incidents" in 2003, more than six times the number it reported 10 years earlier.[18]

Problems Arising from a Mixed Economy

Still other problems have arisen from the post–1979 reform policies that have created a mixed socialist and capitalist economy. For example, with decentralization, industrial enterprises have tried to hide their profits but "socialize" their losses. That is, profitable enterprises (whether state-owned, collective TVEs, or private) try to hide their profits to avoid paying taxes to the state. State-owned enterprises that are losing money ask the state for sub-

sidies to keep them in business. In spite of the dramatic increase in the value of industrial output since 1979, therefore, the amount of profits turned over to the state has actually declined.

Another problem resulting from China's mixed economy is that some enterprises withhold materials normally allocated by the state, such as rolled steel, glass, cement, and timber. They either hoard them as a safeguard against future shortages, or resell them at unauthorized higher prices. By doing so, these enterprises make illegal profits for themselves and deprive other enterprises of the materials they need for production. Thus, with the state controlling the pricing and allocation of some resources, and the free market determining the rest, there are many opportunities for corruption and abuse of the system.

The needs of the centralized state economy remain in tension with the interests of provinces, counties, towns, and individuals, most of which now operate under the dual rules of an economy controlled by both the market and the state. Thus, even as enterprises are relying on market signals to determine whether they will expand production facilities, the state continues to centrally allocate resources based on a national plan. A clothing factory that expands it production, for instance, requires more energy (coal, oil, water) and more cotton. The state, already faced with inadequate energy resources to keep most industries operating at even 70 percent capacity, continues to allocate the same amount to the now-expanded factory. Profitable enterprises want a greater share of centrally allocated scarce resources, but find they cannot acquire them without the help of "middlemen" and a significant amount of under-the-table dealing. Corruption has, therefore, become rampant at the nexus where the capitalist and socialist economies meet.

Widespread corruption in the economic sector has led the Chinese government to wage a series of campaigns against economic crimes such as embezzlement and graft. An increasing number of economic criminals are going to prison, and serious offenders are sometimes executed. Until energy and transportation bottlenecks and the scarcity of key resources are dealt with, however, it will be extremely difficult to halt the bribery, smuggling, embezzlement, and extortion now pervasive in China. The combination of relaxed centralized control, the mandate for the Chinese people to "get rich," and a mixed economy have exacerbated what was already a problem under the socialist system. In a system suffering from serious scarcities but controlled by bureaucrats, it is political power, not the market, that determines who gets what. This includes not only goods, but also opportunities, licenses, permits, and approvals.

Although the Chinese may now purchase in the market many essential products previously distributed through bureaucratically controlled channels, there are still many goods available only through the "back door"—that is, through people they know and with whom they exchange favors or money. Scarcity, combined with bureaucratic control, has led to "collective corruption": Individuals engage in corrupt practices, even cheat the state, in order to benefit the enterprise for which they work. Since nonstate-owned "collectives" and private enterprises will not be bailed out by the state if they suffer losses, the motivation for corrupt activities is stronger than under the previous system.

Liberalization of the economy is providing a massive number and variety of goods for the marketplace. The Chinese people may buy almost any basic consumer goods in stores or the open markets. But the nexus

between continued state control and the free economy still fuels a rampant corruption that threatens the development of China's economy. The ability of corrupt individuals to abscond to a foreign country has made capital flight an increasingly worrisome issue. China's Ministry of Commerce reported that, since 1984, thousands of corrupt officials (many of whom are said to be the children and relatives of political leaders) have illegally transferred a total of US$5 billion to companies registered in offshore tax shelters and financial havens.[19]

Unequal Benefits

Not all Chinese have benefited equally from the last 20 years of economic reforms. Those who inhabit cities in the interior, and peasants living far from cities and transportation lines or tilling less arable land, have reaped far fewer rewards. And although the vast majority of Chinese have seen some improvement in their standard of living, at the same time short-term gains in income are threatened in the long term by the deterioration of the infrastructure for education and medical care in large parts of China's hinterland. This deterioration is due to the elimination of the commune as the underlying institution for funding these services. Wealthier peasants send their children into the larger towns and cities for schooling, and their family members travel to the more comprehensive health clinics and hospitals farther away. In some areas, however, the wealthier peasants have actually built local schools and private hospitals. Furthermore, in a remarkable move away from dependency on the state for benefits, Chinese citizens now support charitable organizations, some of which have as their purpose improving education and health care for children.

DEVELOPMENT

China has become an active participant in the international economic system. China has a large surplus in its trade relationships with most countries, and produces a large variety of consumer and light industrial products. The world's largest dam has been built on the Yangtze River. In 2002, China began a massive transfer of water from the south to the north to supply the needs of the dry northern plains, including Beijing. Rapid development has raised serious questions concerning environmental sustainability.

In the cities, employees of state-run enterprises suffer from fixed state salaries that are barely adequate to buy goods whose prices are no longer state-controlled. By the mid-1990s, however, the state managed to bring inflation under con-

trol. Inflation is now at a very low level in China, but many urban workers feel they must have two jobs in order to make ends meet. In such an environment, there is also a greater temptation to engage in corruption in order to lead a better life. Without an adequate system of regulations and laws to replace the state's regulatory system, white-collar crime is surging.

The visible polarization of wealth, which had been virtually eradicated in the first 30 years of Communist rule, has returned to China with the return of the free market. The creation of a crassly ostentatious wealthy class and a simply ostentatious middle class, in the context of high unemployment, poverty, and a mobile population, is breeding the very class conflict that the Chinese Communists fought a revolution to eliminate. When reforms began some 20 years ago, street crime was almost unheard of in China's cities. Now it is a serious problem, one that the government fears may erode the legitimacy of CCP rule if left uncontrolled.

Mortgaging the Future

One of the most damaging aspects of the capitalist "get-rich" atmosphere prevailing in China is the willingness to sacrifice the future for short-term profits. The environment has been rapidly degraded by uncontrolled pollution; the rampant growth of new towns, cities, and highways; the building of houses on arable land; the overuse of nonrenewable water resources; and the destruction of forests. Some state institutions have turned the open spaces within their walls into parking lots in China's crowded cities to provide spaces for the rapidly growing number of privately owned cars. For example, some middle schools have turned playgrounds and basketball courts into parking lots, and have used state funds allocated for education to build commercial shops and offices along their outside walls. Teachers and administrators deal themselves the profits; but in the meantime, classroom materials and facilities are deteriorating.

Nevertheless, the tensions generated among the central state, the provinces, the cities, the towns, and enterprises have been generally beneficial to economic growth. Greater economic autonomy has also led to growth in political autonomy. Thus, provinces, especially those that are producing significant revenues, can now challenge central economic policy, and even refuse to carry it out. For example, the wealthier east-coast provinces have successfully challenged the central government's right to collect more taxes on provincial revenues. Similarly, cities now have greater au-

Settling civil disputes in China today still usually involves neighborhood mediation committees, family members, or friends.

tonomy vis-à-vis the formerly all-encompassing power of the provinces in which they are located. In short, this sort of economic power, won through the free market, has brought with it the political power necessary to challenge China's leaders. Indeed, some commentators wonder whether the unwillingness of the provinces to dutifully obey Beijing will lead to the breakup of China in the twenty-first century. In this sense also, the Chinese must be wondering whether they are mortgaging the future for short-term gains.

THE CHINESE LEGAL SYSTEM

Before legal reforms began in 1979, the Chinese cultural context and the socialist system were the primary factors shaping the Chinese legal system. Since 1979, reforms have brought about a remarkable transformation in Chinese attitudes toward the law, with the result that China's laws and legal procedures—if not practices and implementation—look increasingly like those in the West. This is particularly true for laws that relate to the economy, including contract, investment, property, and commercial laws. China's legal system has evolved to accommodate its increasingly market-oriented economy and privatization. Chinese citizens have discovered that the legal system can protect their rights, especially in economic transactions. In 2004, 4.4 million civil cases were filed, double the number 10 years earlier. This is a strong indication that the Chinese now believe they can use the law to protect their rights, and to hold others (including officials) accountable for their actions.[20] Criminal law and procedure have also undergone reforms. In civil law (such as disputes with neighbors and family members), however, the Chinese are still more inclined to rely on mediation and traditional cultural values in the settling of disputes.

Ethical Basis of Law

In imperial China, the Confucian system provided the basis for the traditional social and political order. Confucianism posited that good governance should be based on maintaining correct personal relationships among people, and between people and their rulers. Ethics were based on these relationships. A legal system did exist, but the Chinese resorted to it in civil cases only in desperation, for the inability to resolve one's problems oneself or through a mediator usually resulted in considerable "loss of face"—a loss of dignity or standing in the eyes of others. (In criminal cases, the state normally became involved in determining guilt and punishment.)

This perspective on law carried over into the period of Communist rule. Until legal reforms began in 1979, most Chinese preferred to call in CCP officials, local neighborhood or factory mediation committees, family members, or friends to settle disputes. Lawyers were rarely used, and only when mediation failed did the Chinese resort to the courts. By contrast, the West lacks both this strong support for the institution of mediation and the concept of "face." So Westerners have difficulty understanding why China has had so few lawyers, and why the Chinese have relied less on the law than on personal relationships when problems arise.

Like Confucianism, Marxism-Leninism is an ideology that embodies a set of ethical standards for behavior. After 1949, it easily built on China's cultural predisposition toward ruling by ethics instead of law. Although Marxism-Leninism did not completely replace the Confucian ethical system, it did establish new standards of behavior based on socialist morality. These ethical standards were embodied in the works of Marx, Lenin, and Mao, and in the Chinese Communist Party's policies; but in practice, they were frequently undercut by preferential treatment of party officials. Today, socialist ethics are gradually giving way to more universal ethics, particularly the principle of equality before the law.

Law and Politics

From 1949 until the legal reforms that began in 1979, Chinese universities trained few lawyers. Legal training consisted of learning law and politics as an integrated whole; for according to Marxism, law is meant to reflect the values of the "ruling class" and to serve as an instrument of "class struggle." The Chinese Communist regime viewed law as a branch of the social sciences, not as a professional field of study. For this reason, China's citizens tended to view law as a mere propaganda tool, not as a means for protecting their rights. They had never really experienced a law-based society. Not only were China's laws and legal education highly politicized, but politics also pervaded the judicial system.

With few lawyers available, few legally trained judges or prosecutors in the courts, and even fewer laws to refer to for standards of behavior, inevitably China's legal system has been subject to abuse. China has been ruled by people, not by law; by politics, not by legal standards; and by party policy, not by a constitution. Interference in the judicial process by party and local state officials has been all too common.

After 1979, the government moved quickly to write new laws. Fewer than 300 lawyers, most of them trained before 1949 in Western legal institutions, undertook the immense task of writing a civil code, a criminal code, contract law, economic law, law governing foreign investment in China, tax law, and environmental and forestry laws. One strong motivation for the Chinese Communist leadership to formalize the legal system was its growing realization, after years of a disappointingly low level of foreign investment, that the international business community was reluctant to invest further in China without substantial legal guarantees.

Even China's own potential entrepreneurs wanted legal protection against the *state* before they would assume the risks of developing new businesses. Enterprises, for example, wanted a legal guarantee that if the state should fail to supply resources contracted for, it could be sued for losses issuing from its nonfulfillment of contractual obligations. Since the leadership wanted to encourage investment, it has had to supplement economic reforms with legal reforms. Codification of the legal system has fostered a stronger basis for modernization and has helped limit party/state abuse of the people's rights.

In addition, new qualifications have been established for all judicial personnel. Demobilized military officers who became judges and prosecutors during the Cultural Revolution can no longer serve in the judiciary. Judges, prosecutors, and lawyers must now have formal legal training and pass a national judicial examination. It is hoped that requiring judicial personnel to have higher qualifications, paying them more, and making judicial systems financially autonomous from local governments will diminish judicial corruption and enhance the autonomy of judicial decisions.[21]

Criminal Law

Procedures followed in Chinese criminal courts have differed significantly from those in the United States. Although the concept of "innocent until proven guilty" was introduced in China in 1996, it is still presumed that people brought to trial in criminal cases were guilty. This presumption is confirmed by the judicial process itself. That is, after a suspect is arrested by the police, the procuratorate (*procuracy*, the investigative branch of the judiciary system) will spend considerable time and effort examining the evidence gathered by the police and establishing whether the suspect is indeed guilty. This is important to understand when assessing the fact that 99 percent of all the accused brought to trial in China are judged guilty. Theoretically, had the facts not substantiated their guilt, the procuracy would have dismissed their cases before going to trial.

Ideally, then, those adjudged to be innocent are never brought to trial in the first place. For this reason, court trials function mainly to present the evidence upon which the guilty verdict is based—not to weigh the evidence to see if it indicates guilt—and to remind the public that criminals are punished. A trial is a "morality play" of sorts: the villain is punished, justice is done, the people's interests are protected. In addition, the trial process continues to emphasize the importance of confessing one's crimes, for those who confess and

appear repentant in court will usually be dealt more lenient sentences. Criminals are encouraged to turn themselves in, on the promise that their punishment will be less severe than if they are caught. Those accused of crimes are encouraged to confess rather than deny their guilt or appeal to the next level, all in hopes of gaining a more lenient sentence from the judge. This is a system weighted against the innocent.

In short, the system tends to focus on confession, not on fact finding. The result is that police are more inclined to use brutality in order to exact a confession; but of course, police in Western liberal-democratic countries also have been known to use brutality, and even torture, to get a confession from the accused.

Another serious problem with the Chinese system was that the procuracy, which investigated the case, also prosecuted the case. Once the procuracy established "the facts," they were not open to question by the lawyer or representative of the accused. (In China, a person may be represented by a family member, friend, or colleague, largely because there are not enough lawyers to fulfill the guarantee of a person's "right to a defense.") The lawyer for the accused was not allowed to introduce new evidence, make arguments to dismiss the case based on technicalities or improper procedures (such as wire tapping), call witnesses for the defendant, or make insanity pleas for the client Instead, the lawyer's role in a criminal case was simply to represent the person in court and to bargain with the court for a reduced sentence for the repentant client. The 1996 legal reforms were aimed at improving the rights of the accused: They may now call their own witnesses and introduce their own evidence, they cannot be held for more than 30 days without being formally charged with a crime, and they are supposed to have access to a lawyer within several days of being formally arrested. But many suspects are still not accorded these rights.

The accused have the right to a defense, but it has always been presumed that a lawyer will not defend someone who is guilty. Most of China's lawyers are still employed and paid by the state. As such, a lawyer's obligation is first and foremost to protect the state's interests, not the individual's interests at the expense of the state. When lawyers have done otherwise, they have risked being condemned as "counterrevolutionaries" or treasonous. Small wonder that after 1949, the study of law did not attract China's most talented students.

Today, however, the law profession is seen as potentially lucrative and increasingly divorced from politics. Lawyers can now enter private practice or work for for-

CENTRAL GOVERNMENT ORGANIZATION OF THE PEOPLE'S REPUBLIC OF CHINA

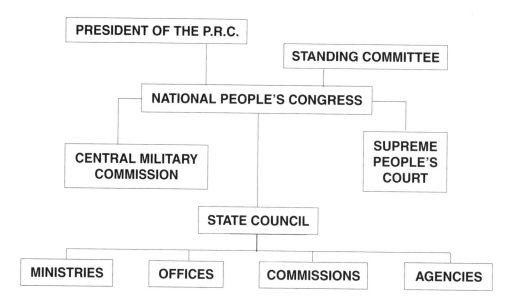

This central government organization chart represents the structure of the government of the People's Republic of China as it appears on paper. However, since all of the actions and overall doctrine of the central government must be reviewed and approved by the Chinese Communist Party, political power ultimately lies with the party. To ensure this control, virtually all top state positions are held by party members.

THE CHINESE COMMUNIST PARTY (CCP)

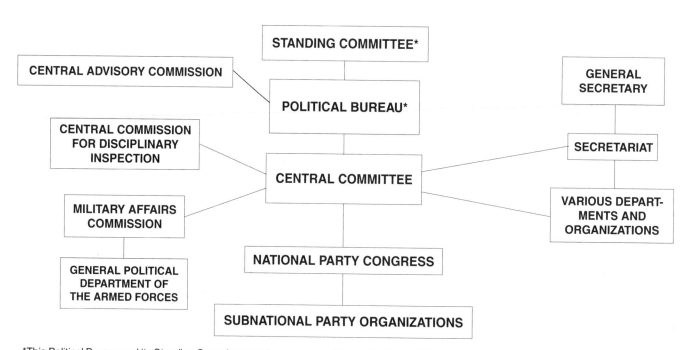

*This Political Bureau and its Standing Committee are the most powerful organizations within the Chinese Communist Party and are therefore the real centers of power in the P.R.C.

eigners in a joint venture. The All-China Lawyers Association, established in 1995 by the Ministry of Justice to regulate the legal profession, also functions as an interest group to protect the rights of lawyers against the state. Thus, when in recent years lawyers have found themselves in trouble with the law because they have defended the political rights of their clients against the state, the association tries to protect them.

Lawyers in Civil and Commercial Law

In the areas of civil and commercial law, the role of the lawyer has become increasingly important since the opening of China's closed door to the outside world. Because foreign trade and investment have become crucial to China's development, its goal is to train at least one lawyer for every state, collective, or private organization and enterprise. Chinese judicial personnel are learning that in today's China, upholding the law is no longer simply a matter of correctly understanding the party "line" and then following it in legal disputes.

China's limited experience in dealing with economic, liability, corporate, and contractual disputes in the courts and the insistence by foreign investors that Chinese courts be prepared to address such issues have forced the leadership to train lawyers in the intricacies of Western law and to draft literally thousands of new laws and regulations. To protect themselves against what is difficult to understand in the abstract, the Chinese used to refuse to publish their newly written laws. Claiming a shortage of paper or the need to protect "state secrets," they withheld publication of many laws until their actual impact on China's state interests could be determined. This practice frustrated potential investors, who dared not risk capital investment in China until they knew exactly what the relevant laws were. The rising complexity of foreign-investment issues as well as the entrepreneurial activities of their own citizens have led the Chinese government to publish most of its laws as quickly as possible.

THE POLITICAL SYSTEM

The Party and the State

In China, the Chinese Communist Party is the fountainhead of power and policy. But not all Chinese people are party members. Although the CCP has 60 million members, this number represents less than 5 percent of the population. Joining the CCP is a competitive, selective, rigorous process. Some have wanted to join out of a commitment to Communist ideals, others in hopes of climbing the ladder of success,

still others to gain access to limited goods and opportunities. Ordinary Chinese are generally suspicious of the motives of those who do join the party. By the late 1980s, so many well-educated Chinese had grown cynical about the CCP that they refused to join. Still, those who travel to China today are likely to find that many of the most talented people they meet are party members. Party hacks who are ideologically adept but incompetent at their work are gradually being squeezed out of a party desperate to maintain its leading position in a reforming China.

Today's CCP wants the best and brightest of the land as members, and in 2001, Jiang Zemin, then general-secretary of the CCP, announced a new "theory of the three represents": The Party represents not just the workers and peasants, but *all* the people's interests, including both intellectuals and capitalists. Not only does this addition to party theory indicate that the Chinese Communist Party recognizes the important role that intellectuals and business people have played—and will play—in modernizing China; but it also acknowledges the reality that many individuals already within the party have become capitalists. By enshrining this theory in the party constitution at the 16th Party Congress in 2002, the party was in effect attempting to shore up its legitimacy as China's ruling party. At the same time, it was in effect announcing that it had relinquished its role as a revolutionary party in favor of its new position as the *governing* party of China.

The CCP is still China's ultimate institutional authority. Although in theory the state is distinct from the party, in practice the two overlapped almost completely from the late 1950s to the early 1990s. Efforts by the reformers to get the party out of the day-to-day work of the government and out of the management of economic enterprises have had some effect; but in recent years, more conservative leaders within the CCP have exerted considerable pressure to keep the party in charge.

The state apparatus consists of the State Council, headed by the premier. Under the State Council are the ministries and agencies and "people's congresses" responsible for the formulation of policy. The CCP has, however, exercised firm control over these state bodies through interlocking organizations. For example, CCP branches exist within all government organizations, and all key state personnel are also party members. At every administrative level from the central government in Beijing down to the villages, almost everyone in a leadership position is also a party member.

China's political system is subject to enormous abuses of power. The lines of au-

thority within both the CCP and the state system are poorly defined, as are the rules for success to the top leadership positions. In the past, this allowed individuals like Mao Zedong and the Gang of Four to usurp power and rule arbitrarily. By the late 1980s, China's bureaucracy appeared to have become more corrupt than at any time since 1949. Anger at the massive scale of official corruption was, in fact, the major factor unifying ordinary citizens and workers with students during the antigovernment protests in the spring of 1989.

Although campaigns to control official corruption continue, the problem appears to be growing even worse. Campaigns do little more than scratch the surface. This is in part because with so many opportunities to make money in China, especially for party and state officials whose positions give them the inside track for making profitable deals, the potential payoff for corruption can be huge—and the risks of getting caught appear small. Chinese institutions lack the transparency, acquired through financial checks within the system and open access to accounting books that could help rein in corruption. The situation is exacerbated by a society that by its complicity encourages official corruption.

Individuals may write letters to the editors of the country's daily newspapers or to television stations to expose corruption. Many Chinese, especially those living in the countryside, feel that the only way in which local corruption will be addressed is if the media send reporters to investigate and publicly expose criminality. The China Central Television station has a popular daily program in prime time that records the successes of China's public-security system in cracking down on official corruption and crime. The press also devotes substantial space to sensational cases of official corruption, in part because it helps sell newspapers. The media only has the resources, however, to address comparatively few of the numerous cases begging for investigation.

So far, most efforts to control official corruption have had little effect. Officials continue to use their power to achieve personal gain, trading official favors for others' services (such as better housing, jobs for their children, admission to the right schools, and access to goods in short supply). Getting things done in a system that requires layers of bureaucratic approval still depends heavily upon a complex set of personal connections and relationships, all reinforced through under-the-table gift giving. This stems in part from the still heavily centralized aspect of Chinese governance, and in part from the overstaffing of a bureaucracy that is plagued by red

Although the Chinese Communist Party (shown here at its 11th National Congress) has hand-picked one candidate for each public office in the past, it is now allowing more than one person to run for the same post.

tape. Countless offices must sign off on requests for anything from installing a telephone to processing a request for a license or additional electrical outlets. This gives enormous power to individual officials who handle those requests, allowing them to ask for favors in return, or to stonewall if the payoff is inadequate.

In today's more market-oriented China, officials have lost some of their leverage over the distribution of goods in short supply. Now, instead of waiting for a work-unit official to decide whose turn it is to purchase a bicycle, anyone with adequate funds may buy. But because all electrical service is controlled by the government, people who purchase something like an air conditioner must then pay off an official to allow the electrical service to their living units to be upgraded so they can actually use it. Similarly, brothels can be run in the open, virtually without interference from the police, who are bribed to look the other way. In short, officials may have lost control over the distribution of many consumer goods, but they have kept their ability to facilitate or obstruct access or the provision of many services.

Controlling the abuse of official privilege is difficult in part because of the large discretionary budgets that officials have,

and in part because the Chinese have made an art form out of going around regulations. For example, the government issued a regulation stipulating that governmental officials doing business could order only four dishes and one soup at a meal. But as most Chinese like to eat well, especially at the government's expense, the restaurants accommodated them by simply giving them much larger plates on which they put many different dishes, and write it up as if it were just one dish!

The definition of what is corrupt behavior has also become more complex as the country moves from a socialist economy to a market economy. For example, in the initial stages of introducing a market economy, importing goods and then selling them in China for high profits was considered corrupt, as was paying middlemen to arrange business transactions. In the 1980s, some businesspeople were arrested, and even executed, for such behavior. Now, the important role of middlemen is recognized, and government regulations allow a middleman to keep 5 percent of the total value of the transaction as a "fee." The Chinese have also adopted the custom in other countries of permitting tour guides who take tourists to a shop to receive a percent-

age of the total sales, behavior earlier considered corrupt.

Reform of the party/state has been an important avenue through which the government has attempted to curb corruption; but its overall goal in reforming the party/state has been to improve the quality of China's leadership. Reforms have encouraged, if not demanded, that the Chinese state bureaucracy reward merit more than mere seniority, and expertise more than political activism. And in 1996, the government's practice of allowing officials to stay in one ministry during their entire career was replaced by new regulations requiring officials from divisional chiefs up to ministers and provincial governors to be rotated every five years. Restrictions on tenure in office have brought a much younger generation of leaders into power; and they have placed a time limit on any one individual's access to power, including the "back door" to wealth. In addition, no high official may work in the same office as his or her spouse or direct blood relative. It is hoped that such regulations will curtail the building of power bases that undercut effective leadership and foster corruption within both the party and the state. Finally, the emphasis on a collective leadership since reforms began in 1979 has made it virtually impossi-

Villagers wait patiently to cast their votes during the September 2001 election in a small village outside Shanghai.

ble for a leader to develop a personality cult, such as that which reached fanatical proportions around CCP chairman Mao Zedong during the Cultural Revolution.

ENVIRONMENT FOR DEMOCRACY

The Chinese political system reflects a history, political culture, and values entirely different from those in the West. For millennia, Chinese thought has run along different lines, with far less emphasis on such ideals as individual rights, privacy, and limits on state power. The Chinese political tradition is weighed down with a preference for authoritarian values, such as obedience and subordination of individuals to their superiors in a rigidly hierarchical system, and a belief in the importance of moral indoctrination. China's rulers have throughout history shown greater concerns for establishing their authority and maintaining unity in the vast territory and population they control than in protecting individual rights. Apart from China's intellectuals in the twentieth century, the overwhelming majority of the Chinese people have appeared to be more afraid of chaos than rule by an authoritarian despot. As a result, even today the Chinese people seem

more concerned that their rulers have *enough* power to control China than that the rights of citizens vis-à-vis their rulers be protected.

Cultural and Historical Authoritarianism

This is not to suggest that Confucianism and China's other traditions did not consider such rights. They did; but the *dominant* strand of Chinese political culture, which survived into the late twentieth century, was authoritarian. It was critical in shaping the development of today's political system. As a result, when the Communists came to power in 1949, they were trying to operate within the context of an inherited patriarchal culture, in which the hierarchical values of superior–inferior and subordination, loyalty, and obedience prevailed over those of equality; and in which there was a historical predisposition toward official secrecy; a fear of officials and official power; and a traditional repugnance for courts, lawyers, and formal laws that protected individual rights. Thus, when Western democratic values and institutions were introduced, China's political culture and institutions were ill prepared to embrace them.

China's limited experience with democracy in the twentieth century was bitter. Virtually the entire period from the fall of China's imperial monarchy in 1911 to the Communist victory in 1949 was marred by warlordism, chaos, war, and a government masking brutality, greed, and incompetence under the label of "democracy." Although it is hardly fair to blame this period of societal collapse and externally imposed war on China's efforts to practice "democracy" under the "tutelage of the Kuomintang," the Chinese people's experience of democracy was nevertheless negative.

Moreover, the Chinese Communists believed that Western liberal democracy had been too weak a political system to prevent the Great Depression and two world wars in the twentieth century. And, in any event, foreign values were always suspect in China, as foreigners had repeatedly declared war on China in order to advance their own national interests. China was inclined to view the propagation of liberal democratic values as just one more effort by Western countries to enhance their own national power.

China's experience with the Western powers and their own efforts to implement democracy before 1949, together with China's traditional political culture, help explain the people's reluctance to embrace Western liberal democratic ideals. During the period of the Republic of China (1912–1949), China's democratic political institutions (in form) and legal system (on paper) proved inadequate to guarantee the nation's welfare, much less the protection of individual rights. But even under Communist rule, the one period described as "democratic mass rule" (the "10 bad years" or Cultural Revolution, from 1966 to 1976) was in fact a period of mass tyranny. For the Chinese, the experience of relinquishing power to "the masses" turned into the most horrific period of unleashed terrorism and cruelty they had experienced since the Communist takeover in 1949.[22]

Socialist Democracy

When the CCP came to power in 1949, it inherited a country torn by civil war, internal rebellion, and foreign invasions for more than 100 years. The population was overwhelmingly illiterate and desperately poor, the economy in shambles. The most urgent need was for order. Despite some serious setbacks and mistakes under the leadership of Mao Zedong, China made great strides in securing its borders, establishing institutions of government, and enhancing the material well-being of its people. But in the name of order and stability, China's leaders also severely limited the development of "democracy" in its Western liberal-democratic sense.

FREEDOM

China has made significant progress in opening up the media and allowing pluralization of interests and ideas. Today, there are more than 200,000 associations and interest groups, and elections of leaders in rural villages and urban districts.

The Chinese people are accustomed to "eating bitterness," not to standing up to authority. The traditional Confucian emphases on the group rather than the individual and on respect for authority continue to this day, although they are now being undercut by the effects of modernization, internationalization, and disenchantment with the CCP leadership. Rapid modernization has similarly undermined Marxist-Leninist-Maoist values, the glue that, along with traditional values, has helped hold China together. None of this, however, necessarily bodes well for the propagation of democratic values; for the destabilizing social effects of a loss of values, and the concomitant rise of nationalism and aggressive materialism, hardly provide a receptive environment for liberal democratic values.

Nevertheless, since 1949, and especially since the reform period began in 1979, there has been a gradual accretion of individual rights for Chinese people, including far greater freedom of speech; access to more information and a diversity of perspectives; the right to vote in local elections; and the development of the rights to privacy, to choose one's own work (as opposed to being assigned by the government), to move, to private property, and many more. Moreover, the impersonal market forces of supply and demand have undercut the power of officials to control the distribution of resources and opportunities in the society, with the result that the Chinese people now have greater equality of access to material goods. They no longer need to be as beholden to officials as in the past. As a result, they normally do not have to cultivate relationships with officials and others by gift giving, banqueting, and outright bribery to acquire goods. This equality of access in the marketplace contributes to a greater sense of control by the people over their daily lives.

Unfortunately, even as such rights grow, other very important rights previously enjoyed by the Chinese people—such as a job, health care, and education—are being eroded. Such "welfare rights" have provided the context in which other rights have gained meaning. At the same time that it has led to greater political liberalization, economic liberalization has contributed to the polarization and destabilization of Chinese society.

Patterns of political participation are also changing. Participation in the political process at the local level is already leading to greater official responsiveness to the common people's needs. Some village officials eagerly seek out advice for improving the economic conditions in their localities, and now incompetent officials are usually unable to gain reelection. In spite of the election laws and other efforts to advance village democracy, however, villagers in many parts of China do not believe that elections have made much of a difference. Even in those areas where there has been extraordinary economic development, villagers do not necessarily see the connection of elections and democratization with prosperity.[23]

China has experienced only limited open popular demand for democracy. When the student-led demonstrations in Beijing began in 1989, the demands for democratic reforms were confined largely to the small realm of the intellectual elite—that is, students and well-educated individuals, as well as some members of the political and economic ruling elite. The workers and farmers of China remained more concerned about bread-and-butter issues—inflation, economic growth, unemployment, and their own personal enrichment—not democratic ideals.

By the mid-1990s, many Chinese had discovered that they could get what they wanted through channels other than mass demonstrations, because of the development of numerous alternative groups, institutions, and processes. Many of these groups are not political in origin, but the process by which they are pressing for policy changes in the government is highly political. Literally hundreds of thousands of interest groups and associations have sprung up over the past decade. While many of these interest groups are organized and controlled by the state (such as the Women's Association and the Communist Party Youth League), others are not. It is perhaps ironic that the Chinese Communist Party's penchant for organizing people resulted in teaching them organizational skills that they now use to pressure the government to change policy.[24] They work through these organizations to protect and advance their members' interests within the framework of existing law and regulations.

Even those Chinese not working through officially organized associations ban together to petition local officials using the organizational techniques they learned from the CCP. For example, urban neighborhoods join forces to stop local noise pollution emanating from stereos blasting on the street where thousands of Chinese couples learn ballroom dancing, or from the cymbals, tambourines, gongs, and drums of old ladies doing "fan dancing" on the city streets.[25] In turn, the dancers petition the local officials to maintain their "right" to express themselves through dance in the streets, some of the few public spaces available to them in a crowded urban environment.

The tendency to organize around issues and interests in China today is more than a reflection of the decline of the role of Communist ideology in shaping policy. It also reflects the government's focus on problem solving and pragmatic concerns in policymaking. Many constituencies take advantage of this approach to issues and policy. Although China has never been homogeneous and uncomplicated, it is certainly a far more complex economy, society, and polity today than it was before 1980. Today's China has far more diverse needs and interests than was the case previously, and specialized associations and interest groups serve the need of articulating these interests.

Compared to Western liberal democratic systems, Chinese citizens are politically passive. But is this a sign of satisfaction with the CCP regime? An indication that the Chinese are a "nation of sheep"? The outcome of their fear of the repercussions of being critical of the regime? Is Chinese submissiveness a sign of "collusion" with their oppressors? One could argue that, like Eastern Europeans, the Chinese have participated in their own political oppression simply by complying with the demands of the system. As the Czech Republic's then-president Vaclav Havel stated, "All of us have become accustomed to the totalitarian system, accepted it as an unalterable fact and therefore kept it running.... None of us is merely a victim of it, because all of us helped to create it together.[26]

Can we say that the Chinese, any more than the Czechs, collaborated with their authoritarian rulers if they did not flee into exile under Communist rule, or did not refuse to work? Is anyone who does not actively revolt against an oppressive system necessarily in collusion with it? In the case of China, one cannot assume that the major reason why people are not challenging the Communist system is out of fear of punitive consequences—with the exception of those who want to directly challenge CCP rule through the formation of new political parties or trade unions, or through the explicit public criticism of China's top political leaders. The more logical explanation is that the Chinese have developed a completely different style from that of citizens in Western liberal democratic countries for getting what they want. This style is largely based on cultivating personal relationships rather than more formal and open institutionalized forms of political participation. More "democratic" behavioral skills can, of course, be acquired through practice; and as the Chinese political system gradually adopts more liberal democratic practices, the political culture is likely to evolve—indeed, it *is* evolving—in a more democratic direction.

Democratization in China has been hampered by the people's inability to envision an alternative to CCP rule. What form would it take? How would it get organized? And if the CCP were overthrown, who would lead a new system? These questions are still far from being answered. So far, no dissident leadership capable of offering an alternative to CCP leadership, and laying claim to popular support, has formed. Even the mass demonstrations in 1989 were not led by either a worker, a peasant, or even an intellectual with whom the common people could identify:

[C]ompared with the intellectuals of Poland and Czechoslovakia, for example, Chinese intellectuals have little contact with workers and peasants and are not sensitive to their country's worsening social crisis; they were caught unawares by the democratic upsurge of 1989, and proved unable to provide the people with either the theoretical or practical guidance they needed.[27]

In fact, during the Tiananmen protests in 1989, students were actually annoyed by the workers' participation in the demonstrations. They wanted to press their own political demands—not the workers' more concrete, work-related issues. Some Chinese believe that the students' real interest in demanding respect for their own goals from China's leadership was because they wanted to enhance their own prestige and their power vis-à-vis the regime: The students' major demands were for a "dialogue" with the government as "equals," and for free speech. These issues were of primary interest to them, but of little interest to the workers of China.

Many Chinese believe that had the leaders of the 1989 demonstrations suddenly been catapulted to power, their behavior would have differed little from the ruling CCP elite. The student movement itself admitted to being authoritarian, of kowtowing to its own leaders, and of expecting others to obey rather than to discuss decisions. As one Beijing University student wrote during the 1989 Tiananmen Square protests:

The autonomous student unions have gradually cut themselves off from many students and become a machine kept constantly on the run in issuing orders. No set of organizational rules widely accepted by the students has emerged, and the democratic mechanism is even more vague.[28]

In any event, few of those who participated in the demonstrations in 1989 are interested in politics or political leadership today. Most have thrown themselves into business and making money.

Apart from students and intellectuals, some of the major proponents of democratic reform today hail from China's newly emerging business circles; but these groups have not united to achieve reform, as they neither like nor trust each other. Intellectuals view venture capitalists "as uncultured, and business people as driven only by crass material interests." The latter in turn regard intellectuals and students as "well-meaning but out of touch with reality and always all too willing and eager to serve the state" when it suits their needs.[29] Moreover, although the business community is interested in pushing such rights as the protection of private property and the strengthening of the legal system in the commercial and economic spheres, it tends to be more supportive of the regime's "law and order" values than of democratic values; for an unstable social and political environment would not be conducive to economic growth.

Those dissidents who left China and remain abroad have lost their political influence. Apart from everything else, it is difficult for dissidents abroad to make themselves heard in China, even if their articles are published or appear on the Web. Although they may keep in touch with the dissident movement in China, their influence is largely limited to their ability to supply it with funds. Doing so, however, often gets the recipients in China in trouble, and may even lead to lengthy prisons sentences.

The Impact of Global Interdependency on Democratization

Since the late 1970s, the cultural context for democracy in China has shifted. The expansion of the international capitalist economy and increasing cultural and political globalization have led to a social and economic transformation of China. For the first time in Chinese history, a significant challenge to the "we–they" dichotomy—of China on the one hand, against the rest of the world on the other—is occurring. This in turn has led many Chinese to question the heretofore assumed superiority of Chinese civilization.

Such an idea does not come easily for a people long-accustomed to hearing about China's greatness. Hence the fuss caused by *River Elegy*, a TV series first shown on Chinese national television in 1988. In this series, the film producers argued that the Chinese people must embrace the idea of global interdependency—technological, economic, and cultural. To insist at that time in history on the superiority of Chinese civilization, with the isolation of China from the world of ideas that this implied, would only contribute to China's continued stagnation. The series suggested that the Chinese must see themselves as equal, not superior to, others; and as interdependent with, not as victims of, others. Such concepts of equality, and the opening up of China to ideas from the rest of the world, have led to a remarkable transformation of the Chinese political, cultural, and economic landscape.

STOCK MARKETS, GAMBLING, AND LOTTERIES

China has had two stock markets since just before the Tiananmen Square demonstrations of 1989. One is in the Special Economic Zone (SEZ) of Shenzhen; the other is in Shanghai. With only seven industries originally registered on them, strict rules about how much daily profit or loss (1.2 percent for the Shanghai exchange until July 1992) a stock could undergo, and deep public suspicion that these original issues of stocks were worthless, these markets got off to a slow start. But when these same stocks were worth five times their original value just a year later, the public took notice. Rumors—as important in China as actual news—took over and exaggerated the likelihood of success in picking a winning stock. The idea of investors actually losing money, much less a stock-market crash, did not seem to be an idea whose time had come.

Soon there were so many Chinese dollars chasing so few stocks that the government began a lottery system: Anyone who wanted to buy a stock had first to buy a coupon, which was then put into a national lottery. The supply/demand ratio for stocks was so out of proportion that an individual had only a one in 100 chance of having a coupon chosen from the lottery. The coupon would, in turn, enable its bearer to buy a mere five shares of a stock that might or might not make a profit. Today, thanks to the rapid increase in stocks registered on the two stock exchanges, there is a 70 in 100 chance of getting the right to buy a stock.

As China is still largely a cash economy, making a stock-market transaction does not resemble what happens in a Western country. Instead of simply telephoning a stockbroker and giving an order, with a simple bank transfer or check to follow shortly, most Chinese must still appear in person, stand in line, and pay cash on the spot. Taiwan has added its own angle to the stock-mania by selling to the Mainlanders small radios that are tuned in to only one frequency—stock-market news.

Issuing, buying, and selling stocks has become a near-national obsession. Not only do ordinary companies selling commercial goods, such as computers and clothing, issue stocks.

So do taxicab companies and even universities. Thus far, few such stocks are actually listed on the national stock exchanges; but employees of these work units are eager to purchase the stocks. In most cases, the original issues are sold at far higher prices than their face value, as employees (and even nonemployees) eagerly buy up fellow employees' rights to purchase stocks, at grossly inflated prices. Presumably, the right of employees to own stock in their own work unit will make them eager to have it do well, and thus increase efficiency and profits.

Companies that are owned and run by state agencies and ministries, such as the Ministry of Defense, have used state funds to invest in stocks and real estate, including international stock markets and real estate abroad. Regrettably, they have sometimes lost millions of yuan in the process. As a result, the government has now prohibited state-run units from investing state money in stocks and real estate. Nevertheless, some still do it through the companies that they control, especially those with foreign connections.

Learning from Western practices and catering to a national penchant for gambling (illegal, but indulged in nevertheless, in mahjong and cards), the Chinese have also begun a number of lotteries. Thus far, most of these have been for the purpose of raising money for specific charities or causes, such as for victims of floods and for the disabled. Recently, because the items offered, such as brand-new cars, have been so desirable, the lotteries have generated billions of yuan in revenue. The government has found these government-controlled lotteries to be an excellent way of addressing the Chinese penchant for gambling while simultaneously generating revenues to compensate for those it seems unable to collect through taxes.

Finally, companies following such Western marketing gimmicks for increasing sales as putting Chinese characters on the inside of packages or bottle caps to indicate whether the purchasers have won a prize. With a little Chinese ingenuity, the world could witness never-before-imaged realms of betting and competitive business practices that appeal to people's desire to get something for nothing.

The Press and Mass Media

At the time that the student-led demonstrations for democracy began in Beijing's Tiananmen Square in the spring of 1989, China's press had grown substantially and become increasingly liberalized. With some 1,500 newspapers, 5,000 magazines, and 500 publishing houses, the Chinese were able to express a wider variety of viewpoints and ideas than at any time since the CCP came to power in 1949. The production of millions of television sets, radios, short-wave radios, cassette recorders, and VCRs also facilitated the growth of the

mass media in China. They were accompanied by a wide array of foreign audio and video materials. The programs of the British Broadcasting Corporation (BBC) and the Voice of America, the diversification of domestic television and radio programs (a choice made by the Chinese government and facilitated by satellite communication), and the importation and translation of foreign books and magazines—all contributed to a more pluralistic press in China. In fact, by 1989, the stream of publications had so overwhelmed the CCP Propaganda Department that it was simply no longer able to monitor their contents.

During the demonstrations in Tiananmen Square, the international press in Beijing, unlike the Chinese press, was freely filming the events and filing reports on the demonstrations. The Chinese press then took a leap into complete press freedom. With camera operators and microphones in hand, Chinese reporters covered the student hunger strike that began on May 13 in its entirety; but with the imposition of martial law in Beijing on May 20, press freedom came to a crashing halt.

In the immediate aftermath of the crackdown on Tiananmen Square demonstrators in June 1989, the CCP imposed a ban on a variety of books, journals, and magazines. The government ordered the "cleansing" of media organizations, with "bad elements" to be removed and not permitted to leave Beijing for reporting. All press and magazine articles written during the prodemocracy movement, as well as all television and radio programs shown during this period, were analyzed to see if they conformed to the party line. If they did not, those individuals responsible for editing were dismissed. And, as had been the practice in the past, press and magazine articles once again had to be on topics specified by the editors, who were under the control of the Propaganda Department of the CCP. In short, press freedom in China suffered a significant setback because of the pro-democracy demonstrations in the spring of 1989.

In the new climate of experimentation launched by Deng Xiaoping in 1992, however, the diversity of television and radio programs soared. China's major cities now have multiple channels carrying a broad range of programs from Hong Kong, Taiwan, Japan, and the West. These programs—whether soap operas about the daily life of Chinese people living in Hong Kong and Taiwan, or art programs introducing the world of Western religious art through a visual art-history tour of the Vatican, or in news about protests and problems faced by other nations in the world—expose the Chinese to values, events, ideas, and standards of living previously unknown to them. They are even learning about the American legal process through China's broadcasting of American television dramas that focus on police and the judicial system. Chinese are fascinated that American police, as soon as they arrest suspects, inform them that they have "the right to remain silent" and "the right to a lawyer." Such programs may do more to promote reform of the Criminal Procedure Code than anything human-rights groups do.

Today, television ownership is widespread. In addition to dozens of regular channels, China has numerous cable stations. There are more than 100 million cable-television subscribers, with millions of new subscribers being added each year. And virtually all families have radios. Round-the-clock all-news radio stations broadcast the latest political, economic, and cultural news and conduct live interviews; and radio talk shows take phone calls from anonymous listeners about everything from sex to political and economic corruption. There are even blatant critiques of police brutality,[30] and analyses of failed government policies on everything from trade policy to health care and unemployment. There is also far better coverage of social and economic news in

all the media than previously, and serious investigative reporting on corruption and crime. A popular half hour program during evening primetime looks at cases of corruption and crime that are being investigated, or have been solved. Figures indicate the exponential growth of the media. The number of books published annually in China rose from about 5,000 in 1970 to about 104,000 in 1995. The number of newspapers grew from 42 in 1970 to 2,202 in 1995—an overall increase of more than 4,000 percent in 25 years. Similarly, the number of journals and magazines grew from a handful in 1970 to 8,135 by 1995—an official number that excludes the large number of nonregistered magazines and illegal publications.[31] The number of radio stations nearly doubled from 635 to 1,210 in 1995; and the number of television stations grew from 509 to 980. Close to 90 percent of the population have access to television.[32] And, CCTV-9, the "CNN" of China, is a slick, English language program that is watched in many parts of the world, and actually has broader viewership in Asia than does CNN. By 1999, there were 750 cable-television stations and an estimated 100 million customers. The government's National Cable Company is forming a countrywide network through which to expand services geographically and to include new Internet and telecommunications connections.[33]

This effort to expand Internet availability indicates the government's dilemma: It wants China to modernize rapidly, and to have the scientific, economic, commercial, and educational resources on the Internet available to as many as possible. It has demanded that all government offices be updated to use the Internet in order to improve communication, efficiency, and transparency. At the same time, it wants to control which Web sites can be accessed, and what type of material may be made available to China's citizens on those sites. In spite of the government's best efforts, the Internet is virtually uncontrollable. Angry and inflammatory commentary on events in China, and even criticism of China's leadership, appear with increasing regularity. Some sites are closed down because they violate the unstated boundaries of acceptable commentary; but it is difficult in a computer-savvy world to block access to all sites that the government might find irritating, if not downright seditious. China's Internet users know what is going on in the world; and they can spend their time chatting to dissidents abroad and at home on e-mail if they so choose. But most prefer to use the Internet for business, news, sports, games, music, socializing, and, of course,

pornography. With print and electronic media so prolific and diverse, much of it escapes any monitoring whatsoever, especially newspapers, magazines, and books. Even the weekend editions of the remaining state-run newspapers print just about any story that will sell. Often about a seamier side of Chinese life, they undercut the puritanical aspect of communist rule and expand the range of topics available for discussion in the public domain. Were the state able to control the media, it would, at the very least, crack down on pornographic literature on the streets.

The party tends, however, to concentrate its limited resources on the largest and most influential journals, magazines, newspapers, and publishing houses. It seems to have written off the rest as the inevitable downside of a commercialized media market—and the part that it no longer supports, nor controls, financially. Funded by advertising and consumer demand, the media must now "march to the market." As for China's film industry, because it produces relatively few films each year, it is more heavily censored than print media. Furthermore, all films are shot in a small number of studios, making control easier. Finally, a film is likely to have a much larger audience than most books, and so the censors are concerned that it be carefully reviewed before being screened.[34]

The sheer quantity of output on television, radio, books, and the press allows the Chinese people to make choices among the types of news, programs, and perspectives they find most appealing. Their choices are not necessarily for the most informative or the highest quality, but for the most entertaining. Because the media (with few exceptions) are market-driven, consumers' preferences, not government regulations or ideological values, shape programming and publishing decisions.

This market orientation is due to economic reforms. By the 1990s, the government had cut subsidies to the media, thereby requiring that even the state-controlled media had to make money or be shut down. This in turn meant that the news stories it presented had to be more newsworthy in order to sell advertising and subscriptions. Even in the countryside, the end of government subsidies to the media has spurred publication. Thinking there is money to be made, township and village enterprises, as well as private entrepreneurs, have set up thousands of printing facilities during the last 10 to 15 years.[35] In short, "Even though China's media can hardly be called free, the emergence of divergent voices means the center's ability to control people's minds has vanished."[36]

Human Rights and Democracy: The Chinese View

Surveys indicate that most urban Chinese citizens believe the government has adopted policies that have greatly improved their daily lives. Many have seen the government's law-and-order campaigns—which sometimes involve crackdowns on perceived dissidents (such as Falun Gong activists)—as necessary to China's continued economic prosperity and political stability. They have tended to be far more interested in the prospect of a higher standard of living than in the rights of dissidents.

Even China's intellectuals no longer seem interested in protest politics. They do not "love the party," but they accept that status quo. Some just want a promotion and to make money. Others have become advisers to the government's think tanks and advisory committees. Many have gone into business. As one university professor put it, it is easy to be idealistic in one's heart; but to be idealistic in action is a sign of a true idealist, and there haven't been many of those in China since 1989. Today in China, it is difficult to find any student or member of the intellectual elite who demonstrated in Tiananmen Square in 1989 doing anything remotely political today.

Still, those who are discontent with party rule have far more outlets today for their grievances: the mass media, and journals, as well as think tanks, policy advisory committees, and professional associations that actually influence policy. Street protests are no longer considered the best way for the educated and professional classes to change policy, although ordinary workers and peasants resort to them with increasing frequency.

Many members of China's elites are committed to reform, but the number of idealists committed to democracy—or communism—is limited. Few Chinese, including government officials, want to discuss communis ideology, and even fewer agree on what democracy means. They prefer to talk about business and development, and do so in terms familiar to capitalists throughout the world, but also in terms of the overall objective of strengthening China as a nation. In this respect, they are appealing to the strong nationalism that has virtually replaced communism as the normative glue holding the country together.

Apart from their changed perspectives on what really matters, many Chinese feel that they do not know enough to challenge government policy on human rights issues. They assume that the government lies to them, and on such issues as the treatment of dissidents, most Chinese know no more than what the government tells them. Why should they risk their careers to fight for the rights of jailed dissidents about whom they know almost nothing, they ask. They know of the derogation of human rights in Western liberal democracies, such as the killing of student protesters at Kent State by the National Guard during the Vietnam War and the many deaths attributed to the British forces in Northern Ireland. They have heard of the unseemly behavior of several student leaders of the Tiananmen Square demonstrations, both during the movement and after it. They wonder aloud if in the treatment of criminal suspects and those suspected of treason or efforts to harm the country—especially suspected terrorists—Western democratic states exhibit more virtuous behavior than does China.

Some Chinese intellectuals argue that the recent difficulties in the United States and other Western democracies indicate that their citizens frequently elect the wrong leaders, people who not only make bad policies but who are also increasingly involved in corrupt money politics—the very issue of most concern to the Chinese. This, they suggest, indicates that democracy is no more able than socialism to produce good leaders. Furthermore, many support the view that the Chinese people are inadequately prepared for democracy because of a low level of education. The U.S. Congress's impeachment hearings of President Bill Clinton made ordinary Chinese wonder why any country would want to imitate American democracy. And the blatant political maneuverings and other problems surrounding the 2000 U.S. presidential election hardly offered reassurance as to the virtues of American democracy.

Within China itself, moreover, it is frequently the people, not the government, who demand the harshest penalties for common criminals, if not political dissidents. And it is China's own privileged urbanites who often demand that the government ignore the civil rights of other citizens. For example, urban residents in Beijing have repeatedly demanded that the government remove migrant squatters and their shantytowns, asserting that they are breeding grounds for criminality in the city. Many ordinary people now seem to accept the government's overall assessment of the events of the spring of 1989, which is that the demonstrations in Tiananmen Square posed a threat to the stability and order of China and justified the military's crackdown. To many Chinese people, no less than to their government, stability and order are critical to the continued economic development of China. Advancement toward democracy and protection of human rights take a back seat.

INTERNATIONAL RELATIONS

From the 1830s onward, foreign imperialists nibbled away at China's territorial sovereignty, subjecting China to one national humiliation after another. As early as the 1920s, both the Nationalists and the Communists were committed to unifying and strengthening China in order to rid it of foreigners and resist further foreign incursions. When the Communists achieved victory over the Nationalists in 1949, they vowed that they would never again allow foreigners to tell China what to do. This historical background is essential to understanding China's foreign policy in the period of Chinese Communist rule.

From Isolation to Openness

By the early 1950s, the Communists had forced all but a handful of foreigners to leave China. The target of the United States' Cold War policy of "isolation and containment," China under Chairman Mao Zedong charted an independent, and eventually an isolationist, foreign policy. Even China's relations with socialist bloc countries were suspended after the Cultural Revolution began in 1966. China took tentative steps toward re-establishing relations with the outside world with U.S. president Richard Nixon's visit to China in 1972, but it did not really pursue an "open door" policy until more pragmatic "reformers" gained control in 1978. By the 1980s, China was hosting several million tourists annually, inviting foreign investors and foreign experts to help with China's modernization, and allowing Chinese to study and travel abroad. Nevertheless, inside the country, contacts between Chinese and foreigners were still affected by the suspicion on the part of ordinary Chinese that ideological and cultural contamination comes from abroad and that association with foreigners might bring trouble.

These attitudes have moderated considerably, to the point where some Chinese are more willing to socialize with foreigners, invite them to their homes, and even marry them. However, this greater openness to things foreign sits uncomfortably together with a new nationalism that has emerged since the West so heavily criticized, and punished, China for the military crackdown on Tiananmen Square demonstrators in June 1989. The broad masses of Chinese people remain suspicious, even disdainful of foreigners. A strong xenophobia (dislike and fear of foreigners) and an awareness of the history of China's victimization by Western and Japanese imperialism mean that the Chinese are likely to rail at any effort by other countries to tell them what to do. The Chinese continue to exhibit this sensitivity on a wide variety of issues, from human rights to

China's policies toward Tibet and Taiwan; from intellectual property rights to prison labor and environmental degradation.

Ordinary Chinese people tended to concur with the government's anger over the U.S. threat of economic sanctions to challenge China's human rights policies (something that is no longer possible since China joined the World Trade Organization (WTO) in 2001). They were enraged by the American bombing of the Chinese Embassy in Belgrade during the war in Yugoslavia, which the U.S. government insisted was by error, not intent; the crash of a U.S. spy plane with a Chinese jet in 2001, which the Chinese believed to be in their own airspace; efforts to prevent China from entering the World Trade Organization or becoming a site for the Olympics; American accusations of Chinese spying and stealing of American nuclear-weapons secrets, charges whose motivations appeared even more as part of an anti-China campaign because they were ultimately dismissed for lack of evidence; American accusations of illegal campaign funding by the Chinese; the on-going human-rights barrage; the strengthening of U.S. military ties with Japan and Taiwan; and continued American interference in China's efforts to regain control of Taiwan.

Only a narrow line separates a benign nationalism essential to China's unity, however, and a popular nationalism that has militant overtones and threatens to careen out of control.[37] These aspects of nationalism worry China's government, whose primary concerns remain economic growth and stability. The government does not want to be forced into war by a militant nationalism. Nevertheless, China's xenophobia continues to show up in its efforts to keep foreigners isolated in certain living compounds; to limit social contacts between foreigners and Chinese; to control the import of foreign literature, films, and periodicals; and in general to keep foreign political and cultural values out of China. In 1996, in an effort to protect China's culture, the government ordered television stations to broadcast only Chinese-made programs during prime time. The government has also attempted to enhance national pride through economic success; and by participation in international events, including the Olympic and Asian Games, music competitions, and film festivals.

Yet, in some respects, it has been a losing battle, with growing numbers of foreigners in China socializing with Chinese; television swamped with foreign programs; Kentucky Fried Chicken, MacDonald's, and Pizza Hut stores proliferating; "Avon calling" at several million homes;[38] and bodybuilding and disco becoming part of the

culture. Further, with millions of Chinese tourists and officials tromping through the world, huge foreign investors in China, China's hosting of trade fairs, important international meetings, and the 2008 Olympics, as well as a heavy reliance on foreign experts to help China reform its economic, legal, financial, and banking systems, and set up a commodities and futures market, China is awash with values that contend with traditional Chinese ones. China is now a full participant in the international economic and financial system. It is an important participant in the war on terrorism, and a key player in Interpol and other efforts to control international crime syndicates, smuggling of weapons and drugs, and to stop international trafficking in children and women. China today is seen by most countries as a partner rather than an adversary, a part of the solution, not the problem. Its powerful economy and investments abroad have gained it respect (leavened with fear) worldwide. Today, China is a key international actor. It cannot be dismissed as a poor country without the wherewithal to enter the modern world.

China is a much more open country than at any time since 1949. This is in spite of the efforts by the party's more conservative wing to limit the impact of China's "open door" policy on the political system (the influx of ideas about democracy and individual rights), on economic development (a market economy, corruption, and foreign control and ownership), and on Chinese culture ("pollution" from foreign literature, pornography, and such values as individualism). At the same time, China's flourishing business community and growing middle class have little interest in disrupting the emphasis on stability by calling for greater political democracy. Foreign investment in China (total foreign direct investment in 2003 was US$57 billion; and more than US$60 billion in 2004) which far outstrips foreign investment in any other developing country, is due not just to China's potential market size but also to the favorable investment climate created by the party/state. The government has, in short, seemed less worried about the invasion of foreign values than it is anxious to attract foreign investment. The government's view now seems to be: If it takes nightclubs, discos, exciting stories in the media, stock markets, rock concerts, the Internet, and consumerism to make the Chinese people content and the economy flourish under CCP rule, then so be it.

THE SINO–SOVIET RELATIONSHIP

While forcing most other foreigners to leave China in the 1950s, the Chinese Com-

munist regime invited experts from the Soviet Union to China to give much-needed advice, technical assistance, and aid. This convinced the United States (already certain that Moscow controlled communism wherever it appeared) that the Chinese were puppets of the Soviets. Indeed, until the Great Leap Forward in 1958, the Chinese regime accepted Soviet tenets of domestic and foreign policy along with Soviet aid. But China's leaders soon grew concerned about the limits of Soviet aid and the relevance of Soviet policies to China's conditions—especially the costly industrialization favored by the Soviet Union. Ultimately, the Chinese questioned their Soviet "big brother" and turned to the Maoist model of development, which aimed to replace expensive Soviet technology with human labor. Soviet leader Nikita Khrushchev warned the Chinese of the dangers to China's economy in undertaking the Great Leap Forward; but Mao Zedong interpreted this as evidence that the Soviet "big brother" wanted to hold back China's development.

The Soviets' refusal to use their military power in support of China's foreign-policy objectives further strained the Sino–Soviet relationship. First in the case of China's confrontation with the United States and the forces of the "Republic of China" over the Offshore Islands in the Taiwan Strait in 1958, and then in the Sino-Indian border war of 1962, the Soviet Union backed down from its promise to support China. The Soviets also refused to share nuclear technology with the Chinese. The final blow to the by-then fragile relationship came with the Soviet Union's signing of the 1963 Nuclear Test Ban Treaty. The Chinese denounced this as a Soviet plot to exclude China from the "nuclear club" that included only Britain, France, the United States, and the Soviet Union. Subsequently, Beijing publicly broke party relations with Moscow.

The Sino–Soviet relationship, already in shambles, took on an added dimension of fear during the Vietnam War, when the Chinese grew concerned that the Soviets (and Americans) might use the war as an excuse to attack China. China's distrust of Soviet intentions was heightened in 1968, when the Soviets invaded Czechoslovakia in the name of the "greater interests of the socialist community"—which, they contended, "overrode the interests of any single country within that community." Soviet skirmishes with Chinese soldiers on China's northern borders soon followed.

Ultimately, it was the Chinese leadership's concern about the Soviet threat to China's national security that, in 1971, caused it to reassess its relationship with the United States and led ultimately to the establishment of diplomatic relations with China in 1979. Indeed, the real interest of China and the United States in each other was as a "balancer" against the Soviet Union. Thus, in the midst of the Cold War, which began in 1947 and did not end until the late 1980s, China had already started to move out of the Soviet-led camp after the debacle of the 1958 Great Leap Forward; yet China did not begin benefiting from friendship with Western countries in the power balance with the Soviet Union until after it became a member of the United Nations in 1971 and Nixon's visit in 1972.

The Sino–Soviet relationship moved toward reconciliation only near the end of the Cold War. In 1987, the Soviets began making peaceful overtures: They reduced troops on China's borders and withdrew support for Vietnam's puppet government in neighboring Cambodia. Beijing responded positively to the new *glasnost* ("open door") policy of the Soviet Communist Party's general-secretary, Mikhail Gorbachev. Border disputes were settled and ideological conflict between the two Communist giants abated; for with the Chinese themselves shelving Marxist dogma in their economic policies, they could hardly continue to denounce the Soviet Union's "revisionist" policies and make self-righteous claims to ideological orthodoxy. With both the Soviet Union and China abandoning their earlier battle over who should lead the Communist camp, they shifted away from conflict over ideological and security issues to cooperation on trade and economic issues.

GUNS AND RICE

With the collapse of Communist party rule, first in the Central/Eastern European states in 1989, and subsequently in the Soviet Union, the dynamics of China's foreign policy changed dramatically. Apart from fear that their own reforms might lead to the collapse of CCP rule in China, the breakup of the Soviet Union into 15 independent states removed China's ability to play off the two superpowers against each other: The formidable Soviet Union simply no longer existed. Yet its fragmented remains had to be treated seriously, for the state of Russia still has nuclear weapons and shares a common border of several thousand miles with China, and the former Soviet republic of Kazakhstan shares a border of nearly 1,000 miles.

The question of what type of war the Chinese military might have to fight has affected its military modernization. For many years, China's military leaders were in conflict over whether China would have to fight a high-tech war or a "people's war," in which China's huge army would draw in the enemy on the ground and destroy it. In 1979, the military modernizers won out, jettisoning the idea that a large army, motivated by ideological fervor but armed with hopelessly outdated equipment, could win a war against a highly modernized military such as that of Japan or even the Soviet Union. The People's Liberation Army (PLA) began by shedding a few million soldiers and putting its funds into better armaments. A significant catalyst to further modernizing the military came with the Persian Gulf War of 1991, during which the jettisoning the idea that a large army, motivated CNN news network broadcasts vividly conveyed the power of high-technology weaponry to China's leaders.[39]

China's military believed that it was allocated an inadequate budget for modernization, so it struck out on its own along the capitalist road to raise money. By the late 1990s, the PLA had become one of the most powerful actors in the Chinese economy. It had purchased considerable property in the Special Economic Zones near Hong Kong; acquired ownership of major tourist hotels and industrial enterprises; and invested in everything from golf courses, brothels, and publishing houses to CD factories and the computer industry, as a means for funding military modernization. In 1998, however, President Jiang Zemin demanded that the military relinquish its economic enterprises and return to its primary task of building a modern military and protecting China. The promised payoff was that China's government would allocate more funding to the PLA, making it unnecessary for it to rely on its own economic activities.

In recent years, China's military has purchased weaponry and military technology from Russia as Moscow scales back its own military in what sometimes resembles a going-out-of-business sale; but in doing so, China's military may have simply bought into a higher level of obsolescence, since Russia's weaponry lags years behind the technology of the West. China possesses nuclear weapons and long-distance bombing capability, but its ability to fight a war beyond its own borders is quite limited. Asian countries, torn between wondering whether China or Japan will be a future threat to their territory, do not seem concerned by China's military modernization, except when China periodically makes threatening statements about Taiwan or the Spratly Islands in the South China Sea. Even here, however, Beijing usually relies on economic and diplomatic instruments. In the case of Taiwan, it is essentially tying Taiwan's economy to the

mainland by welcoming economic investment and trade; the hope is ultimately to bring Taiwan under the control of Beijing without a war. In the case of the Spratlys, Beijing has reached tentative agreement with the five governments involved in competing claims to the Spratlys to avoid the possibility of armed clashes.

Nevertheless, in spite of China's remarkable economic and diplomatic gains since the reform period began in 1979, the leadership continues to modernize China's military capabilities. Beijing is ever alert to threats to its national security, but there are no indications that it is preparing for aggression against any country. China's military modernization is primarily aimed at defensive capabilities and maintaining its deterrent capability against an American nuclear attack. It has also increased the number of missiles it aims at Taiwan as Taiwan's President, Chen Shui-bian, has talked increasingly of declaring Taiwan independence. With the arrival of the George W. Bush administration in Washington in early 2001, the American leadership began to seriously consider the possibility of deploying a limited "national missile defense" in the United States, and even to deploy a "theater missile defense" (TMD) around Japan—and possibly Taiwan. Were a missile-defense system successfully deployed, it would limit the ability of China to prevent Japan or the United States from attacking it—or to prevent Taiwan from declaring independence; for a defensive system would deny China the ability to cause "mutual assured destruction" in a retaliatory strike on the aggressor, as China's limited number of ICBMs (estimated to be fewer than a dozen) would not be able to penetrate even a limited national missile-defense system. In the case of Taiwan, TMD would allow it to declare independence with impunity. Denying China the ability to retaliate is, then, perceived by Beijing as an aggressive move by the United States, and helps explain China's efforts to substantially increase the number of missiles it aims at Taiwan in order to overcome any defensive system (including "theater missile defense") that might be installed.

National missile defense and theater missile defense hardly provide the basis for confidence-building, peace, and security in Asia. Nevertheless, because the United States officially adopted a decision to deploy NMD within the context of international terrorism, and because the Chinese see themselves on the same side with the Americans in the war on terrorism, the Chinese have felt less threatened by NMD than they otherwise might.

China's leadership is, in any event, primarily concerned with economic development. China is working to become an integral part of the international economic, commercial, and monetary systems. It has rapidly expanded trade with the international community, and it won membership in the World Trade Organization in 2001. Today, China is focusing its efforts primarily on infrastructure development and investment, not just in China but throughout the world. With the exponential growth in per capita income for more than 200 million Chinese, China is pressed to acquire natural resources to satisfy rocketing increases in consumerism. It is investing heavily in resources for the future in Latin America, Southeast Asia, Africa, and the Middle East. It is ironic that China, considered "the sick man of Asia " in the early 20th century, should now in the 21st century be buying up companies not just in the developing world, but also in Europe and the United States; and that it has, along with Japan, become the "banker" for the United States, buying American debt and keeping the U.S. dollar from declining still further in value. For these reasons, China holds substantial bargaining power vis-à-vis the United States and many other countries in the world.

THE SINO–AMERICAN RELATIONSHIP

China's relationship with the United States has historically been an emotionally turbulent one.[40] It has never been characterized by indifference. During World War II, the United States gave significant help to the Chinese, who at that time were fighting under the leadership of the Nationalist Party, headed by General Chiang Kai-shek. When the Americans entered the war in Asia, the Chinese Communists were fighting together with the Nationalists in a "united front" against the Japanese, so American aid was not seen as directed against communism.

After the defeat of Japan at the end of World War II, the Japanese military, which had occupied much of the north and east of China, was demobilized and sent back to Japan. Subsequently, civil war broke out between the Communists and Nationalists. The United States attempted to reconcile the two sides, but to no avail. As the Communists moved toward victory in 1949, the KMT leadership fled to Taiwan. Thereafter, the two rival governments each claimed to be the true rulers of China. The United States, already in the throes of the Cold War because of the "iron curtain" falling over Eastern Europe, viewed communism in China as a major threat to its neighbors.

Korea, Taiwan, and Vietnam

The outbreak of the Korean War in 1950 helped the United States to rationalize its decision to support the Nationalists, who had already lost power on the mainland and fled to Taiwan. The Korean War began when the Communists in northern Korea attacked the non-Communist south. When United Nations troops (mostly Americans) led by American general Douglas Mac-Arthur successfully pushed Communist troops back almost to the Chinese border and showed no signs of stopping their advance, the Chinese—who had frantically been sending the Americans messages about their concern for China's own security, to no avail—entered the war. The Chinese forced the UN troops to retreat to what is today still the demarcation line between North and South Korea. Thereafter, China became a target of America's Cold War isolation and containment policies.

With the People's Republic of China condemned as an international "aggressor" for its action in Korea, the United States felt free to recognize the Nationalist government in Taiwan as the legitimate government to represent all of China. The United States supported the Nationalists' claim that the people on the Chinese mainland actually wanted the KMT to return to the mainland and defeat the Chinese Communists. As the years passed, however, it became clear that the Chinese Communists controlled the mainland and that the people were not about to rebel against Communist rule.

Sino–American relations steadily worsened as the United States continued to build up a formidable military bastion with an estimated 100,000 KMT soldiers in the tiny islands of Quemoy and Matsu in the Taiwan Strait, just off China's coast. Tensions were exacerbated by the steady escalation of U.S. military involvement in Vietnam from 1965 to the early 1970s. China, fearful that the United States was really using the war in Vietnam as the first step toward attacking China, concentrated on civil-defense measures: Chinese citizens used shovels and even spoons to dig air-raid shelters in major cities such as Shanghai and Beijing, with tunnels connecting them to the suburbs. Some industrial enterprises were moved out of China's major cities in order to make them less vulnerable in the event of a massive attack on concentrated urban areas. The Chinese received a steady barrage of what we would call propaganda about the United States "imperialist" as China's number-one enemy; but it is important to realize that the Chinese leadership actually *believed* what it told the people, especially in the context of the United States' steady escalation of the war in Vietnam toward the Chinese

border, and the repeated "mistaken" over-flights of southern China by American planes bombing North Vietnam. Apart from everything else, it is unlikely that China's leaders would have made such an immense expenditure of manpower and resources on civil-defense measures had they not truly believed that the United States was preparing to attack China.

Diplomatic Relations

By the late 1960s, China was completely isolated from the world community, including the Communist bloc. In the throes of the Cultural Revolution, it had withdrawn its diplomatic staff from all but one of its embassies. It saw itself as surrounded on all sides by enemies—the Soviets to the north and west, the United States to the south in Vietnam as well as in South Korea and Japan, and the Nationalists to the east in Taiwan. Internally, China was in such turmoil from the Cultural Revolution that it appeared to be on the verge of complete collapse.

In this context, Soviet military incursions on China's northern borders, combined perhaps with an assessment of which country could offer China the most profitable economic relationship, led China to consider the United States as the lesser of two evil giants and to respond positively to American overtures. In 1972, President Richard Nixon visited China, the first official American contact with China since breaking diplomatic relations in 1950. When the U.S. and China signed the Shanghai Communique at the end of his visit, the groundwork was laid for reversing more than two decades of hostile relations.

Thus began a new era of Sino–American friendship, but it fell short of full diplomatic relations, This long delay in bringing the two states into full diplomatic relations reflected not only each country's domestic political problems but also mutual disillusionment with the nature of the relationship. Although both sides had entered the relationship with the understanding of its strategic importance as a bulwark against the Soviet threat, the Americans had assumed that the 1972 opening of partial diplomatic relations would lead to a huge new economic market for American products; the Chinese assumed that the new ties would quickly lead the United States to end its diplomatic relations with Taiwan. Both were disappointed. Nevertheless, pressure from both sides eventually led to full diplomatic relations between the United States and the People's Republic of China on January 1, 1979.

The Taiwan Issue in U.S.–China Relations

Because the People's Republic of China and the Republic of China both claimed to be the legitimate government of the Chinese people, the establishment of diplomatic relations with the former necessarily entailed breaking them with the latter. Nevertheless, the United States continued to maintain extensive, informal economic and cultural ties with Taiwan. It also continued the sale of military equipment to Taiwan. Although these military sales are still a serious issue, American ties with Taiwan have diminished, while China's own ties with Taiwan have grown steadily closer since 1988. Taiwan's entrepreneurs have become one of the largest groups of investors in China's economy. More than one million people from Taiwan live on the mainland, with 300,000 living in Shanghai alone. Taiwan used to have one of the cheapest labor forces in the world; but because its workers now demand wages too high to remain competitive, Taiwan's entrepreneurs have dismantled many of its older industries and reassembled them on the mainland. With China's cheap labor, these same industries are now profitable, and both China's economy and Taiwan's entrepreneurs benefit. Taiwan's businesspeople are also investing in new industries and new sectors, and they are competing with other outside investors for the best and brightest Chinese minds, so the relationship has already moved beyond simply exploiting the mainland for cheap labor and raw materials.

Ties with the mainland have also been enhanced since the late 1980s by the millions of tourists from Taiwan, most of them with relatives in China. They bring both presents and good will. Family members who had not seen one another since 1949 have reestablished contact, and "the enemy" now seems less threatening.

China hopes that its economic reforms and growth, which have substantially raised the standard of living, will make reunification more attractive to Taiwan. This very positive context has, however, been disturbed over the years by events such as the military crackdown on the demonstrators in Tiananmen Square in 1989, Taiwan president Lee Teng-hui's visit to the United States in 1995, and subsequent efforts by Taiwan's leaders to declare independence. In 1996, China responded to such efforts by "testing" its missiles in the waters around Taiwan. High-level talks to discuss eventual reunification were broken off and did not resume for several years. The 2000 election of the Democratic Progressive Party candidate, Chen Shui-bian as president (and his re-election in 2004), led to still more diplomatic-crises with Beijing. President Chen, who has campaigned on the platform of an independent Taiwan, has refused acknowledge Beijing's "one China" principle, and

he has insisted that Taiwan negotiate with the P.R.C. as "an equal." This has further strained the relationship and led to raising the bellicosity decibel level in Beijing. Nevertheless, both sides recognize it is in their interests for the foreseeable future to maintain the status quo—a peaceful and profitable relationship in which Taiwan continues to act as an independent state, but does not declare its independence.

So far, the battle between Taipei and Beijing remains at the verbal level. At the same time, massive investments by Taiwan in the mainland and the 2000 opening of Taiwan's Offshore Islands of Quemoy and Matsu for trade with the P.R.C. are bringing the two sides still closer together. Their two economies are becoming steadily more intertwined, and both sides benefit from their commercial ties. This does not mean that they will soon be fully reunified in law. Furthermore, there remains the black cloud of Beijing possibly using military force against Taiwan if it declares itself an independent state. Beijing refuses to make any pledge never to use military force to reunify Taiwan with the mainland, on the grounds that what it does with Taiwan is China's internal affair. Hence, in Beijing's view, no other country has a right to tell China what to do about Taiwan.

Human Rights in U.S.–China Relations

Since U.S. president Jimmy Carter established diplomatic relations with China in 1979, each successive American president has campaigned on a platform that decried the abuses of human rights in China and vowed, if elected, to take strong action, including economic measures, to punish China. The Chinese people have been confused and distraught at this prospect. They do not see the point in punishing hundreds of millions of Chinese for human rights abuses committed not by the people, but by their leadership. Nor do they necessarily believe that their own government has been more abusive of human rights than other states not so condemned. In any event, within a few months (if not sooner) of being sworn in, each successive president has abandoned his campaign platform and chosen to break the linkage between "most-favored-nation" (MFN) trade status for China and its human rights record.

Why was this? Once inauguration day was over, it was quickly explained to the new president that the United States dare not risk jeopardizing its relations with an increasingly powerful state containing one quarter of the world's population through measures that would probably simply give Japan and other countries a better trading position while undercutting the opportunity for Americans to do business with

China. By 2000, President Bill Clinton had managed to get Congress to vote for "permanent normal trading relations" (PNTR). No longer would normal trade relations with China be subjected each spring to a congressional review of its human-rights record. This in turn cleared the way for China to join the World Trade Organization with an American endorsement in 2001. Under WTO rules, one country may not use trade as a weapon to punish another for political reasons.

Clinton's China policy was also shaped by a new strategy of "agreeing to disagree" on certain issues such as human rights, while efforts continued to be made to bring the two sides closer together. This strategy came out of a belief that China and the United States had so many common interests that neither side could afford to endanger the relationship on the basis of a single issue. The American policy of "engagement" with China, which began with the Clinton administration, was based on the belief that isolating China had proven counterproductive. The administration argued that human rights issues could be more fruitfully addressed in a relationship that was more positive in its broader aspects. "Engagement" allowed the two countries to work together toward shared objectives, including the security of Asia.

Upon entering office in January 2001, George W. Bush appeared intent on ending this policy of engagement, and treating China as a "strategic adversary." Yet the Bush administration soon abandoned this policy—an act made complete by the September 11, 2001, terrorist attacks on the United States. After those events, President Bush told the world, "You are either with us or against us." China, not wishing to needlessly bring trouble on itself, immediately sided with the United States in the war on terrorism. This had important implications for the role that human rights could play in the U.S.–China relationship; for with the focus on terrorism, human rights took a back seat, even in the United States itself. With libertarians raising countless questions about the American government's treatment of suspected terrorists, the United States was hardly in a position to be pressing for the amelioration in the practice of human rights in China. Just as important, the United States did not want to raise gratuitous questions concerning China's alleged derogation of human rights when it needed China on its side, not just in the war on terrorism, but on almost every issue of international significance.

In spite of the White House's tendency to be pro-China and avoid the issue of human rights, the U.S. Congress has been a different matter. It was Congress that

Timeline: PAST

1842
The Treaty of Nanking cedes Hong Kong to Great Britain

1860
China cedes the Kowloon Peninsula to Great Britain

1894–1895
The Sino–Japanese War

1895–1945
Taiwan is under Japanese colonial control

1898
China leases Northern Kowloon and the New Territories (Hong Kong) to Great Britain for 99 years

1900–1901
The Boxer Rebellion

1911
Revolution and the overthrow of the Qin Dynasty

1912–1949
The Republic of China

1921
The Chinese Communist Party (CCP) is established

1931
Japanese occupation of Manchuria (the northeast province of China)

1934–1935
The Long March

1937–1945
The Japanese invasion and occupation of much of China

1942–1945
The Japanese occupation of Hong Kong

1949
Civil war between the KMT and CCP

The KMT establishes the Nationalist government on Taiwan

The People's Republic of China is established

1950
The United States recognizes the Nationalist government in Taiwan as the legitimate government of all China

1958
The "Great Leap Forward"; the Taiwan Strait crisis (Offshore Islands)

1963
The public Sino–Soviet split

1966–1976
The "Great Proletarian Cultural Revolution"

1971
The United Nations votes to seat the P.R.C. in place of the R.O.C.

1972
U.S. president Richard Nixon visits the P.R.C.; the Shanghai Communique

1976
Mao Zedong dies; removal of the Gang of Four

1977
Deng Xiaoping is restored to power

1979
The United States recognizes the P.R.C. and withdraws recognition of the R.O.C.; the Sino–Vietnamese War

1980s–1990s
Resumption of arms sales to Taiwan

The Shanghai Communique II: the United States agrees to phase out arms sales to Taiwan

China and Great Britain sign an agreement on Hong Kong's future

Sino–Soviet relations begin to thaw

Student demonstrations for democracy are widespread in China's cities; political liberalization is condemned

China sells Silkworm missiles to Iran and Saudi Arabia

Student demonstrations in Tiananmen Square; military crackdown; political repression follows

Deng encourages "experimentation" and the economy booms

The United States bombs the Chinese Embassy in Belgrade; says "an accident"

Den Xiaoping dies; Jiang Zemin assumes power

PRESENT

2000s
China bans the Falun Gong religion

Terrorist attacks of 9/11 lead to stronger ties between China and the United States, but China opposes the war on Iraq

SARS (Severe Acute Respiratory Syndrome) outbreak

Hu Jintao and Wen Jiabao take over the reins of the party and government from outgoing Jiang Zemin in a peaceful transition of power

---

pressed the human rights agenda, especially under President Clinton. Indeed, during his second term in office, Congress used Clinton's favorable treatment of China as one more reason for trying to force him out of office. All that changed with the inauguration of President Bush in 2001. With a Republican in the White House and a Republican-controlled House of Representatives, there was no longer any need for Congress to use the China issue to attack the president.[41] Coupled with 9/11, the Chinese human-rights issue has virtually disappeared from the congressional agenda. U.S.–China relations are no longer plagued by a sense that anti-China policies have powerful backers in Congress, with the result that U.S.–China relations are smoother, more predictable, and less emotional than they have been for a long time.

THE FUTURE

Since 1979, China has moved from being a relatively closed and isolated country to one that is fully engaged in the world. China's agenda for the future is daunting. It must avoid war; maintain internal political stability in the context of international pressures to democratize; continue to carry out major economic, legal, and political reforms without destablizing society and endangering CCP control; sustain economic growth while limiting environmental destruction; and limit population growth.

After the death of Deng Xiaoping in 1997, China carried out a smooth leadership transition. Although his successor, Jiang Zemin, was not as powerful as Deng had been, he did not need to be in the new context of a collective leadership. Under Jiang, those committed to reform retained dominance within the leadership. With the naming of Hu Jintao as the new head of the Chinese Communist Party in 2002 and president of China in 2003, it looks as if the party/state will continue along the road of reform. This is in spite of growing economic polarization and social instability caused by economic reforms and rapid modernization.

As long as economic growth continues, the CCP leadership will probably remain in power. The party may ultimately change its name to one more reflective of its princi-

ples. That is, it is no longer a party promoting communism but, rather, one promoting capitalist economic principles and socialist social policies. It may imitate European leftist parties, which tend to label themselves as "democratic socialist" or "socialist democratic" parties. Whatever name it adopts, it is unlikely to allow the creation of a multiparty system that could challenge its leadership.

Governing the world's largest population is a formidable task, one made even more challenging by globalization. The integration of China into the international community has heightened the receptivity of China's leaders to pressures from the international system on a host of specific issues: human rights, environmental protection, intellectual property rights, prison labor, arms control, and legal codes. China's leadership insists on moving at its own pace and in a way that takes into account China's culture, history, and institutions; but China is now subject to globalizing forces, as well as internal social and economic forces, that have a momentum of their own.

In the meantime, China, like so many other developing countries, must worry about the growing gap between the rich and the poor, high levels of unemployment, uncontrolled economic growth, environmental degradation, and the strident resistance by whole regions within China against following economic and monetary policies formulated at the center. It is also facing a major HIV/AIDS epidemic that may within 7 years lead to as many as 20 million Chinese being HIV-positive,[42] a potential collapse of the entire banking and financial system, the need to finance a social safety net and retirement pensions, and a looming threat to its state-owned enterprises as a result of China's entry into the WTO. To wit, the international community is pressing China to revalue the Chinese currency, the yuan, so that Chinese goods will be priced higher, thus making them less competitive internationally. Common criminality, corruption, and social instability provide additional fuel that could one day explode politically and bring down Chinese Communist Party rule.

At this time there is no alternative leadership waiting in the wings to take up the burden of leading China and ensuring its stability. An unstable China would not be in anyone's interest, neither that of the Chinese people, nor of any other country. An insecure and unstable China would be a more dangerous China, and it would be one in which the Chinese people would suffer immeasurably.

NOTES

1. Pam Woodall, "The Real Great Leap Forward," *The Economist,* October 2, 2004, p. 6.
2. A concern about "spiritual pollution" is not unique to China. It refers to the contamination or destruction of one's own spiritual and cultural values by other values. Europeans are as concerned about it as the Chinese and have, in an effort to combat spiritual pollution, limited the number of television programs made abroad (i.e., in the United States) that can be broadcast in European countries.
3. The essence of the "get civilized" campaign was an effort to revive a value that had seemingly been lost: respect for others and common human decency. Thus, drivers were told to drive in a "civilized" way—that is, courteously. Ordinary citizens were told to act in a "civilized" way by not spitting or throwing garbage on the ground. Students were told to be "civilized" by not stealing books or cheating, keeping their rooms and bathrooms clean, and not talking loudly.
4. The turmoil that ensued after his death had also ensued after the death of the former beloved premier, Zhou Enlai. This helps explain why the central leadership was almost paralyzed when, in January 2005, Zhao Ziyang, the Premier of China at the time of the Tiananmen demonstrations in 1989, died. Zhao, who was dismissed from his position because in the end he supported the students' demands for political reform, had been accused of trying to "split" the party. He spent the next 15 years, until his death, under house arrest in Beijing.
5. "Campaign to Crush Dissent Intensifies," *South China Morning Post* (August 9, 1989).
6. Chinese student (anonymous) in the United States, conversation (Summer 1990).
7. "China Intends to Check Unbalanced Sex Ratio by Amending Criminal Law," *People's Daily Online* (January 10, 2005).
8. Most of these young women tend to be sold as brides to men who live in remote villages where there are not enough women.
9. For example, Chinese women (at least in state-owned enterprises) have had paid maternity leave and child care in the workplace.
10. Susan K. McCarthy, "The State, Minorities, and Dilemmas of Development in Contemporary China," *The Fletcher Forum of World Affairs,* Vol. 26:2 (Summer/Fall 2002).
11. June Teufel Dreyer, "Economic Development in Tibet under the People's Republic of China," *Journal of Contemporary China*, Vol. 12, no. 36 (August 2003), pp. 411–430.
12. That Pachen Lama died in 1989.
13. For excellent detail on Chinese religious practices, see Robert Weller, *Taiping Rebels, Taiwanese Ghosts, and Tiananmen* (Seattle: University of Washington Press, 1994); and Alan Hunter and Kim-kwong Chan, *Protestantism in Contemporary China* (Cambridge: Cambridge University Press, 1993). The latter notes that Chinese judge gods "on performance rather than theological criteria" (p. 144). That is, if the contributors to the temple in which certain gods were honored were doing well financially and their families were healthy, then those gods were judged favorably. Furthermore, Chinese pray as individuals rather than as congregations. Thus, before the Chinese government closed most temples, they were full of individuals praying randomly, children playing inside, and general noise and confusion. Western missionaries have found this style too casual for their own more structured religions (p. 145).
14. Professor Rudolf G. Wagner (Heidelberg University). Information based on his stay in China in 1990.
15. Richard Madesen, "Understanding Falun Gong," *Current History* (September 2000), Vol. 99, No. 638, pp. 243–247.
16. For a better understanding of how Chinese characters are put togther, see John DeFrancis, *Visible Speech: The Diverse Oneness of Writing Systems* (Honolulu: University of Hawaii Press, 1989); and Bob Hodge and Kam Louie, *The Politics of Chinese Language and Culture: The Art of Reading Dragons* (New York: Routledge, 1998.)
17. "Number of University Students Recruited Doubles in Six Years," *People's Daily Online* (December 7, 2004).
18. David J. Lynch, "Discontent in China Boils into Public Protest," *USA Today*, Sept. 14, 2004.
19. Jonathan Watts, "Corrupt Officials Have Cost China 330 Million Pounds in 20 Years," *The Guardian* (August 20, 2004).
20. Philip P. Pan, "In China, Turning the Law into the People's Protector," *The Washington Post Foreign Service* (Dec. 28, 2004), p. A1.)
21. Suzanne Ogden, *Inklings of Democracy in China* (Cambridge: Harvard University Asia Center and Harvard University Press, 2002), pp. 234–236.
22. Of course, during the "10 Bad Years," the Chinese people were really manipulated by power-hungry members of China's elite, an ever-shifting nouveau elite who were in a desperate competition with other pretenders to power.
23. Based on the author's trip to interview village leaders in 2000 and the author's visit with President Carter to monitor elections in a Chinese village in 2001. See Ogden, pp. 183–220.
24. For an excellent analysis of how the "patterns of protest" in China have replicated the "patterns of daily life," see Jeffrey N. Wasserstrom and Liu Xinyong, "Student Associations and Mass Movements," in Deborah S. Davis, Richard Kraus, Barry Naughton, Elizabeth J. Perry, eds., *Urban Spaces in Contemporary China: The Potential for Autonomy and Community in Post-Mao China* (Cambridge: Cambridge University Press and Woodrow Wilson Center Press, 1995), pp. 362–366, 383–386. The authors make the point that students learned how to organize, lead, and follow in school. This prepared them for organizing so masterfully in Tiananmen Square. The same was true for the workers who participated in the 1989 protests "not as individuals or members of 'autonomous' unions but as members of *danwei* delegations, which were usually organized with either the direct support or the passive approval of work-group leaders, and which were generally led onto the streets by people carrying flags emblazoned with the name of the unit." p. 383.
25. In 1996–1997, the citizens of Beijing who were unable to sleep through the racket finally forced the government to pass a noise ordinance that lowered the decibel level allowed on streets by public performers, such as the fan and ballroom dancers.
26. Vaclav Havel, as quoted by Timothy Garton Ash, "Eastern Europe: The Year of Truth," *New York Review of Books* (February 15, 1990), p. 18, referenced in Giuseppe De Palma, "After Leninism: Why Democracy Can Work in Eastern Europe," *Journal of Democracy,* Vol. 2, No. 1 (Winter 1991), p. 25, note 3.
27. Liu Binyan, "China and the Lessons of Eastern Europe," *Journal of Democracy,* Vol. 2, No. 2 (Spring 1991), p. 8.
28. Beijing University student, "My Innermost Thoughts—To the Students of Beijing Universities" (May 1989), Document 68, in Suzanne Ogden, et al., eds., *China's Search for Democracy,* pp. 172–173.
29. Vivienne Shue, in a speech to a USIA conference of diplomats and scholars, as quoted and summarized in "Democracy Rating Low in Mainland," *The Free China Journal* (January 24, 1992), p. 7.

30. Joyce Barnathan, et al., "China: Is Prosperity Creating a Freer Society?" *Business Week* (June 6, 1994), p. 98.

31. These figures are a composite of figures from Information Office of the State Council of the People's Republic of China, "The Progress of Human Rights in China" (December 1995), *Beijing Review,* Special Issue (1996), pp. 11–12; and Shaoguang Wang, "The Politics of Private Time: Changing Leisure Patterns in Urban China," in Davis, et al., *Urban Spaces in Contemporary China,* from charts on pp. 162–163.

32. Information Office of the State Council of the People's Republic of China, "The Progress of Human Rights in China" (December 1995), *Beijing Review,* Special Issue (1996), pp. 11–12.

33. "Lexis-Nexis Country Report, 1999: China," http://www.lexis-nexis.com.

34. Wang Meng (former minister of culture and a leading novelist in China), speech at Cambridge University (May 23, 1996). An example of a movie banned in China is the famous producer Chen Kaige's *Temptress Moon.* This movie, which won the Golden Palm award at the Cannes Film Festival in 1993, is, however, allowed to be distributed abroad. The government has adopted a similar policy of censorship at home but distribution abroad for a number of films, including *Farewell My Concubine,* by China's most famous film directors.

35. Wang, in Davis, pp. 170–171.

36. Barnathan, et al., pp. 98–99.

37. For more on Chinese nationalism, see Suzanne Ogden, "Chinese Nationalism: The Precendence of Community and Identity Over Individual Rights," *Asian Perspective.* Vol. 25, N. 4 (2001), pp. 157–185.

38. In 1998, Avon was, at least temporarily, banned from China, as were other companies that used similar sales and marketing techniques. Too many Chinese found themselves bankrupted when they could not sell the products that they had purchased for resale.

39. It is rumored today that China has acquired Patriot missiles, used in the Gulf War with such vaunted success, from Israel.

40. For excellent analyses of the Sino–American relationship from the nineteenth century to the present, see Warren Cohen, *America's Response to China: A History of Sino–American Relations,* 3rd ed. (New York: Columbia University Press, 1990); Richard Madesen, *China and the American Dream: A Moral Inquiry* (Berkeley, CA: University of California Press, 1995); Michael Schaller, *The United States and China in the Twentieth Century,* 2nd ed. (New York: Oxford University Press, 1990); David Shambaugh, ed., *American Studies of Contemporary China* (Armonk, NY: M.E. Sharpe, 1993); and David Shambaugh, *Beautiful Imperialist: China Perceives America, 1972–1990* (Princeton, NJ: Princeton University Press, 1991).

41. Kenneth Lieberthal, the Charles Neuhauser Memorial Lecture, Harvard University (November 2002).

42. "Special Report: The SARS Epidemic: China Wakes Up," *The Economist* (April 26, 2003), p. 18.

Hong Kong Special Administration Region Map

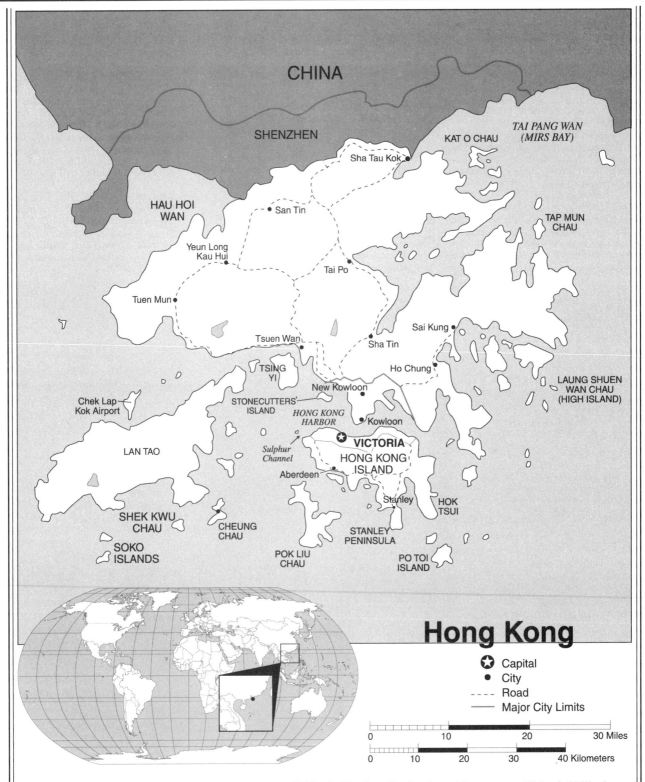

Hong Kong is comprised of the island of Hong Kong (1842), the Kowloon Peninsula and Stonecutters' Island (1860), the New Territories (1898) that extend from Kowloon to the Chinese land border; and 230 adjacent islets. Land is constantly being reclaimed from the sea, so the total land area of Hong Kong is continually increasing by small amounts. All of Hong Kong reverted to Chinese sovereignty on July 1, 1997. It was renamed the Hong Kong Special Administrative Region.

Hong Kong Special Administrative Region

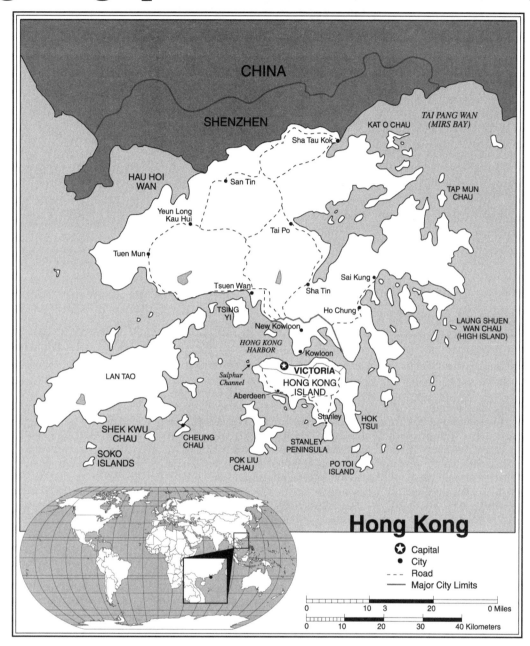

Hong Kong Statistics

GEOGRAPHY

Area in Square Miles (Kilometers): 671 (1,054) (about 6 times the size of Washington, D.C.)

Capital (Population): Victoria (na)

Environmental Concerns: air and water pollution

Geographical Features: hilly to mountainous, with steep slopes; lowlands in the north

Climate: tropical monsoon

PEOPLE

Population

Total: 6,810,000

Annual Growth Rate: 0.2%

Rural/Urban Population Ratio: 9/91

Major Languages: Chinese (Cantonese); English

Ethnic Makeup: 95% Chinese (mostly Cantonese); 5% others

Religions: 90% a combination of Buddhism and Taoism; 10% Christian

Health

Life Expectancy at Birth: 78.6 years (male); 84.3 years (female)

Infant Mortality: 2.4/1,000 live births

Physicians Available: 1.3/1,000 people

HIV/AIDS Rate in Adults: 0.1%

Education
Adult Literacy Rate: 93.5%

COMMUNICATION
Telephones: 3,801,300 main lines
Internet Users: 3,212,800

TRANSPORTATION
Highways in Miles (Kilometers): 1,135 (1,831)
Railroads in Miles (Kilometers): 22 (34)
Usable Airfields: 4

GOVERNMENT
Type: Special Administrative Region (SAR) of China
Head of State/Government: Chief Executive Donald Tsang
Political Parties: Democratic Alliance for the Betterment of Hong Kong; Democratic Party; Association for Democracy and People's Livelihood; Hong Kong Progressive Alliance; Citizens Party; Frontier Party; Liberal Party
Suffrage: direct elections universal at 18 for residents who have lived in Hong Kong for at least 7 years

MILITARY
Military Expenditures (% of GDP): defense is the responsibility of China
Current Disputes: none

ECONOMY
Currency ($ U.S. Equivalent): 7.79 Hong Kong dollars = $1
Per Capita Income/GDP: $28,800/$180 billion $22,988 (GDP)
GDP Growth Rate: 3.2%
Inflation Rate: -2.6%
Unemployment Rate: 7.9%
Natural Resources: outstanding deepwater harbor; feldspar
Agriculture: vegetables; poultry; fish; pork
Industry: textiles; clothing; tourism; electronics; plastics; toys; clocks; watches
Exports: $225.9 billion (primary partners China, US, Japan)
Imports: $230.3 billion (primary partners China, Taiwan, US, Japan, Singapore, South Korea)

SUGGESTED WEB SITES
http://www.info.gov.hk/
 sitemap.htm
http://www.odci.gov/cia/
 publications/factbook/geos/
 hk.html
http://www.state.gov
http://www.worldBANK.org

Hong Kong Report

Hong Kong, the "fragrant harbor" situated on the southeastern edge of China, was under British rule characterized by such epithets as the "pearl of the Orient," a "borrowed place living on borrowed time," and a "den of iniquity." The British colonial administration supported a market economy in the context of a highly structured and tightly controlled political system. This allowed Hong Kong's dynamic and vibrant people to shape the colony into one of the world's great success stories. The history of Hong Kong's formation and development, its achievements, and its handling of the difficult issues emanating from a "one country, two systems" formula since it was returned to Chinese rule in 1997 provide one of the most fascinating stories of cultural, economic, and political transition in the world. Hong Kong's "borrowed time" has ended; but its efforts to shape itself into the "Manhattan of China" are in full swing.

HISTORY

In the 1830s, the British sale of opium to China was creating a nation of drug addicts. Alarmed by this development, the Chinese imperial government banned opium; but private British "country traders," sailing armed clipper ships, continued to sell opium to the Chinese by smuggling it (with the help of Chinese pirates) up China's coast and rivers. In an effort to enforce the ban, the Chinese imperial commissioner, Lin Zexu, detained the British in Canton (Guangzhou) and forced them to surrender their opium. Commissioner Lin took the more than 21,000 chests of opium and destroyed them in public.[1] The British, desperate to establish outposts for trade with an unwilling China, used this siege of British warehouses as an excuse to declare war on the Chinese. Later called the Opium War (1839–1842), the conflict ended with China's defeat and the Treaty of Nanjing.

Great Britain did not wage war against the Chinese in order to sell an addictive drug that was banned by the Chinese government. Rather, the Chinese government's attack on the British opium traders, whose status as British citizens suddenly proved convenient to the British government, provided the necessary excuse for Great Britain getting what it really wanted: free trade with a government that had restricted trade with the British to one port, Canton. It also allowed London to assert Great Britain's diplomatic equality with China, which considered itself the "Central Kingdom" and superior to all other countries. The Chinese imperial government's demand that all "barbarians," including the British, kowtow to the Chinese emperor, incensed the British and gave them further cause to set the record straight.

The China trade had been draining the British treasury of its gold and silver species; for the British purchased large quantities of Chinese porcelain, silk, tea, and spices, while the Chinese refused to purchase the products of Great Britain's nineteenth-century Industrial Revolution. Smug in their belief that their cultural and moral superiority was sufficient to withstand any military challenge from a "barbarian" country, the Chinese saw no need to develop a modern military or to industrialize. An amusing example of the thought process involved in "Sinocentrism"—the belief that China was the center of the world and superior to all other countries—was Imperial Commissioner Lin's letter to Queen Victoria. Here he noted "Britain's dependence on Chinese rhubarb, without which the English would die of constipation."[2] China's narrow world view blinded it to the growing power of the West and resulted in China's losing the opportunity to benefit from the Industrial Revolution at an early stage. The Opium War turned out to be only the first step in a century of humiliation for China—the step that led to a British foothold on the edge of China.

For their part, the British public did not generally see the sale of opium as a moral issue, or that large-scale addiction was a possible outcome for China. Opium was available for self-medication in Britain, was taken orally (not smoked as it was in China), was administered as a tranquilliser for infants by the working class, and was not considered toxic by the British medical community at that time.[3] The Hong Kong colonial government remained dependent on revenues from the sale of opium until Hong Kong was occupied by Japan during WWII.[4]

The Treaty of Nanjing gave the British the right to trade with the Chinese from

New York Public Library

THE SECOND ANGLO/ CHINESE CONVENTION CEDES THE KOWLOON PENINSULA TO THE BRITISH

The second Anglo/Chinese Convention, signed in 1898, was the result of a string of incidents and hostilities among the Chinese, the British, and the French. Although the French were involved in the outbreak of war, they were not included in the treaty that resulted from conflict.

The catalyst for the war was that, during a truce, the Chinese seized the chief British negotiator and executed 20 of his men. In reprisal, the English destroyed nearly 200 buildings of the emperor's summer palace and forced the new treaty on the Chinese. This called for increased payments ("indemnities") by the Chinese to the English for war-inflicted damages as well as the cession of Kowloon Peninsula to the British.

five Chinese ports; and Hong Kong, a tiny island off the southern coast of China, was ceded to them "in perpetuity." In short, according to the practices of the colonizing powers of the nineteenth century, Hong Kong became a British colony forever. The Western imperialists were still in the acquisition phase of their history. They were not contemplating that one day the whole process of colonization might be reversed. As a result, Great Britain did not foresee that it might one day have to relinquish the colony of Hong Kong, either to independence or to Chinese rule.

Hong Kong Island's total population of Chinese villagers and people living on boats then numbered under 6,000. From 1842 onward, however, Hong Kong became the primary magnet for Chinese immigrants fleeing the poverty, chaos, and cruelty of China in favor of the relatively peaceful environment of Hong Kong under British rule. Then, in 1860, again as a result of a British victory in battle, the Chinese ceded to the British "in perpetuity" Stonecutters' Island and a small (3 1/2 square miles) but significant piece of land facing the island of Hong Kong: Kowloon Peninsula. Just a few minutes by ferry (and, since the 1970s, by tunnel) from Hong Kong Island, it became an important part of the residential, commercial, and business sectors of Hong Kong. The New Territories, the third and largest

part (89 percent of the total area) of what is now known as "Hong Kong," were not granted "in perpetuity" but were merely leased, to the British for 99 years under the second Anglo–Chinese Convention of Peking in 1898. The New Territories, which extended from Kowloon to the Chinese land border, comprised the major agricultural area supporting Hong Kong.

The distinction between those areas that became a British colony (Hong Kong Island and Kowloon) and the area "leased" for 99 years (the New Territories) is crucial to understanding why, by the early 1980s, the British felt compelled to negotiate with the Chinese about the future of "Hong Kong"; for although colonies are theoretically colonies "in perpetuity," the New Territories were merely leased, and would automatically revert to Chinese sovereignty in 1997. Without this large agricultural area, the rest of Hong Kong could not survive; the leased territories had, moreover, become tightly integrated into the life and business of Hong Kong Island and Kowloon.

Thus, with the exception of the period of Japanese occupation (1942–1945) during World War II, Hong Kong was administered as a British Crown colony from the nineteenth century onward. After the defeat of Japan in 1945, however, Britain almost did not regain control over Hong Kong because of the United States, which

insisted that it did not fight World War II in order to return colonies to its allies. But Britain's leaders, both during and after World War II, were determined to hold on to Hong Kong because of its symbolic, economic, and strategic importance to the British Empire. India, Singapore, Malaya, Burma—all could be relinquished, but not Hong Kong.[5] Moreover, although during World War II, a U.S. presidential order had stated that at the end of the war, Japanese troops in Hong Kong were to surrender to Chiang Kai-shek, the leader of the Republic of China, it did not happen. Chiang, more worried about accepting surrender of Japanese troops in the rest of China before the Chinese Communist could, did not rush to Hong Kong. Meanwhile, a British fleet moved rapidly to Hong Kong to pre-empt Chiang occupying Hong Kong, even though Chiang averred he would not have stayed. The British doubted this, and argued that Hong Kong was still British sovereign territory and would itself accept the surrender of the Japanese.[6]

At the end of the civil war that raged in China from 1945 to 1949, the Communists' Red Army stopped its advance just short of Hong Kong. Beijing never offered an official explanation. Perhaps it did not want to get into a war with Britian in order to claim Hong Kong, or perhaps the Chinese Communists calculated that Hong

(United Nations photo)

Hong Kong's economy is supported by a hardworking and dynamic population. The stall keepers at this outdoor market typify the intense entrepreneurial tendency of Hong Kong's citizens.

Kong would be of more value to them if left in British hands. Indeed, at no time after their victory in 1949 did the Chinese Communists attempt to force Great Britain out of Hong Kong, even when Sino-British relations were greatly strained, as during China's "Cultural Revolution."[7]

This did not mean that Beijing accepted the legitimacy of British rule. It did not. After coming to power on the mainland in 1949, the Chinese Communist Party held that Hong Kong was a part of China stolen by British imperialists, and that it was merely "occupied" by Great Britain. Hence the notion of Hong Kong as "a borrowed place living on borrowed time." The Peo-

ple's Republic of China insisted that Hong Kong *not* be treated like other colonies; for the process of decolonization has in practice meant sovereignty and freedom not for a former colony's people.[8] China was about to allow Hong Kong to become independent. After the People's Republic of China gained the China seat in the United Nations in 1971, it protested the listing of Hong Kong and Macau (a Portuguese colony) as colonies by the UN General Assembly's Special Committee on Colonialism. In a letter to the Committee, Beijing insisted they were merely

part of Chinese territory occupied by the British and Portuguese authori-

ties. The settlement of the questions of Hong Kong and Macao is entirely within China's sovereign right and does not at all fall under the ordinary category of colonial territories. Consequently they should not be included in the list of colonial territories covered by the declaration on the granting of independence to colonial countries and peoples. ... The United Nations has no right to discuss these questions.[9]

No doubt the Chinese Communists were ideologically uncomfortable after winning control of China in 1949 in proclaiming China's sovereign rights and spouting

Communist principles while at the same time tolerating the continued existence of a capitalist and British-controlled Hong Kong on its very borders. China could have acquired control within 24 hours simply by shutting off Hong Kong's water supply from the mainland. But China profited from the British presence there and, except for occasional flare-ups, did little to challenge it.

China made it clear that, unlike other colonies, Hong Kong's colonial subjects did not have the option of declaring independence, for overthrowing British colonial rule would have led directly to the re-imposition of China's control. And although there is for the Hong Kong Chinese a cultural identity as Chinese, after 1949 few wanted to fall under the rule of China's Communist Party government. Furthermore, Beijing and London as a rule did not interfere in Hong Kong's affairs, leaving these in the capable hands of the colonial government in Hong Kong. Although the colonial government formally reported to the British Parliament, in practice it was left to handle its own affairs. Still, the colonial government did not in turn cede any significant political power to its colonial subjects.[10]

By 1980, the Hong Kong and foreign business communities had grown increasingly concerned about the expiration of the British lease on the New Territories in 1997. The problem was that all land in the New Territories (which by then had moved from pure agriculture to becoming a major area for manufacturing plants, housing, and commercial buildings) was *leased* to businesses or individuals, and the British colonial government could not grant any land lease that expired after the lease on the New Territories expired. Thus, all land leases—regardless of which year they were granted—would expire three days in advance of the expiration of the main lease on the New Territories on July 1, 1997. As 1997 grew steadily closer, then, the British colonial government had to grant shorter and shorter leases. Investors found buying leases increasingly unattractive. The British colonial government felt compelled to do something to calm investors.[11]

For this reason, it was the British, not the Chinese, who took the initiative to press for an agreement on the future status of the colony and the rights of its people. Everyone recognized the inability of the island of Hong Kong and Kowloon to survive on their own, because of their dependence upon the leased New Territories for food, and because of the integrated nature of the economies of the colonial and leased parts of Hong Kong. Everyone (everyone, that is, except for British prime minister Margaret Thatcher) also knew that Hong Kong was militarily indefensible by the British and that the Chinese were unlikely to permit the continuation of British administrative rule over Hong Kong after it was returned to Chinese sovereignty.[12] So, a series of formal Sino-British negotiations over the future of Hong Kong began in 1982. By 1984, the two sides had reached an agreement to restore all three parts of Hong Kong to China on July 1, 1997.

The Negotiations Over the Status of Hong Kong

Negotiations between the People's Republic of China and Great Britain over the future status of Hong Kong got off to a rocky start in 1982. Prime Minister Thatcher set a contentious tone for the talks when she claimed, after meeting with Chinese leaders in Beijing, that the three nineteenth-century treaties that gave Great Britain control of Hong Kong were valid according to international law; and that China, like other nations, had an obligation to honor its treaty commitments. Thatcher's remarks infuriated China's leaders, who denounced the treaties that resulted from imperialist aggression as "unequal," and lacking legitimacy in the contemporary world.

Both sides realized that Chinese sovereignty over Hong Kong would be reestablished in 1997 when the New Territories lease expired, but they disagreed profoundly on what such sovereignty would mean in practice. The British claimed that they had a "moral commitment" to the people of Hong Kong to maintain the stability and prosperity of the colony. Both the British and the Hong Kong population hoped that Chinese sovereignty over Hong Kong might be more symbolic than substantive and that some arrangement could be worked out that would allow for continuing British participation in the administration of the area. The Chinese vehemently rejected what they termed "alien rule in Chinese territory" after 1997, as well as the argument that the economic value of a Hong Kong *not* under its administrative power might be greater.[13] Great Britain agreed to end its administration of Hong Kong in 1997, and together with China worked out a detailed and binding arrangement for how Hong Kong would be governed under Chinese sovereignty.

The people of Hong Kong itself did not formally participate in these negotiations over the colony's fate. Although the British and Chinese consulted various interested parties in the colony, they chose to ignore many of their viewpoints. China was particularly adamant that the people of Hong Kong were Chinese and that the government in Beijing represented *all Chinese* in talks with the British.

In September 1984, Great Britain and the People's Republic of China initialed the Joint Declaration on the Question of Hong Kong. It stated that, as of July 1, 1997, Hong Kong would become a "Special Administrative Region" (SAR) under the control of the central government of the People's Republic of China. The Chinese came up with the idea of "one country, two systems," whereby, apart from defense and foreign policy, the Hong Kong SAR would enjoy near autonomy. Hong Kong would maintain its current social, political, economic, and legal systems alongside China's systems; would remain an international financial center; and would retain its ability to establish independent economic (but not diplomatic) relations with other countries.

The Sino-British Joint Liaison Group was created to oversee the transition to Chinese rule. Any changes in Hong Kong's laws made during the transition period, if they were expected to continue after 1997, had to receive final approval from the Joint Liaison Group. If there were disagreement within the Liaison Group between the British and Chinese, they were obligated to talk until they reached agreement. This procedure gave China veto power over any proposed changes in Hong Kong's governance and laws proposed from 1984 to 1997.[14] When London's newly appointed governor, Christopher Patten, arrived in 1992 and attempted to change some of the laws that would govern Hong Kong after 1997, China had reason to use that veto power.

The Basic Law

The Basic Law is the crucial document that translates the *spirit* of the Sino-British Joint Declaration into a legal code. Often referred to as a "mini-constitution" for Hong Kong after it became an SAR on July 1, 1997, the Basic Law essentially defines where Hong Kong's autonomy ends and Beijing's governance over Hong Kong begins. The British had no role in formulating the Basic Law, as the Chinese considered it an internal, sovereign matter. In 1985, China established the Basic Law Drafting Committee, under the direction of the National People's Congress (NPC). The Committee had 59 members—36 from the mainland, 23 from Hong Kong. Of the latter, almost all were "prominent figures belonging to high and high-middle strata," with Hong Kong's economic elite at its core. In addition, China established a "Consultative Committee" in Hong Kong of 180 members. Its purpose was to function as a nonofficial representative organ of the people of Hong Kong from all walks of life, an organ that would channel their viewpoints to the Basic Law Drafting Committee. By so including Hong Kong's elite and a Hong Kong-wide civic representative organ in consultations about the Basic Law, China hoped to provide political legitimacy to the Basic Law.[15] Once the Basic Law was approved in April 1990 by China's NPC, the final draft was promulgated.

The Basic Law gave Hong Kong a high degree of autonomy after 1997, except in matters of foreign policy and defense, which fell under Beijing's direct control. The government was to be made up of local civil servants and a chief executive chosen by an "Election Committee" appointed by the Standing Committee of the National People's Congress.[16] The chief executive was given the right to appoint key officials of the Special Administrative Region (subject to Beijing's approval). Provisions were made to allow some British and other foreign nationals to serve in the administration of the SAR, if the Hong Kong government so desired. An elected Legislature was made responsible for formulating the laws.[17] The maintenance of law and order remained the responsibility of local authorities, but China took over from the British the right to station military forces in Hong Kong. The local judicial and legal system were to remain basically unchanged, but China's NPC reserved the right to approve all new laws written between 1990 and 1997.[18]

Thus, the Joint Declaration and Basic Law brought Hong Kong under China's rule, with the National People's Congress in Beijing accorded the right of the final interpretation of the meaning of the Basic Law in case of dispute; but the Basic Law allows Hong Kong considerable independence over its economy, finances, budgeting, and revenue until the year 2047. China is thus committed to preserving Hong Kong's "capitalist system and lifestyle" for 50 years and has promised not to impose the Communist political, legal, social, or economic system on Hong Kong. It also agreed to allow Hong Kong to remain a free port, with its own internationally convertible currency (the Hong Kong dollar), over which China would not exercise authority. The Basic Law states that all Hong Kong residents shall have freedom of speech, press, publication, association, assembly, procession, and demonstration, as well as the right to form and join trade unions, and to strike. Freedom of religion, marriage, choice of occupation, and the right to social welfare are also protected by law.[19]

Beijing agreed to continue to allow the free flow of capital into and out of Hong Kong. It also agreed to allow Hong Kong to enter into economic and cultural agreements with other nations and to participate in relevant international organizations as a separate member. Thus, Hong Kong was not held back from membership in the World Trade Organization (WTO) by China's earlier inability to meet WTO membership qualifications. Similarly, Hong Kong is a separate member of the World Bank, the Asian Development Bank, and the Asian-Pacific Economic Conference (APEC). Hong Kong is also allowed to continue issuing its own travel documents to Hong Kong's residents and to visitors.

FREEDOM

Hong Kong was a British colony, and residents enjoyed some of the civil liberties guaranteed by British law. Under the Basic Law of 1997, the Chinese government has agreed to maintain the capitalist way of life and civic freedoms for 50 years.

When China promulgated the Basic Law in 1990, Hong Kong residents by the thousands took to the streets in protest, burning their copies of the Basic Law. Some of Hong Kong's people saw Britain as having repeatedly capitulated to China's opposition to plans for political reform in Hong Kong before 1997, and as having traded off Hong Kong's interests in favor of Britain's own interests in further trade and investment in China. Hong Kong's business community, however, supported the Basic Law, believing that it would provide for a healthy political and economic environment for doing business. Other Hong Kong residents believed that it was Hong Kong's commercial value, not the Basic Law, that would protect it from a heavy-handed approach by the Chinese government.

The Joint Declaration of China and Great Britain (1984), and the Basic Law (1990) are critical to understanding China's anger in 1992 when Governor Patten proceeded to push for democratic reforms in Hong Kong without Beijing's agreement—particularly since Patten's predecessor, Governor David Wilson, always did consult Beijing and never pushed too hard. After numerous threats to tear up the Basic Law, Beijing simply stated in 1994 that, after the handover of Hong Kong to Chinese sovereignty in 1997, it would nullify any last-minute efforts by the colonial government to promote a political liberalization that went beyond the provisions in the Basic Law. And that is precisely what China did on July 1, 1997. As is noted later in this report, the changes that Patten advocated were largely last-ditch efforts to confer on Hong Kong's subjects democratic rights that they had never had in more than 150 years of British colonial rule. These rights related largely to how the Legislature was elected, the expansion of the electorate, and the elimination of such British colonial regulations as one requiring those who wanted to demonstrate publicly to first acquire a police permit.

The Chinese people were visibly euphoric about the return of Hong Kong "to the embrace of the Motherland." The large clock in Beijing's Tiananmen Square counted the years, months, days, hours, minutes, and even seconds until the return of Hong Kong, helping to focus the Chinese people on the topic. Education in the schools, special exhibits, the movie *The Opium War* (produced by China), and even T-shirts displaying pride in the return of Hong Kong to China's control reinforced a sense that a historical injustice was at last being corrected. On July 1, 1997, celebrations were held all over China, and the pleasure was genuinely and deeply felt by the Chinese people. In Hong Kong, amidst a drenching rain, celebrations were also held. At midnight on June 30, 1997, 4000 guests watched as the Union Jack was lowered, and China's flag, together with the new Hong Kong Special Administrative Region flag, were raised. President Jiang Zemin, together with Charles, the Prince of Wales, and Tony Blair, Prime Minister of Great Britain, represented their respective countries, but the Hong Kong people were mere spectators, without an official representative at the handover.[20]

THE SOCIETY AND ITS PEOPLE

Immigrant Population

In 1842, Hong Kong had a mere 6,000 inhabitants. Today, it has almost 7 million people. What makes this population distinctive is its predominantly immigrant composition. Waves of immigrants have flooded Hong Kong ever since 1842. Even today, barely half of Hong Kong's people was actually born there. This has been a critical factor in the political development of Hong Kong; for instead of a foreign government imposing its rule on submissive natives, the situation has been just the reverse. Chinese people voluntarily emigrated to Hong Kong, even risking their lives to do so, to subject themselves to alien British colonial rule.

In recent history, the largest influxes of immigrants came as a result of the 1945–1949 Civil War in China, when 750,000 fled to Hong Kong; as a result of the "three bad years" (1959–1962) following the economic disaster of China's Great Leap Forward policy; and from 1966 to 1976, when more than 500,000 Chinese went to Hong Kong to escape the societal turmoil generated by the Cultural Revolution. After the

The refugees who came to Hong Kong and settled in squatter communities such as the one shown above voluntarily subjected themselves to foreign (British) rule. Government-built housing has largely replaced areas like the one pictured above.

Vietnam War ended in 1975, Hong Kong also received thousands of refugees from Vietnam as that country undertook a policy of expelling many of its ethnic Chinese citizens. Many Chinese from Vietnam risked their lives on small boats at sea to attain refugee status in Hong Kong.

Although China's improving economic and political conditions beginning in 1979 greatly stemmed the flow of immigrants from the mainland, the absorption of refugees into Hong Kong's economy and society remained one of the colony's biggest problems. Injection of another distinct refugee group (the Chinese from Vietnam) generated tension and conflict among the Hong Kong population.

Because of a severe housing shortage and strains on the provision of social services, the British colonial government first announced that it would confine all new refugees in camps and prohibit them from outside employment. It then adopted a policy of sending back almost all refu-

gees who were caught before they reached Hong Kong Island and were unable to prove they had relatives in Hong Kong to care for them. Finally, the British reached an agreement with Vietnam's government to repatriate some of those Chinese immigrants from Vietnam who were believed to be economic rather than political refugees. The first few attempts at this reportedly "voluntary" repatriation raised such an international furor that the British were unable to systematize this policy. By the mid-1990s, however, better economic and political conditions in Vietnam made it easier for the British colonial government to once again repatriate Vietnamese refugees.

Before the July 1, 1997, handover, moreover, Beijing insisted that the British clear the camps of refugees. It was not a problem that China wanted to deal with. As it turned out, the British failed to clear the camps, leaving the job to the Chinese after the handover. The last one was closed in

the summer of 2000. Today, China still maintains strict border controls, in an effort to protect Hong Kong from being flooded by Chinese from the mainland who are hoping to take advantage of the wealthy metropolis, or just wanting to look around and shop in Hong Kong.[21]

The fact that China's economy has grown rapidly for the last 25 years, especially in the area surrounding Hong Kong, has diminished the poverty that led so many Mainlanders to try illegally emigrating to Hong Kong. Nevertheless, overpopulation is still an important social issue because it stretches Hong Kong's limited resources and has contributed to the high levels of unemployment in recent years. Rulings that have greatly limited the right of Mainlanders with at least one Hong Kong parent (so-called "right-of-abode" seekers) to migrate to Hong Kong have eased concerns somewhat.

Today, the largest number of immigrants to Hong Kong are still Mainlanders; but

(United Nations photo/A. Jongen)

Even in modern-day Hong Kong, Chinese cultural values are still very strong. Here, women in traditional Chinese dress take a work break.

they are more likely than in the past to come from distant provinces. Although Hong Kong is made up almost entirely of immigrants and their descendants, the older immigrants look down upon their new non-Cantonese-speaking country cousins as "uncivilized." They are socially discriminated against and find it difficult to get the better-paying jobs in the economy. This is in striking contrast to past attitudes: From the 1960s to the 1980s, Hong Kong residents generally expressed deep sympathy with Mainlanders, building roof-top schools for them, and throwing food onto the trucks when the British colonial government transported Mainlanders back to China against their will.[22]

Language and Education

Ninety-eight percent of Hong Kong's people are Chinese. The other 2 percent is primarily European and Vietnamese. Although a profusion of Chinese dialects are spoken, the two official languages, English and the Cantonese dialect of Chinese, predominate. Since the Chinese written language is in ideographs, and the same ideographs are usually used regardless of how they are pronounced in various dialects, all literate Hong Kong Chinese are able to read Chinese newspapers—and 95 percent of them do read at least one of the 16 daily newspapers available in Hong Kong.[23] Even before the handover, moreover, the people were intensively studying spoken Mandarin, the official language of China.

Since the handover in 1997, a source of bubbling discontent has been the decision

HEALTH/WELFARE

Schooling is free and compulsory in Hong Kong through junior high school. The government has devoted large sums for low-cost housing, aid for refugees, and social services. Housing, however, tends to be cramped and inadequate because of the continuous influx of immigrants.

of the Executive Council to require all children to be taught in Chinese. The government's rationale was that the students would learn more if they were taught in their own language. This decision caused an enormous furor. Many Hong Kong Chinese, especially from the middle classes, felt that if Hong Kong were going to remain a major international financial and trading center, its citizens must speak English. Many suspected that the real reason for insisting on Chinese was to respond to Beijing's wishes to bind the Hong Kong people to a deeper Chinese identity. In response to strong public pressures, the Hong Kong government finally relented and allowed 100 schools to continue to use English for instruction. Many of the other schools are, Hong Kong parents complain, suffering from a decline in the quality of education generally, and language skills in particular. Those who can afford it now try to get their children into the growing number of private schools.

Chinese cultural values of diligence, willingness to sacrifice for the future, commitment to family, and respect for education have contributed to the success of

Hong Kong's inhabitants. (The colonial government guaranteed nine years of compulsory and free education for children through age 15, helping to support these cultural values.) As a result, the children of immigrants have received one of the most important tools for material success. Combined with Hong Kong's rapid post-World War II economic growth and government-funded social-welfare programs, education improved the lives of almost all Hong Kong residents, and allowed remarkable economic and social mobility. A poor, unskilled peasant who fled across China's border to Hong Kong to an urban life of grinding poverty—but opportunity—could usually be rewarded by a government-subsidized apartment, and by grandchildren who graduated from high school and moved on to white-collar jobs.

ACHIEVEMENTS

Hong Kong has held together a society where the gap between rich and poor is enormous. Refugees have gained education, social accepetance, and opportunities for economic advancement. The real wealth of Hong Kong is its talented, educated, and hardworking people.

Since the 1997 handover, the Hong Kong SAR has continued to support compulsory and free education. But in this rapidly changing society, juvenile delinquency is on the rise, because parents are working long hours and spend little time with their children, and an increasing number of those

(Photo by Lisa Clyde)

A fishing family lives on this houseboat in Aberdeen Harbor. On the roof, strips of fish are hung up to dry.

who finish the basic nine years of school now leave the educational system. Criminal gangs *(triads)* recruit some of them to promote criminal activities.[24]

For those who wish to continue their education beyond high school, access to higher education is limited, so ambitious students work hard to be admitted to one of the best upper-middle schools, and then to one of the even fewer places available in Hong Kong's universities. Hong Kong's own universities are becoming some of the best in the world, and admission is highly competitive. An alternative chosen by many of Hong Kong's brightest students is to go abroad for a college education. This has been important in linking Hong Kong to the West.

Living Conditions

Hong Kong has a large and growing middle class. By 1995, in fact, Hong Kong's per capita income had surpassed that of its colonial ruler, Great Britain.[25] Its people are generally well-dressed; and restaurants, buses, and even subways are full of people yelling into their cellphones. Enormous malls full of fashionable stores can be found throughout Hong Kong. McDonald's is so much a part of the cityscape that most residents do not realize it has American origins. After school, the many McDonald's are full of teenagers meeting with their friends, doing homework, and sharing a "snack" of the traditional McDonald's meal—hamburgers, fries, and Coca-Cola (served hot or cold). The character Ronald McDonald, affectionately referred to by his Cantonese name, Mak

Dong Lou Suk Suk (Uncle McDonald), is recognized throughout Hong Kong.[26]

Nevertheless, Hong Kong's people suffer from extremes of wealth and poverty. The contrast in housing that dots the landscape of the colony dramatically illustrates this. The rich live in luxurious air-conditioned apartments and houses on some of the world's most expensive real estate. They may enjoy a social life that mixes such Chinese pleasures as mahjong, banqueting, and participation in traditional Chinese and religious rituals and festivals with British practices of horseracing, rugby, social clubs, yacht clubs, and athletic clubs for swimming and croquet[27] (but these practices have faded greatly with the exit of the British). The Chinese have removed the name "Royal" from all of Hong Kong's old social clubs, including the Royal Jockey Club, which today is neither royal nor British. Times have changed: Hong Kong is no longer a colony, and exclusivity is a thing of the past. The mix of Hong Kong people and foreigners in business clubs reflects the fact that Hong Kong's business elite is now both British and Chinese. Of course, memberships in some clubs are still all-British, like cricket clubs, because the Chinese are not interested.[28]

The wealthy are taken care of by cooks, maids, gardeners, and chauffeurs, most of whom are Filipino. The Hong Kong people—and the government—greatly prefer hiring Filipinos to hiring mainlander Chinese, as the former are better trained, speak English, and are less likely to try to apply for permanent residency in Hong Kong. (The preference for hiring Filipinos adds to the sense the Chinese Mainlanders

have of being discriminated against.) Virtually all members of the Filipino workforce (estimated to be about 200,000) have Sundays off, and they more or less take over the parks, and even the streets in Hong Kong Central, where they camp out for the day with their compatriots. The government dares not intervene to get them off the streets, lest it be accused of racism or violating their civil liberties, and of complaints by Filipinios of exploitation and abuse by Hong Kong employers.

There is a heavier concentration of Mercedes, Jaguars, Rolls Royces, and other luxury cars—not to mention cellular phones, car faxes, and fine French brandy—in Hong Kong than anywhere else on Earth.[29] Some Hong Kong businessmen spend lavishly on travel, entertainment, and mistresses. Their mistresses, however, now tend to live in Shenzhen, the "Special Economic Zone" (SEZ) just 40 minutes by train from Hong Kong Central, because purchasing an apartment for them and paying for their upkeep are cheaper there, as are the karaoke bars, golf courses, bars, and massage parlors. Hong Kong is now for the wealthier and more sedate businessmen, while Shenzhen is more attractive to Hong Kong's young professionals.

The wealthiest have at least one bodyguard for each member of their families. Conspicuous consumption and what in the West might be considered a vulgar display of wealth are an ingrained part of the society. Some Hong Kong businesspeople, who have seemingly run out of other ways to spend their money, ask for restaurants to decorate their food with gold leaf. Others spend several hundred thousand dollars just to buy a lucky number for their car license plate. One Hong Kong businessman built a mansion in Beijing that replicates the style of the Qing Dynasty and features countless brass dragons on the ceilings, decorated with three pounds of gold leaf.[30]

In stark contrast to the lifestyles of the wealthy business class, the vast majority of Hong Kong's people live in crowded high-rise apartment buildings, with several poor families sometimes occupying one apartment of a few small rooms,[31] and having inadequate sanitation facilities. Beginning in the mid-1950s, the British colonial government built extensive low-rent public housing, which today accommodates about half of the population. These government-subsidized housing projects easily become run down and are often plagued by crime. But without them, a not-insignificant percentage of the new immigrant population would have continued to live in squalor in squatter villages with no running water, sanitation, or electricity.

When he took office in 1997, Hong Kong's chief executive, Tung Chee-hwa, made a significant commitment to provide more government-funded housing and social-welfare programs—some 85,000 new apartments were to be built each year. But in 2000, Tung suddenly announced that he had scrapped that policy back in 1998 because of falling real-estate prices (owing to the Asian economic crisis)—without telling anyone. The result is a housing scarcity, and the rents for the apartments in the few new residential buildings are far beyond the means of the average Hong Kong resident.

The Economy

A large part of the allure of Hong Kong has, then, been its combination of a dynamic economy with enlightened social-welfare policies. The latter were possible not just because of the British colonial government's commitment to them but also because the flourishing Hong Kong economy provided the resources for them. Hong Kong had a larger percentage of the gross domestic product (GDP) available for social welfare than most governments, for two reasons. First, it had a low defense budget to support its approximately 12,000 British troops (including some of the famous Gurkha Rifles) stationed in the colony for external defense (only 0.4 percent of the GDP, or 4.2 percent of the total budget available). Second, the government was able to take in substantial revenues (18.3 percent of GDP) through the sale of land leases.[32]

Now, of course, China is in charge of Hong Kong's defense; but although land

leases came to an end in 1997 with the return of Hong Kong's leased territories to China, the sale of land is still the primary source of government revenue. Continuing the system established by the British colonial government, the Hong Kong government makes money by selling land one piece at a time to its handful of real-estate developers, who then build on the land.[33] This is how Hong Kong can continue to finance governmental expenditures without any income tax, sales taxes, or capital-gains tax; and with a low profit tax (16.5 percent) for corporations, and a flat 15 percent tax on salaries that kicks in at such a high level that a large percentage of those working in Hong Kong pay nothing at all.

The result is that Hong Kong is a great place to do business, but the cost of private housing is beyond the means of most of Hong Kong's middle class; and, having made its money by selling land to the land developers, the government must then turn around and use a substantial portion of that money to subsidize public housing so that rent is affordable. Half of Hong Kong's populace receive housing free or at minimal (subsidized) rent, but the middle class cannot afford to buy housing. Many have moved to more affordable housing in the Shenzhen Special Economic Zone across the border in China. This may explain in part why the Hong Kong population has fallen by at least 300,000 people in the last few years.

Hong Kong's real estate system has affected the people's viewpoint on housing. Unlike in the West, people do not view the purchase of an apartment as a place where they might want to reside for an indefinite period. Rather, they see it more like buying a stock, and they might buy and sell it within one year in order to make a profit. Before the Asian financial crisis that hit Hong Kong in 1997, this was a fairly sure bet; but housing prices have fallen since then, and many people have lost their savings by speculating in housing.

Beijing's greatest concern before the handover in 1997 was that the colonial administration dramatically increased welfare spending—65 percent in a mere five years.[34] From Beijing's perspective, the British appeared determined to empty Hong Kong's coffers, leaving little for Beijing to use elsewhere in China. The Chinese believed that the British were setting a pattern to justify Hong Kong's continuing expenditures in the next 50 years of protected autonomy. As it turned out, however, the British did not try to deplete Hong Kong assets, and China's companies were deeply involved even in the vastly expensive new airport that the British insisted on

building before their departure. (It opened in July 1998, one year after the handover.)

No sooner had China taken back Hong Kong than, unexpectedly, the Asian economic financial crisis began to emerge, first in Thailand, and then throughout Asia. It wreaked havoc on Hong Kong's economy and challenged its financial and economic system. Beijing, instead of interfering with the decisions made by Hong Kong to address the crisis, took a hands-off approach, except to offer to support the Hong Kong dollar against currency speculators by using China's own substantial foreign-currency reserves of U.S.$150 billion to sustain the Hong Kong dollar's peg to the U.S. dollar. Furthermore, China did not take the easy route of devaluing the Chinese yuan, which would have sent Asian markets, and Hong Kong's in particular, into a further downward spiral.

The Asian financial and economic crisis (1997–2002) brought a severe downturn in living conditions in Hong Kong. Growing increasingly anxious about the situation, the Hong Kong population heavily pressured their government to step in and do more to ease the pain. They demanded that the government take a more activist role in providing social welfare, controlling environmental degradation, and regulating the economy. In response to such pressure, the government required taxis to stop using diesel fuel and enacted some of the toughest emissions standards in the world. In addition, Chief Executive Tung Chee-hwa announced that the government would commit even more to social welfare than it had under British rule. In addition, the Hong Kong government intervened in the financial markets, purchasing large amounts of Hong Kong dollars to foil attempts by speculators to make a profit from selling Hong Kong dollars. It also intervened in the Hong Kong stock market, using up some 25 percent of its foreign-currency reserves to purchase large numbers of shares in Hong Kong companies in a risky effort to prevent a further slide of the stock market. This strategy worked, and government investments in the stock market had doubled in value by the end of 1998.

Hong Kong's woes were aggravated, however, by the world economic turndown that began in 2000, made worse by the September 11, 2001 terrorist attacks on the United States, and by the outbreak of the SARS (severe acute respiratory syndrome) epidemic in early 2003 in the next-door Chinese province of Guangdong. It quickly spread to Hong Kong and shows how vulnerable Hong Kong's geographical position on the edge of China has made it; for the life-threatening illness (with close to a 5 percent mortality rate) and the initial reluctance of China to furnish information

Land is so expensive in Hong Kong that most residences and businesses today are located in skyscrapers. While the buildings are thoroughly modern, construction crews typically erect bamboo scaffolding as the floors mount up, protected by netting. The skyscrapers that appear darker in color in the center of this photo are being built with this technique.

about the spread of the epidemic in China led to disaster for Hong Kong's tourist industry and the cancellation of countless business trips to Hong Kong.

Before the 1997 return of Hong Kong to China, many analysts predicted that Beijing would undercut Hong Kong's prosperity through various political decisions limiting political freedom, tampering with the legal system, and imposing economic regulations that would endanger growth. Instead, Beijing appears to have left Hong Kong in the hands of Chief Executive Tung. Tung, however, proved unable to save Hong Kong's economy from the Asian financial crisis and rapidly lost popularity. To wit, although the economy has now recovered, its growth rate of 3.3% remains well below half that of the China Mainland. The local press has severely criticized Tung for having taken too literally Beijing's promise to the rest of the world in 1997 that Hong Kong would not change. Because Tung has done so little to change Hong Kong in the years since the handover, many worry that Hong Kong's dominant position in Asia as an entrepôt may be lost to Shanghai and the other ports being built along the coast of China; and Hong Kong's role as a financial and economic center lost to Singapore. Even worse, Taiwan has recently emerged as the Asian leader in computer technology, while Hong Kong has fallen further behind in the technology sweepstakes.

The consensus is that Chief Executive Tung and his government lacked a "vision" of where Hong Kong should go and how Hong Kong should reshape itself.[35] The government is blamed for being beholden to the real-estate "tycoons"—not a "class" of wealthy people but a mere half-dozen individuals—who hold the vast majority of Hong Kong's wealth in their hands[36] and who, having made sky-high profits from real estate in the 1990s, did not see the need to do anything more visionary. It is only because of the collapse of Hong Kong's real estate prices in the wake of the Asian economic crisis that they began to reorient themselves and turn to technology. The Internet is also empowering small entrepreneurs. But Hong Kong's economy is plagued with other problems emanating from monopolistic and duopolistic control of many sectors in the economy.[37] A more cynical view is that Hong Kong businesspeople are sycophants to both the Hong Kong and Beijing governments, because their customary operating procedure is to seek favors in exchange for loyalty. In other words, they make money not through creativity but through their connections—a practice common on the mainland as well.

As of 2005, Hong Kong's real estate and stock markets had recovered a significant amount of the value lost during the Asian financial crisis. For the time being, Hong Kong's greatest protection against substantial turmoil and a prolonged recession is the stability and growth of China's econ-

omy, which continues to bubble along at a healthy average of 8 percent per year.

In truth, however, Hong Kong's and Beijing's best efforts probably will not be adequate to address Hong Kong's recession if Japan, whose economy is larger than all the other Asian economies combined, does not reform its own financial system and pull itself out of its own recession through new policies. Indeed, the major reason why so many of Hong Kong's hotels and restaurants are operating at a low percentage of normal capacity, and so many of its retail shops and businesses in the entertainment sector have gone bankrupt, is because the Japanese are staying at home.

Hong Kong as a World Trade and Financial Center

From the beginning, the British designated Hong Kong as a free port. This has meant that Hong Kong has never applied tariffs or other major trade restrictions on imports. Such appealing trade conditions, combined with Hong Kong's free-market economy, deepwater harbor, and location at the hub of all commercial activities in Asia, have made it an attractive place for doing business. Indeed, from the 1840s until the crippling Japanese occupation during World War II, Hong Kong served as a major center of China's trade with both Asia and the Western world.

The outbreak of the Korean War in 1950 and the subsequent United Nations embargo on exports of strategic goods to

China, as well as a U.S.-led general embargo on the import of Chinese goods, forced Hong Kong to reorient its economy. To combat its diminished role as the middleman in trade with the mainland of China, Hong Kong turned to manufacturing. At first, it manufactured mainly textiles. Later, it diversified into other areas of light consumer goods and developed into a financial and tourist center.

Today, Hong Kong continues to serve as a major trade and financial center, with thousands of companies (including Taiwanese ones) still located in Hong Kong for the purpose of doing business with China. Instead of being a middleman in trade with China, however, Hong Kong has actually shifted its own manufacturing base into China. Back in 1980, when almost half of Hong Kong's workforce labored in factories and small workshops, Hong Kong was on the verge of pricing itself out of world markets because of its increasingly well-paid labor and high-priced real estate. Just then, China, with its large and cheap labor supply, inexpensive land, and abundant resources, initiated major internal economic reforms that opened up the country for foreign investment. As a result, Hong Kong transferred its manufacturing base over the border to China, largely to the contiguous province of Guangdong. By the late 1990s, more than 75 percent of the Hong Kong workforce was in the service sector, with only 13 percent remaining in manufacturing. Now it is some 5 million Chinese laborers who work in Hong Kong factories on the mainland.[38]

DEVELOPMENT

Hong Kong is one of the preeminent financial and trading dynamos of the world. Hong Kong's fine harbor, its recently built $21 billion airport, a new Disney theme park, and information-technology "cyberport" have helped fuel its economy; but the Asian financial crisis that began in 1997, followed by a worldwide economic downturn, has made it difficult for Hong Kong to recover its former strength. The SARS outbreak in 2003 further impaired its economy.

Hong Kong's investment accounts for some two thirds of China's total foreign investment, and a full 80 percent of Guangdong's total. Hong Kong owns 18,000 factories in Guangdong and contracts out processing work to another 20,000 companies.[39] More than one third of China's total trade is through Hong Kong. Whether Hong Kong's transfer of its manufacturing and processing base to China's provinces contributed to its current problems or saved it from greater dif-

ficulties during the Asian financial and economic crisis is a matter of debate.[40]

Hong Kong's many assets, including its hard-working, dynamic people, have made it into the world's largest container port; third-largest center for foreign-exchange trade; seventh-largest stock market; and, until recently, Asia's largest banking and financial center (with 79 of the world's 100 largest banks in Hong Kong). Hong Kong has also been a major processing and re-exporting center, making it the world's 10th-largest trading economy.[41] Considering its tiny size and population, these are extraordinary achievements.

Many residents of Hong Kong worry that Shanghai may soon prove a serious threat to Hong Kong's economic growth and position as a financial and trade center. There is no reason, however, why there would not be room for another major financial and trade center in Asia, or to expect that Shanghai's growth would necessarily come at Hong Kong's expense. Singapore presents the greatest challenge to Hong Kong's position as a financial center. South Korea has also challenged Hong Kong's growth; for once China established full diplomatic relations with its government in 1992, it was allowed to deal directly with China, thereby bypassing Hong Kong as an entrepôt for trade and business with China.

The Special Economic Zones (SEZs)

As part of its economic reform program and "open door" policy beginning in 1979, China created Special Economic Zones in areas bordering or close to Hong Kong in order to attract foreign investment. SEZs, until recent years under far more liberal regulations than the rest of China, have blossomed in the 1980s and 1990s. Various branches of China's government have themselves invested heavily in the SEZs, in hopes of making a profit. In the 1990s, even China's military developed an industrial area catering to foreign investors and joint ventures in one of China's SEZs, Shenzhen, as part of its effort to compensate for insufficient government funding for the military. It called its policy "one army, two systems"—that is, an army involved with both military and economic development.[42] Brushing aside its earlier preference for a puritanical society, China's military was as likely to invest in nightclubs, Western-style hotels, brothels, and health spas in the SEZs as it was in the manufacturing sector. In 1998, however, Beijing ordered the military to divest itself of its economic enterprises, so it no longer runs nonmilitary enterprises in Shenzhen.

The bulk of foreign investment in the SEZs, and in the rest of China, comes from Hong Kong Chinese, either with their own

money or acting as middlemen for investors from Taiwan, the United States, and others. Most direct foreign investment in China, in fact, comes *through* Hong Kong, either by setting up companies in Hong Kong, or using Hong Kong companies as intermediaries. In turn, China is the single largest investor in Hong Kong, and its state-owned enterprises and joint ventures also set up companies in Hong Kong. Thus, this integrated area of south China, encompassing Hong Kong, the SEZs, and the provinces of Guangdong, Fujian, and Hainan Island, has become a powerful new regional economy on a par with other Asian "little dragons."

Indeed, even before China took over Hong Kong in 1997, south China had already become an integral part of Hong Kong's empire, with profound political as well as economic implications. Many Hong Kong people who regularly cross into Shenzhen SEZ (and even own property there) pressure the Shenzhen government to be responsive to their interests. Hong Kong's media coverage of Shenzhen affairs also puts pressure on the SEZ's administrators to be more responsive to public concerns, such as the exploitation of Mainlanders working under contract in Hong Kong–owned firms.[43]

At the same time, Hong Kong's growing ties with the SEZs and cross-border trade generally is causing serious problems, including the restructuring of the workforce as jobs in the manufacturing sector move across the Hong Kong border for cheaper labor. The result is a downward pressure on wages in Hong Kong, although many economists believe this will make Hong Kong more competitive. In any event, the Hong Kong business community is moving to integrate the economy even more fully with the mainland to take advantage of China's future growth.

Sensitivity of the Economy to External Political Events

Hong Kong's economic strength rests on its own people's confidence in their future—a confidence that has fluctuated wildly over the years. When Beijing undertook economic retrenchment policies, partially closed the "open door" to international trade and investment, engaged in political repression, or rattled its sabers over Taiwan, Hong Kong's stock market would gyrate, its property values decline, and foreign investment would go elsewhere. Not knowing what the transition to Chinese sovereignty would bring, Hong Kong's professional classes emigrated at the rate of about 60,000 people per year between 1990 and 1997. This drain of both talent and money out of the colony

Even under China's control, the strong work ethic of Hong Kong's small entrepreneurs continues.

was as serious a concern for China as it was for Hong Kong.

London's refusal to allow Hong Kong citizens to emigrate to the United Kingdom contributed to a sense of panic among the middle and upper classes in Hong Kong—those most worried about their economic and political future under Communist rule. Other countries were, however, more than willing to accept these well-educated, wealthy immigrants, who came ready to make large deposits in their new host country's banks. Once emigrants gained a second passport (a guarantee of residency abroad in case conditions warrant flight), however, they tended to return to Hong Kong, where there are still opportunities for entrepreneurs and those in the professions, such as doctors, architects, and engineers.

Beijing's verbal intimidation of Hong Kong dissidents who criticized China in the period following the Tiananmen crackdown in 1989, and again when Governor Christopher Patten began whipping up Hong Kong fervor for greater democratic reforms from 1992 to 1997, also aroused

anxiety in the colony. Earlier, Hong Kong had been afraid that Great Britain would trade the colony's democratic future for good relations with China. Patten's efforts to inject Hong Kong with a heavy dose of democratization before 1997, on the other hand, led to concern that China might clamp down on political freedoms after 1997. A significant portion of Hong Kong public opinion turned against Patten out of such concerns. Many in Hong Kong (especially the businesspeople who had invested heavily in both Hong Kong and China) wondered whether more democracy was worth the risks.[44] In the event, most of those concerns never materialized.

Thus far, China's sovereignty over Hong Kong does not seem to have had a negative effect on its economy. Occasionally China's statements concerning Hong Kong's economy, judicial system, or politics have sent shock waves throughout the colony, causing the Hang Sang stock market to take a nose dive out of fears that China would ignore the principles in the Basic Law guaranteeing Hong Kong's 50 years of autonomy. Simi-

larly, whether due to the Chinese government's incompetence or an intentional effort to conceal information about the spread of SARS on the mainland, the 2003 SARS epidemic had a catastrophic impact on Hong Kong's economy. Such volatility demonstrates just how sensitive Hong Kong is to Beijing's policies and actions. Nevertheless, in the years since the handover, close to 2 million people from Hong Kong have relinquished their British passports for Hong Kong Special Administrative Region passports; so it appears that an increasing number of people feel that their future lies with Hong Kong and participation in China's overall growth.

China is deeply concerned that Hong Kong not be destabilized by its policies or by misinterpretation of its statements. Indeed, since the handover, Beijing has exercised unusual restraint so as to avoid being seen to interfere in Hong Kong's affairs. For example, Beijing no longer permits Chinese ministry officials to visit or oversee their counterparts in Hong Kong without clearance, lest it be interpreted as interference. Furthermore, it is the Hong Kong government, not the Ministry of Foreign Affairs office in Hong Kong, that deals with all of the foreign consulates in Hong Kong. And, unlike the British Commonwealth Office, which always sent copies of government documents to the Hong Kong government, China's Ministry of Foreign Affairs does not, again to avoid being accused of interference.[45] For example, if the ministry were to send out copies of documents on Beijing's position on matters relating to Taiwan or Tibet, or even about Beijing's views on WTO issues or relationships with the United States and other countries, inevitably this would be perceived as pressuring the Hong Kong government to adopt the same position.

Because Beijing has tread lightly in Hong Kong, most businesses already located there have remained. In fact, many foreign corporations rushed to establish themselves in Hong Kong before the handover in order to avoid the unpredictable, lengthy, and expensive bureaucratic hassle of trying to gain a foothold in the China mainland lying beyond Hong Kong. Even Taiwan's enterprises in Hong Kong are standing firm; for without direct trade and transport links between China and Taiwan, Hong Kong is still the major entrepôt for trade between the two places. Many Chinese mainland corporations also establish footholds in Hong Kong to ease the problem of foreign hard-currency transfer and to avoid a host of other difficulties that plague mainland businesses.

Nevertheless, the overall profile of the foreign business community has changed substantially since 1997. The percentage of Americans and Europeans doing business in Hong Kong has increased significantly, while many British have gone home, leaving their companies in the hands of capable Hong Kong Chinese managers, a reflection of the end of the colonial era in the business community as well as in the political system.

Crime

Although Hong Kong is still characterized by a high level of social stability a high crime rate continues to plague society. For more than a decade, ordinary criminality has been steadily augmented by crime under the control of competing Chinese triads. This is in part because the housing and community and mutual-aid groups of the 1970s and 1980s, which used to help the police track down criminals, have disappeared. Their disappearance has also contributed to increasing juvenile delinquency.[46] Opium, largely controlled by the triads, continues to be used widely by the Chinese. As a commentator once put it:

Opium trails still lead to Hong Kong … and all our narcotic squads and all the Queen's men only serve to make the drug more costly and the profits more worthwhile. It comes in aeroplanes and fishing junks, in hollow pipes and bamboo poles and false decks and refrigerators and pickle jars and tooth paste tubes, in shoes and ships and sealing wax. And even cabbages.[47]

Today, Hong Kong is still one of the largest entrepôts for drugs, and the number of drug addicts is skyrocketing. This is in no small part because social and economic liberalization on the mainland has allowed its people to move about freely. Hong Kong triads work in collaboration with triads across the border. Young people cross the border to Shenzhen to buy drugs (which usually originated in Myanmar (Burma), which they then smuggle back across Hong Kong's border. In turn, those recruited in Shenzhen provide a base in the mainland for Hong Kong triads to deal in drugs, prostitutes, and guns, and to set up underground banks and transport illegal immigrants across the border.[48]

Hong Kong and Chinese mainland drug squads cooperate, but their working together is complicated by the fact that Hong Kong's legal system still differs from the legal system of the rest of China. The critical difference is that Hong Kong does not have capital punishment. And China, committed to not changing Hong Kong's legal system for 50 years, has not pressured Hong Kong to change its law on the death penalty. Dozens of crimes such as drug dealing, punishable by execution in the rest of China, will result at most in a life sentence in Hong Kong.

Before the handover, when Hong Kong investigators asked the Chinese to turn over drug dealers to the Hong Kong authorities, the Chinese expended significant resources to find the criminals and turn them over. But when the Chinese asked the Hong Kong drug authorities to do the same, they went so far as to arrest the suspects, but refused to turn them over to China's public security office because of the fairly strong chance that a person convicted on charges of selling drugs in China would be executed. At first China wanted to copy the Singapore model of executing drug dealers, but it soon realized that would mean the execution of thousands. Now only the biggest drug dealers are executed. (The problems emanating from cross-border crime are discussed further in this report, under the topic "The Judiciary, Law, and Order.")

Hong Kong's organized crime has long been powerful in the areas of real estate;

extortion from massage parlors, bars, restaurants, and clubs; illegal gambling; smuggling; the sale of handguns (illegal for ordinary people to purchase); prostitution; and drugs. And, as is common in other Asian countries, gangs are often hired by corporations to deal with debtors and others who cause them difficulties. Triads have also expanded into kidnapping for ransom, and taken on some unexpected roles. As an example, when the British governor in Hong Kong, Christopher Patten, upset Beijing with his proposals for further democratization of Hong Kong before 1997, the Chinese Communist regime (by way of its estimated 60,000 supporters working in Hong Kong) allegedly recruited triad members to begin harassing those within the Hong Kong government who were supporting Patten's proposals. (And, when Patten's dog disappeared one day in 1992 during the crisis stage of Sino-British relations, one rumor had it that the Chinese Communists had kidnapped the dog and were going to ransom it in exchange for halting political reform in Hong Kong. The other rumor was that Patten's pet had been flown into China to be served up for breakfast to Deng Xiaoping. Of course, neither rumor was true.)

POLITICS AND POLITICAL STRUCTURE

Politics and the political structure have changed greatly since the days of Hong Kong's colonial government, when the British monarch, acting on the advice of the prime minister, would appoint a governor, who presided over the Hong Kong government's colonial administration. Colonial rule in Hong Kong may be characterized as benevolent, consultative, and paternalistic, but it was nonetheless still colonial. Although local people were heavily involved in running the colony and the colonial government interfered very little in the business activities and daily lives of Hong Kong Chinese, the British still controlled the major levers of power and filled the top ranks in the government.

The colony's remarkable political stability until the handover in 1997 was, then, hardly due to any efforts by the British to transplant a form of Western-style democracy to Hong Kong. But, the colonial Hong Kong government did seek feedback from the people through the hundreds of consultative committees that it created within the civil service. Similarly, although the British ultimately controlled both the Legislative Council (LegCo) and Executive Council (ExCo), these governmental bodies allowed Hong Kong's socioeconomic elites to participate in the administration of the colony, even if they were unable to participate in the *formulation* of policy. Some 300 additional advisory groups as well as numerous partly elected bodies—such as the municipal councils (for Hong Kong Island and Kowloon), the rural committees (for the New Territories), and district boards—also had considerable autonomy in managing their own affairs. This institutionalized consultation among Chinese administrators and the colonial government resulted in the colony being governed by an elite informed by and sensitive to the needs of the Hong Kong people. As was common to British colonial administration elsewhere, the lower levels of government were filled with the local people. Rarely was political dissent expressed outside the government.[50]

The relatively high approval rating of British colonial rule helps explain why only a small portion of the mere 6 percent of registered voters actually voted. With the government assuring both political stability and strong economic growth, the people of Hong Kong spent most of their time and energy on economic pursuits, not politics. In any event, given the limited scope of democracy in Hong Kong, local people had little incentive to become politically involved. For this reason, as the handover came nearer, Hong Kong residents grew increasingly concerned that there were few competent and trustworthy leaders among the Hong Kong Chinese to take over.[51] They also worried that a government controlled by leaders and bureaucrats who held foreign passports or rights of residence abroad would not be committed to their welfare. Although Beijing withdrew its demand before the handover that all governmental civil servants swear an oath of allegiance to the government of China and turn in their British passports, many did so anyway (including the incoming chief executive, Tung Chee-hwa).

The colonial government remained stable, then, because it was perceived to be trustworthy, competent, consultative, and capable of addressing the needs of Hong Kong's people. Most Hong Kong citizens also believed that a strong political authority was indispensable to prosperity and stability, and they worried that the formation of multiple political parties could disrupt that strong authority. Thus, what is seen in the West as a critical aspect of democracy was viewed by the people in Hong Kong as potentially destabilizing.

Nevertheless, by the late 1980s, many Hong Kong Chinese began to demand that democratic political reforms be institutionalized before the Chinese Communists took over in 1997. The ability of the departing colonial government to deal with these increased pressures to democratize

Hong Kong was, however, seriously constricted by the 1982 Joint Declaration and the Basic Law of 1990, which required Beijing's approval before the British could make any changes in the laws and policies governing Hong Kong. The people of Hong Kong awoke to the fact that their interests and those of the colonial government were no longer compatible. Britain's policy toward Hong Kong had become a mere appendage of British policy toward China, and the status quo was frozen. The Hong Kong colonial government had, essentially, lost its independence to Beijing and London.[52]

What was "handed over" on July 1, 1997, was sovereign control of Hong Kong. Hong Kong became a Special Administrative Region of China, with Beijing guaranteeing autonomy for 50 years in the political, legal, economic, and social realms. But Hong Kong would be governed by its new "constitution," the Basic Law, written by China's leaders. This document provided for certain changes to be made *after* the handover. Notably, Article 23 required the Hong Kong Legislative Council to outlaw treason, succession, subversion, and sedition, as well as other activities that could endanger China's national security. That is, Hong Kong was expected to outlaw, and punish, those individuals and organizations operating in Hong Kong who in the view of Beijing might pose a threat to China's security; but for five years, nothing was done to define how Article 23 could be implemented (see below).

Similarly, for five years after the 1997 handover, there was no real change in the government—except, of course, that Hong Kong's chief executive reported to Beijing, not London. The structure of the post-1997 government, outlined in great detail in the Basic Law, was to be, like Hong Kong's colonial government: structured on a separation of powers among the executive, legislative, and judicial branches of government, serving to check the arbitrary use of power by any single individual or institution of the government. This separation of powers, however, never did exist within the framework of a representative democracy. Thus, the government today remains similar in many respects to what it was under colonial rule, with power continuing to be centralized in the executive branch. LegCo cannot even initiate substantive legislation without the approval of the chief executive, or hold the chief executive accountable for his or her actions. In effect, Hong Kong under the Basic Law has retained the colonial model put in place by the British in the nineteenth century.

China has made some changes in Hong Kong's government to bring it into greater

correspondence with its own government structure, even if in some cases this is merely a matter of changing names. In 2002, under a new "ministerial" system, Beijing changed the Basic Law to allow the chief executive to appoint all 14 policy secretaries in the cabinet of 20. (The other 6 are leading politicians, including two heads of progovernment parties, and close personal advisers.) Previously, cabinet secretaries were senior civil servants who were, at least in theory, politically neutral. Now that they are appointed, they serve at the chief executive's discretion, which means that their political views weigh heavily in their selection. This has raised further questions concerning the accountability of the government to the Legislature.[53]

The Executive Council is run by Hong Kong's chief executive. Fortunately, continuity was maintained in the first years after the handover, when most of the cabinet heads under British colonial rule agreed to serve under Tung Chee-hwa. Sole decision-making authority remains vested in the chief executive, although ultimately Beijing must approve of any of ExCo's policies. So far, Beijing has chosen not to exercise this approval in a way that hampers the chief executive's policy-making authority.

Hong Kong's residents have been remarkably satisfied with Beijing's handling of Hong Kong affairs. The Hong Kong Transition Project has, since 1982, annually taken the pulse of Hong Kong through an extensive survey of public opinion on a variety of issues related to the impact of the transition of Chinese sovereignty. The level of satisfaction has been consistently much higher in the posthandover period than in the 15 years leading up to it.[54] In particular, the percentage of those satisfied with the performance of the PRC government in dealing with Hong Kong affairs remained unusually high. In fact, Hong Kong public opinion in Nov. 2003 continued to give a far higher approval rating to Beijing's leadership than to Hong Kong's chief executive: Those "satisfied" or "very satisfied" with Beijing's dealing with Hong Kong affairs (polled in November 2003) were 72 percent of those polled; whereas for Chief Executive Tung, the approval rating in the same poll was a humiliating 21 percent, a stunning fall from the height of his popularity one year after he took office (49 percent were satisfied with his performance in July 1998). In fact, the satisfaction with the overall performance of the Hong Kong government had fallen from 66 percent approval in June 1997 (still under British rule), to 20 percent by Nov. 2003.[55]

The Hong Kong government has constructed many apartment complexes in an effort to address the severe overcrowding in tenements and squatter communities.

Tung's problems with leadership are understandable: As a colony, Hong Kong had bred strong civil servants, but no real political leaders who had had opportunities to make political decisions. So Beijing had to pick someone whom they felt was safe, but who inevitably lacked real leadership experience. Tung Chee-hwa is a businessman, an elitist who opposes the welfare state and supports "Asian values" of a strong central leadership and obedient citizenry,[56] was quickly eroded by his inability to effectively address a number of problems, including the fallout for Hong Kong of the Asian financial and economic crisis that began within a few months of his entering office. Tung was blamed not only for plummeting land prices and the government's incompetence, but also for his reluctance to intervene in order to relieve the people's suffering that resulted from the crisis. The infrequency of his appearances in front of LegCo, and his refusal to consult with LegCo representatives even in private, caused significant conflict and anger. Although it was the executive branch's role to deal with the financial crisis in Hong Kong, LegCo still believed that it should have been consulted.[57]

Beijing had not been happy with Tung since they had him reappointed for a second term in 2002. Nor were they happy that the people of Hong Kong were dissatisfied with his performance. (Indeed, 58 percent of Hong Kong people polled in late 2003 said they would support China's President Hu Jintao and Premier Wen Jiabao if they were to dismiss Tung for his performance[58] China's leaders criticized his performance, and were particularly concerned with the growing number of demonstrations and protests directed at his government. Suddenly, in March 2005, Tung resigned, apparently at the request of Beijing. Because two years remained in Tung's term as Chief Executive, it caused a major constitutional crisis because the Basic Law had not anticipated how to deal with a mid-term replacement. As this book went to press, it appeared that Beijing would choose Donald Tsang, a member of the Executive Council and an impressive and highly competent leader, to fill out the term that ends in 2007.

Tung's efforts to get the economy back on track had, as noted above, been stymied greatly by events beyond his control—the Asian financial crisis, the global recession

that followed the terrorist attacks on the United States on September 11, 2001; and the SARS epidemic that erupted in 2003. Nevertheless, the public and the press continued to fault the Hong Kong government, not external forces (or Beijing), for Hong Kong's difficulties. Because Tung had tended to be reactive, not proactive, he was seen as a weak leader. He was also blamed for the dearth of good advisers in ExCo, since he made these appointments for political reasons. They engaged in what is known locally as "small-circle politics"—that is, small elites interlocked through networks that can mobilize quickly. Members of this network were simultaneously Hong Kong's economic and political elite, and they tended to believe that only they understood what would make Hong Kong successful and stable.[59] However, more than half of those polled in 2002 expressed dissatisfaction with the government's performance on reducing unemployment (79 percent), implementing educational reform (65 percent), and improving medical services (53 percent).[60]

Hong Kong residents generally believe, then, that Hong Kong's problems arise not from Beijing's control but, rather, from the incompetence of their own government. So the critical question for the next Chief Executive, as it was for Tung, is no longer how autonomous he is of Beijing's control. Instead, it is whether he can establish his legitimacy as chief executive. In any event, most Hong Kong people support direct election of the chief executive over the current system, in which the person is appointed by the Beijing-controlled Election Committee and would prefer direct elections for all LegCo members and district council members.[61]

Article 23 of the Basic Law

As noted earlier, Article 23 outlaws treason, secession, subversion, and sedition. It also prohibits the theft of "state secrets," political activities of foreign political organizations in Hong Kong, and the establishment of ties with foreign political organizations by political organizations in Hong Kong. But while it had been left up to the Hong Kong government to give teeth to Article 23 so that it could be implemented, it had not done so. The purpose of Article 23 is to protect the national security of China as a whole by defining the types of activities in Hong Kong that would be punishable by law. It was one thing to define terms such as "subversion" as it applied to the rest of mainland China; it was another thing to define these terms for residents of Hong Kong. This had to be done without jeopardizing the fundamental rights given to Hong Kong people in the Basic Law, some of which have not been given to the rest of the Chinese.

The terrorist attacks of 9/11 spurred the Hong Kong government's Security Bureau to submit a bill to amend Article 23. In early 2003, after a year-long period of public consultation, Chief Executive Tung submitted a draft bill to LegCo. This predictably caused a public uproar, as it brought back fears that Beijing might have pressured the government to limit the guaranteed rights of people in Hong Kong. The public was particularly worried about a constriction of freedom of the media to report news; and about their rights to demonstrate and protest, to associate with foreign organizations or with organizations banned on the mainland, to access Internet sites, and to organize pressure groups and political groups. There was also concern that the new definitions of "treason," "subversion," and "sedition" might mean that journalists who managed to get hold of unpublished government documents, or refused to reveal their anonymous sources, could be charged with "theft of state secrets."[62]

The concerns about the potential limits on individual freedom by Article 23 spilled over into a concern for the rule of law in

Hong Kong, which is the basis for protecting all other freedoms. The possibility that the Hong Kong government might resort to secret trials on cases involving proscription of certain local organizations, and that "mere association between a local organization and a proscribed one on the Mainland" might be grounds for arrest, also caused anxiety. On July 1, 2003, 500,000 people demonstrated in Hong Kong against the draft bill. Finally, after further demonstrations, and Beijing's realization that the suggested amendments were destabilizing Hong Kong and causing hostility toward Beijing, the bill to amend Article 23 was withdrawn without any date for reconsideration. The impact of this is dramatically illustrated by polls which asked about the performance of China's government in dealing with Hong Kong affairs: In June 2003, before the July 1st demonstrations, there was a 57 percent satisfaction rate with Beijing; 6 months later, after the withdrawal of the legislation, the satisfaction rate shot up to 72 percent.[63] The fact that Beijing in this case catered to the will of the Hong Kong people seems to have won enormous good will for China.

Political Parties and Elections

On July 1, 1997, China immediately repealed Governor Patten's expansion of the franchise for the 1995 elections, which had lowered the voting age to 18 and extended the vote to all of Hong Kong's adult population (adding 2.5 million people to the voting rolls)—without consulting Beijing. As the Joint Declaration required that China had to agree to even a small increase in democracy in the Hong Kong colony, Beijing was within its rights when it repealed the changes and replaced the Legislature elected under those rules. A 400-member "Provisional Legislature" was then selected by China's Preparatory Committee to replace it on July 1, 1997.

What were Beijing's concerns when it canceled the results of the 1995 elections? China's leaders viewed the last-ditch colonial government's efforts to develop a representative government in Hong Kong as part of a conspiracy to use democracy to undercut China's rule in Hong Kong after 1997. They argued, and no doubt truly believed, that the last-minute political reforms could, in fact, jeopardize Hong Kong's prosperity and stability by permitting special interests and political protest to flourish; and that Hong Kong's social problems—narcotics, violence, gangs, prostitution, an underground economy—required that Hong Kong be controlled, not given democracy. China therefore adopted a status quo approach to Hong Kong. After all, Hong Kong's political system under

British colonial control had been imposed from the outside. This system worked well and kept Hong Kong stable and prosperous. Beijing merely wanted to replace a colonial ruler with a Chinese Communist Party ruler.[64]

Beijing did not want a situation in which those citizens living in Hong Kong enjoyed substantially more rights than the rest of China's population. This could easily lead to pressure on Beijing to extend those rights to all Chinese. If Beijing is going to give greater political rights to its people, it would rather do so on its own timetable.

In the 1995 elections, Hong Kong's Democratic Party, whose political platform called for major changes to the Basic Law in order to give Hong Kong people more democratic rights, won two thirds of the vote. When Beijing canceled the results of that election, many commentators thought the message was clear. But as promised, Beijing rescheduled the elections for May 1998; and when those elections led to the reelection of the same number of Democratic Party members as before, Beijing made no effort to remove them from the legislature.[65]

In the fall 2004 LegCo elections, the Democratic Party and their allies won 25 of the 60 seats, meaning that the pro-government/pro-Beijing parties still control the legislature after several rounds of elections. In this last election, however, only 30 of the seats were chosen by direct election from the citizenry. The other 30 were chosen by what are called "functional constituencies," which tend to be more conservative, pro-business, and pro-Beijing. So, even though the democratic camp won the majority of the electoral vote, they still were not able to gain a majority of the seats.

It was, nevertheless, a disappointing showing for the democracy camp, which has been pressing for advancing the timetable for direct elections, when the terms of the Chief Executive (2007) and members of LegCo (2008) expire. Many analysts attribute the electoral results to the preference of the electorate for better local governance over deadlines for democratization. The democracy camp lacks a meaningful platform of political, social, and economic policies that would improve conditions in Hong Kong and be attractive to the electorate. Analysts also point to the disarray within the democracy camp, which seems unable to agree on anything other than that more democracy would be a good thing, and that they should stand in opposition to government policies.

Beijing has warned the pro-democracy group not to move forward with any plan to hold a referendum on direct elections before the original timetable called for them. Combined with this warning, however, have been significant efforts by China to enhance economic integration between Hong Kong and the mainland provinces bordering it, as well as making it easier for Chinese tourists to go to Hong Kong—a potential boon to the flagging tourist industry in Hong Kong.[66] This may help explain why the pro-Beijing forces retained control of LegCo.

Under British colonial rule, ordinary citizens were rarely involved in politics. Today, there is much greater involvement, but a lingering suspicion of and lack of enthusiasm for politics and political parties in Hong Kong. The lack of a coherent platform and strong leadership in the democracy camp, as well as its failure to bring about change in government policies, contributes to a sense of political inefficacy—a sense that there is little ordinary citizens can do to shape policy. This view is also influenced by a belief that there is a not altogether healthy alliance between big business and government, a view reinforced when Tung appointed still more business people and those working in the private sector to his second term cabinet.

On the other hand, the government has the support of some remarkably strong pro-Beijing parties, especially the Democratic Alliance for the Betterment of Hong Kong (DAB). (It may not be appropriate to call the DAB pro-Beijing, but it does make more efforts to work with China for common goals and opposes Beijing far less frequently than many in the democracy camp do.) The DAB has joined up with other "patriotic" (that is, pro-China/pro-government) parties in a loose coalition.[67] The coalition works *with* the government to make policies focused on environmental, social, and economic issues. Many are suspicious of the DAB ties to the government, and concerned about the funding of the DAB by the Chinese government; but it and Hong Kong's "patriotic" organizations are highly effective at mobilizing people at the local level and responding to their concerns on issues important to their daily lives.

It is difficult for the Democratic Party to make inroads on the seats assigned to functional constituencies, which usually go to pro-government business people. It does seem unfair, indeed undemocratic, that a mere 180,000 individuals could vote for the 30 functional constituency seats, while 3.5 million voted for the other 30 seats.

The complicated electoral system virtually guarantees business groups dominance in the Legislature. This is one factor accounting for both Beijing and the Hong Kong government's tolerating the Democratic Party and its political protests, even when the protesters publicly denounce Beijing's leaders. That is, with rare exceptions (such as the protests led by the democracy camp against

the bill to amend Article 23), their protests are without consequences. Besides, LegCo is a rather powerless political body. It is poorly funded, so legislators cannot afford the kind of research support that is essential to making effective policy proposals. And while in session, LegCo meets only once a week, at which time legislators give speeches written for them by their staffs.[68] This is hardly the stuff of legislative debate and policy making.

Beijing has tolerated the gadfly role of the Democratic Party, no doubt hoping that this hands-off policy will help bring Taiwan into negotiations for unification with the mainland sooner. The Hong Kong government has also kept channels open to Democratic Party leaders. The Democratic Party has, in turn, adopted a more moderate stance toward both Beijing and the government than it might otherwise have done—a position facilitated by the departure of several of the party's more radical leaders—and the public's concerns that it not stir up problems that could destabilize Hong Kong.[69]

The Democratic Party's minority position within the Legislature condemns it to the position of a critic and complainer.[70] In the post-transition period, the Hong Kong people believe that its demands for greater democracy have not been appropriate to the problems at hand. These problems include the "bird flu," which threatened public health and forced the government to kill the entire stock of chickens in Hong Kong; the red tide, which killed hundreds of thousands of fish; a disastrous opening of the new airport in 1998; the right-of-abode seekers in Hong Kong; and, the drop in the stock market, and real-estate values, increased unemployment, bankruptcies of retail stores, and a dramatic decline in tourism. These were not the sorts of problems for which the Democratic Party's call for changes in the Basic Law, a timetable for democratization, and opposition to Beijing have been relevant. Indeed, the vast majority of the public have wanted government intervention to address these problems. Thus, the many street demonstrations that occur tend not to be directed to broad demands for democracy but, rather, to issues such as the above. They reflect the frustration of the population in the legislature's ineffectiveness in shaping government policy.

The Legal System and the Judiciary

Under colonial rule, Hong Kong's judiciary was independent, and it should remain so under the "one country, two systems" envisioned in the Basic Law. Judges are appointed and serve for life. English common law, partly adapted to ac-

commodate Chinese custom, has been at the heart of the legal system. Much of the confidence in Hong Kong as a good place to live and do business has been based on the reputation of its independent judiciary for integrity and competence, and the stability of the legal and constitutional system, and Hong Kong's adherence to the rule of law.

China has promised to continue to allow Hong Kong's legal system to rely on such legal concepts as habeas corpus, legal precedent, and the tradition of common law, which do not exist in China. Beijing has said that it will not subject Hong Kong's legal system to the Communist Party Politburo's guidelines for the rest of China. But on the political matters such as legislation, human rights, civil liberties, and freedom of the press, the Basic Law offers inadequate protection. For example, the Basic Law provides for the Standing Committee of the National People's Congress (NPC) in Beijing, not the Hong Kong courts, to interpret the Basic Law and to determine whether future laws passed by the Hong Kong Legislature conflict with the Basic Law.

Furthermore, although Beijing had promised Hong Kong that the chief executive will be accountable to the Legislature, the Basic Law gives the chief executive— *to be appointed by China's NPC until at least the year 2007*—the power to dissolve the Hong Kong Legislature and veto bills. Of even greater concern, Beijing has yet to state the relationship of Hong Kong's Basic Law to China's own Constitution. The fundamental incompatibility between the British tradition (in which the state's actions must not be in conflict with the laws) and China's practice of using law as a tool of the state, as well as China's conferring and withdrawing rights at will, is at the heart of the concern about Chinese rule over Hong Kong.[71]

The first case to be tried by the Hong Kong courts after July 1, 1997, was that of an American streaker who ran nude in crowded downtown Hong Kong and was easily apprehended. In this rather amusing case, the court fined and released the defendant. Serious questions soon arose, however, about the handling of cross-border crime. As noted earlier, the source of the problem is that Hong Kong does not have capital punishment, and China does. In the past, when Hong Kong has arrested a P.R.C. citizen wanted for a criminal activity that is punishable in China by execution, it has refused to turn the person over to China's judicial authorities.

So it is not surprising that when, in 1998, the Chinese arrested a major Hong Kong criminal who, like so many, tried to seek refuge from Hong Kong authorities across the border, the Chinese took custody of the case. The gangster, known as "Big Spender," was alleged to have committed a number of crimes, including armed robbery, smuggling, and kidnapping of two Hong Kong tycoons. China averred that some of the crimes were actually committed in China, and that the victims of the crimes reported them to China's public security bureau, not to the Hong Kong authorities. Hong Kong did not believe that Big Spender, a Hong Kong citizen, could be tried in China for *any* crimes he committed in Hong Kong, because of its judicial independence under the "one country, two systems" structure. In spite of pressures on the government, after a brief trial in China, Big Spender was found guilty and summarily executed. Another case, in which a Chinese citizen allegedly murdered five people in Hong Kong and was then arrested after he crossed the border back into China, raised no questions of jurisdiction.

The problem is, in short, that Hong Kong and Beijing need to come together on the difficult problem of cross-border crime when two different criminal codes may have to be used to deal with a crime.[72] The more porous borders between Hong Kong and the rest of China, the greater mobility of the Chinese people, and the open market economy have seemed like an open invitation to Hong Kong's triads not only to expand their crime rings within China, but also to take refuge there.

Polls have indicated that residents are satisfied with the Basic Law and the rule of law in Hong Kong, though many did not actually know about the provisions of the Basic Law.[73] In addition, most residents appear to think that Beijing has adhered to the Basic Law, but they have not been pleased that the Hong Kong government has several times dragged Beijing into Hong Kong affairs. The most notable example was the Hong Kong government's request to China's National People's Congress in 1999 to rule on the "right of abode" for Mainlanders with at least one Hong Kong parent. (At the time of the NPC's ruling, 1.4 million Mainlanders were claiming eligibility.) The Hong Kong judiciary's decision was overturned, a result that came as a great relief to Hong Kong residents who were more afraid of overcrowding than they were of Beijing's interference. Nevertheless, as of 2005, the right to abode is still an uncomfortable issue; for although there is sympathy for these settlers' plight (many have lived and worked in Hong Kong for most of their lives), the settlers are usually the fruit of relationships between mainland Chinese women and married Hong Kong men who

frequently go to the mainland for business, a point that does not sit well with many citizens, especially Hong Kong women.

Public Security

Under the "one country, two systems" model, Hong Kong continues to be responsible for its own public security. The British colonial military force of about 12,000 has been replaced by a smaller Chinese People's Liberation Army (PLA) force of about 9,000. The military installations used by the British forces have been turned over to the PLA. In efforts to reassure the Hong Kong populace that the primary purpose of the military remains the protection of the border from smuggling and illegal entry, and for general purposes of national security, soldiers are mostly stationed across the Hong Kong border in Shenzhen. Some are also stationed, however, in heavily populated areas, to serve as a deterrent to social unrest and mass demonstrations against the Chinese government. No doubt the high cost of keeping soldiers in metropolitan Hong Kong was also a factor in the PLA decision to move most of its troops to the outskirts.

Apart from diminishing the interactions between the PLA soldiers and the Hong Kong people, China has done much to ensure that the troops do not become a source of tension. To the contrary, China wants them to serve as a force in developing a positive view of Chinese sovereignty over Hong Kong. As part of their "charm offensive," the soldiers must be tall (at least 5 feet 10 inches); have "regular features" (that is, be attractive); be well read (with all officers having college training, and all ordinary soldiers having a high school education); know the Basic Law; and be able to speak both the local dialect of Cantonese and simple English.[74] In short, the PLA, which has no such requirements for its regular soldiers and officers, has trained an elite corps of soldiers for Hong Kong. No doubt the hope is that a well-educated military will be less likely to provoke problems with Hong Kong residents. In fact, PLA troops are even permitted to date and marry local Hong Kong women!

Press, Civil Rights, and Religious Freedom

China's forceful crackdown on protesters in Beijing's Tiananmen Square in 1989 and subsequent repression traumatized the Hong Kong population. China warned the Hong Kong authorities that foreign agents might use such organizations as the Hong Kong Alliance in Support of the Patriotic Democratic Movement in China (a coalition of some 200 groups) to advance their intelligence activities on the mainland, and even

accused that group of "playing a subversive role in supporting the pro-democracy movement."[75] China also announced that it would not permit politically motivated mass rallies or demonstrations against the central government of China after the handover. As it turns out, Beijing has done little to stop the rallies and protests by such diverse groups as doctors, students, property owners, social workers, and civil servants, Mainlanders demanding the right to abode in Hong Kong, or even protests against the bill to amend Article 23. Such protests occur regularly in Hong Kong. Indeed, even Jiang Zemin, during one of his final trips to Hong Kong as president of China, had to face potential humiliation by thousands of Falun Gong demonstrators protesting Beijing's crackdown on the religious group in China. (Rather than prohibiting the group members from protesting, Hong Kong authorities took Jiang along a route where they were not so visible.) Beijing has, however, used subtle and not-so-subtle intimidation to discourage Hong Kong from supporting prodemocracy activities, suggesting that such acts would be "treasonous." In addition, those individuals who assisted the Tiananmen demonstrators in Beijing in 1989, and those leaders of the annual demonstrations in Hong Kong to protest China's use of force to crack down on Tiananmen demonstrators, are blacklisted in China. Were they to try to go to the mainland, they could be arrested for sedition. Only when the chief executive himself has intervened on their behalf have they been allowed to cross the Hong Kong border.

One freedom that Hong Kong subjects under British rule had was freedom of the press. Hong Kong had a dynamic press that represented all sides of the political spectrum. Indeed, it even tolerated the Chinese Communist Party's sponsorship of both its own pro-Communist, pro-Beijing newspaper, and its own news bureau (the New China News Agency) in Hong Kong, which, until the handover, also functioned as China's unofficial foreign office in Hong Kong.

Today, concerns remain that China may one day crack down on Hong Kong's press. In 2000, the Chinese government did admonish Hong Kong journalists and editors (albeit through a second-level official) not to comment favorably on Taiwan independence.[76] Nevertheless, with few exceptions, the possibilities of China seriously constricting Hong Kong's press freedom have declined because of the significant progress made in advancing press freedom in China in the last 25 years; and because members of the Hong Kong press have mastered the art of knowing where to draw the line so as not to offend Beijing. Self-censorship is nothing new for Hong Kong's (or China's) press; and even under the British, it was necessary. How Beijing will respond to a Hong Kong press that openly challenges the Chinese Communist Party remains to be seen; but in China itself, political analyses are no longer confined to parroting the Communist Party's line, and investigative reportage is encouraged.

In short, a major roll back of press freedom in Hong Kong seems unlikely. Nor is it likely that information coming into Hong Kong will be cut off, for to do so would jeopardize Hong Kong's economic success—something China would not want to do. Although a scandal erupted in 2000 over attempted censorship of a University of Hong Kong professor who conducted polls indicating that Tung Chee-hwa was becoming increasingly unpopular,[77] generally the press does not hesitate to criticize the chief executive.

As for religious freedom, Beijing has stated that as long as a religious practice does not contravene the Basic Law, it will be permitted. Given the fact that China's tolerance of religious freedom on the mainland has expanded dramatically since it began liberalizing reforms in 1979, this does not seem to be a likely area of tension. However, the problems raised by the activities of Falun Gong practitioners on the mainland, and the strong support for the sect in Hong Kong and across the border, could one day lead to a confrontation on the issue.

THE FUTURE

The future is, of course, unpredictable; but so far, there is a "business as usual" look about Hong Kong. Hong Kong still appears much as it did before its return to China's sovereign control. And although some analysts believe that power is imperceptibly being transferred to Beijing, Deng Xiaoping's promise that "'dancing and horseracing' would continue unabated" has been kept. "U.S. aircraft carriers still drop anchor and disgorge their crews into the Wanchai district's red-light bars. Anti-China demonstrators continue their almost weekly parades through the Central district. ... Meanwhile, the People's Liberation Army has made a virtue of being invisible."[78] It may well be that fears of Hong Kong becoming just another Chinese city were misplaced.

China's leaders have stated that for 50 years after 1997, the relationship between China and the Hong Kong Special Administrative Region will be "one country, two systems." Moreover, China's leaders clearly want their cities to look more like Hong Kong, not the other way around. In fact, with China's commercial banks starting to act much the same as banks in any capitalist economy, the phasing out of its state-run economy in favor of a market economy, a budding stock market, billions of dollars in foreign investment, and entry into the World Trade Organization, China's economy is looking more like Hong Kong's and less like a centrally planned socialist economy. China's leaders have repeatedly stated the importance of foreign investment, greater openness, and experimentation; and they are doing everything possible to integrate the country into the global economy. The imposition of a socialist economy on Hong Kong is, at this point, unthinkable.

China's political arena is also changing so profoundly that the two systems, which just 20 years ago seemed so far apart, are now much closer. China has undergone some political liberalization, increasing electoral rights at the local level, permitting freedom in individual lifestyles, mobility, and job selection, and according greater freedom to the mass media. The rapid growth of private property and business interests is also bringing significant social change to China, including the proliferation of interest groups and associations to promote their members' interests.

Southern China's extraordinary economic boom has made Hong Kong optimistic about the future. Many of its residents see a new "dragon" emerging, one that combines Hong Kong's technology and skills with China's labor and resources. Others, however, do not see Hong Kong happily working as one unit with China. Instead, they see Hong Kong as a rival competing with Shanghai and with the many new ports that China is building; and even with Shenzhen, which borders Hong Kong and also now has a deepwater container port. Shenzhen is developing a high-quality pool of labor that costs just one tenth that of Hong Kong, so many corporations are moving their operations—and their wealth—across the border to China. Moreover, the P.R.C. no longer needs Hong Kong as an entrepôt to export its products. Not only can it now do so through its own ports, but it is also building plants in other countries, such as India, where it will manufacture and export goods that otherwise would have gone through Hong Kong. In addition, China's membership in the WTO since 2001, as well as its increasingly direct contacts and trade with Taiwan, could easily lead to a partial eclipse of Hong Kong.

Many Hong Kong analysts believe that the greatest threat to Hong Kong's success is not political repression and centralized control from Beijing; rather, it would take the more insidious form of China's bureaucracy and corruption simply smothering Hong Kong's economic vitality. One concern is that mainland companies operating in Hong Kong may be allowed to stand above the law or to use their Hong Kong ties to exert inappropriate influence on behalf of pro-China business interests and corrupt the Hong Kong economy.[79] So far, Hong Kong has escaped this fate, and is still considered one of the least corrupt business environments in the world. Another concern is that China may bring the features of the Singapore political model to Hong Kong, whereby it would be ruled by pro-China business tycoons who are insensitive to the political, social, and economic concerns of other groups, and where democratic parties would find it impossible to gain a majority in a legislative system stacked against them. Political rights and freedom would also be restricted by moving toward a Singapore political model.[80] Yet another worry is that the dangers lie *within* Hong Kong, notably the minimalist efforts of the Hong Kong government to reshape it in a way that allows it to remain competitive.

Beijing has an important stake in its takeover of Hong Kong not being perceived as disruptive to its political or economic system, and that Hong Kong's residents and the international business community believe in its future prosperity. Policies and events that threatened that confidence in the 1980s led to the loss of many of Hong Kong's most talented people, technological know-how, and investment. China does not want to risk losing still more. Beijing also wants to maintain Hong Kong as a major free port and the regional center of trade, financing, shipping, and information—although it is also doing everything possible to turn Shanghai into a competitive center.[81]

William Overholt has labelled the major underlying sources of tension between Hong Kong and Beijing as the "Three Confusions:" Beijing's "confusion of Hong Kong, where there is virtually no separatist sentiment, with Taiwan;" Beijing's confusion due to a failure to distinguish between the types of lawful demonstrations that have traditionally taken place in Hong Kong on a regular basis "with disruptive demonstrations in the mainland;" and Beijing's confusion because of Beijing's failure to distinguish between some of the older leaders of the democracy movement "with the moderate loyal sentiments of the overwhelming majority of the democratic movement."[82] To the degree that China eliminate such confusion, it will be able to avoid many potential problems with Hong Kong.

Finally, regardless of official denials by the government in Taiwan, Beijing's suc-

Timeline: PAST

A.D. 1839–1842
The first Opium War

1842
The Treaty of Nanjing cedes Hong Kong to Britain

1856
The Chinese cede Kowloon and Stonecutter Island to Britain

1898
England gains a 99-year lease on the New Territories

1911
A revolution ends the Manchu Dynasty; the Republic of China is established

1941
The Japanese attack Pearl Harbor and take Hong Kong; Hong Kong falls under Japanese control

1949
The Communist victory in China produces massive immigration into Hong Kong

1980s
Great Britain and China agree to the return of Hong Kong to China

1990s
China resumes control of Hong Kong on July 1, 1997

PRESENT

2000s
Hong Kong's efforts to recover from the Asian economic and financial crisis were hindered by a worldwide economic downturn and, in 2003, by the outbreak of SARS

2003
After massive demonstrations, government withdraws bill to amend the Basic Law, Article 23, on sedition from further consideration

cessful management of "one country, two systems" in Hong Kong will profoundly affect how Taiwan feels about its own peaceful integration with the mainland. If Beijing wants to regain control of Taiwan by peaceful means, it is critical that it handle Hong Kong well.

NOTES

1. R. G. Tiedemann, "Chasing the Dragon," *China Now,* No. 132 (February 1990), p. 21.

2. Jan S. Prybyla, "The Hong Kong Agreement and Its Impact on the World Economy," in Jurgen Domes and Yu-ming Shaw, eds., *Hong Kong: A Chinese and International Concern* (Boulder, CO: Westview Special Studies on East Asia, 1988), p. 177.

3. Tiedemann, p. 22.

4. Steven Tsang, *A Modern History of Hong Kong* (New York: I.B.Tauris, 2004), p. 271.

5. Siu-kai Lau, "The Hong Kong Policy of the People's Republic of China, 1949–1997," *Journal of Contemporary China* (March 2000), Vol. 9, No. 23, p. 81.

6. Tsang, pp. 133–135.

7. Robin McLaren, former British ambassador to China, seminar at Cambridge University, Centre for International Relations (February 28, 1996).

8. Ambrose Y. C. King, "The Hong Kong Talks and Hong Kong Politics," in Domes and Shaw, p. 49.

9. Hungdah Chiu, Y. C. Jao, and Yual-li Wu, *The Future of Hong Kong: Toward 1997 and Beyond* (New York: Quorum Books, 1987), pp. 5–6.

10. Siu-kai Lau, "Hong Kong's 'Ungovernability' in the Twilight of Colonial Rule," in Zhiling, Lin and Thomas W. Robinson, *The Chinese and Their Future: Beijing, Taipei, and Hong Kong* (Washington, D.C.: The American Enterprise Institute Press, 1994), pp. 288–290.

11. McLaren.

12. McLaren noted that it was not easy to convince Prime Minister Thatcher in her "post-Falklands mood" (referring to Great Britain's successful defense of the Falkland Islands, 9,000 miles away, from being returned to Argentine rule), that Hong Kong could not stay under British administrative rule after 1997.

13. T. L. Tsim, "Introduction," in T. L. Tsim and Bernard H. K. Luk, *The Other Hong Kong Report* (Hong Kong: The Chinese University Press, 1989), p. xxv.

14. Norman J. Miners, "Constitution and Administration," in Tsim and Luk, p. 2.

15. King, pp. 54–55.

16. Annex I, Nos. 1 and 4 of *The Basic Law of the Hong Kong Special Administrative Region of the People's Republic of China* (hereafter cited as *The Basic Law*). Printed in *Beijing Review,* Vol. 33, No. 18 (April 30–May 6, 1990), supplement. This document was adopted by the 7th National People's Congress on April 4, 1990.

17. For specifics, see Annex II of *The Basic Law* (1990).

18. Article 14, *The Basic Law* (1990).

19. Articles 27, 32, 33, and 36 of *The Basic Law* (1990).

20. Tsang, p. 271.

21. Chinese cities on the mainland are overrun by transients. On a daily basis, Shanghai alone has well over a million visitors, largely a transient population of country people. Hong Kong does not feel that it can handle such an increase in its transient population.

22. James Tang and Shiu-hing (Sonny) Lo, University of Hong Kong, interview in Hong Kong (June 2000).

23. Michael E. DeGolyer, director, *1982–2007 Hong Kong Transition Project: Accountability & Article 23* (Hong Kong: Hong Kong Baptist University, December 2002), pp. 43–44. These newspapers include several that are pro-Beijing.

24. Tang and Lo.

25. In 1995, Hong Kong's per capita GDP was U.S.$23,500. Wang Gungwu and Wong Siu-lun, eds., *Hong Kong in the Asian-Pacific Region: Rising to the New Challenge* (Hong Kong: University of Hong Kong, 1997), p. 2.

26. James L. Watson, "McDonald's in Hong Kong: Consumerism, Dietary Change, and the Rise of a Children's Culture," in James L. Watson, ed., *Golden Arches East: Mc-Donald's in East Asia* (Palo Alto, CA: Stanford University Press, 1997), pp. 77–109.

27. Tsim, in Tsim and Luk, p. xx.

28. Howard Gorges (South China Holdings Corporation), interview in Hong Kong (June 2000).

29. Wang and Wong, p. 3.

30. Keith B. Richburg, "Uptight Hong Kong Countdown," *The Washington Post* (July 2, 1996), pp. A1, A12.

31. An average apartment measures 20 feet by 23 feet, or 460 square feet.

32. Tsim, in Tsim and Luk, p. xxi.

33. The purchaser of the property pays tax, develops the property, and then pays a 2 percent tax on every real-estate transaction (renting or selling) that occurs thereafter.

34. Keith B. Richburg, "Chinese Muscle-Flexing Puts Hong Kong Under Pessimistic Pall," *The Washington Post* (December 26, 1996), p. A31.

35. Christine Loh, *Newsletter,* Hong Kong (August 11, 2000).

36. The six largest real-estate companies are among the top 20 companies in the Hang Seng Index of stocks. Others are banks and technology companies. The largest is Cheong Kong Holdings, under Li Ka-shing's control. Companies linked to him and his son accounted for 26 percent of the total market capitalization of the Hang Seng Index! *Asiaweek* (May 26, 2000), pp. 33–36.

37. Al Reyes, journalist (*Asiaweek*), interview in Hong Kong (June 16, 2000).

38. "Is Hong Kong Ripe for a Bit of Central Planning?" *The Economist* (April 12, 1997).

39. Wang and Wong, p. 4.

40. "Is Hong Kong Ripe… "

41. Wang and Wong, pp. 2–4.

42. Tammy Tam, "Shenzhen Industrial Estate Developed to Boost Military Funds," The *Hong-Kong Standard* (September 5, 1989), p. 1.

43. Shiu-hing Lo, "Hong Kong's Political Influence on South China," *Problems of Post-Communism,* Vol. 46, No. 4 (July/August 1999), pp. 33–41.

44. "Sheriff Patten Comes to Town," *The Economist* (November 14, 1992), p. 35.

45. Tang and Lo, interview, (June 2000). Also Christine Loh, LegCo legislator and founder of the Citizens' Party, interview in Hong Kong (June 2000). Loh did not run for reelection in September 2000. She is a businessperson-turned-politician. Many well-placed individuals in Hong Kong think she may become the next chief executive. She has been a self-described "armchair critic" of the British and Hong Kong SAR governments, has not been a radical street demonstrator, and is generally supportive of the business community.

46. Tang and Lo.

47. John Gordon Davies, "Introduction," *Hong Kong Through the Looking Glass* (Hong Kong: Kelly & Walsh, 1969).

48. Tang and Lo.

49. At first China wanted to copy the Singapore model of executing drug dealers, but it soon realized that would mean the execution of thousands. Now only the biggest drug dealers are executed.

50. King, in Domes and Shaw, pp. 45–46.

51. Prybyla, in Domes and Shaw, pp. 196–197; and Lau, in Lin and Robinson, p. 302.

52. Lau, in Lin and Robinson, pp. 293–294, 304–305.

53. Willy Wo-Lap Lam, "New Faces to Star in Hong Kong's New Cabinet," CNN Web site (June 24, 2002).

54. In June 1997, just before the July 1 handover, only 45 percent was "satisfied," and 41 percent were "dissatisfied." De-Golyer, 2002, Table 145, p. 103; http://www.hkbu.edu.hk/~hktp.

55. Hong Kong Transition Project, *Listening to the Wisdom of the Masses:Hong Kong People's Attitudes toward Constitutional Reforms* (Hong Kong: Civic Exchange and Hong Kong Transition Project), January 2004, Tables 38, 135, 9, pp. 32, 68, 14.)

56. Loh, interview (June 2000)

57. "Silent Treatment: Hong Kong's Chief and Its Legislature Aren't Talking," *Far Eastern Economic Review* (September 17, 1998), p. 50. Hong Kong Transition Project, *Listening to the Wisdom*, Table 25, pp. 22-23.

58. Ibid., Table 135, p. 68.

59. Tang and Lo, interview (June 2000); and Micahel DeGolyer, interview in Hong Kong (June 2000).

60. DeGolyer, 2002, Table 132, p. 97.

61. Hong Kong Transition Project, *Listening to the Wisdom*, Table 96, pp. 50-57.

62. Data based on surveys done in November 2002. The government submitted the proposed changes to Article 23 two months earlier. In almost all categories of rights, and across all occupations, ages, education, and so on, Hong Kong people showed an increase in concern about rights because of the potential amendments to Article 23. *Ibid*, pp. 36–39, 49–66.

63. Hong Kong Transition Project, *Listening to the Wisdom,* Table 38, p. 32.

64. King, in Domes and Shaw, pp. 51, 56, 57.

65. In the 1998 elections, 20 members of LegCo were for the first time elected directly. Of these, 13 seats went to the Democratic Party. Of the remaining 40 seats, which were indirectly elected, seven went to the Democratic Party.

66. Mark Magnier, "Hong Kong Warned to Drop Vote Idea," *LA Times*, Nov. 10, 2004.

67. Other patriotic parties include the Liberal Party, the Federation of Trade Unions, and the Hong Kong Progressive Alliance.

68. Loh, interview (June 2000).

69. Alvin Y. So, "Hong Kong's Problematic Democratic Transition: Power Dependency or Business Hegemony? *The Journal of Asian Studies,* Vol. 59, No. 2 (May 2000), pp. 375–376.

70. According to the Hong Kong Transition Project 2000 polls, only 30 percent of the people believed that political parties wielded significant influence on the government, whereas 74 percent thought that Beijing officials did. Michael E. DeGolyer, director, *The Hong Kong Transition Project: 1982–2007* (Hong Kong: Hong Kong Baptist University, 2000), p. 25.

71. James L. Tyson, "Promises, Promises...." *The Christian Science Monitor* (April 20, 1989), p. 2.

72. According to Article 7 of China's Criminal Code, China has the right to prosecute a Chinese national who commits a crime *anywhere* in the world, providing the crime was either planned in, or had consequences in, China. "Another Place, Another Crime: Mainland Trial of Alleged Gangster Puts 'One Country, Two Systems' to Test," *Far Eastern Economic Review* (November 5, 1998), pp. 26–27.

73. DeGolyer, 2000, pp. 3–8. Of students, 54 percent are satisfied, 28 percent neutral.

74. Kevin Murphy, "Troops for Hong Kong: China Puts Best Face on It," *The International Herald Tribune* (January 30, 1996), p. 4.

75. Miu-wah Ma, "China Warns Against Political Ties Abroad," *The Hong Kong Stan-dard* (September 1, 1989), p. 4; and Viola Lee, "China 'Trying to Discourage HK People,'" *South China Morning Post* (August 21, 1989). The article, which originally appeared in a *People's Daily* article in July, was elaborated upon in the August edition of *Outlook Weekly*, a mouthpiece of the CCP.

76. Beijing feels that it can tell Hong Kong what to say about Taiwan because Beijing views Taiwan, like Tibet, as a "national" issue, not a "freedom of press" issue. Hong Kong commentators tend to say that reunification of Taiwan with the mainland is best; but what they are really thinking is "Let's not get involved in this. If Taiwan tries for independence, there will be a war. It is not realistic to push for independence."Loh, interview (June 2000).

77. While Tung Chee-hwa denied any responsibility for telling the university's provice-chancellor to admonish the professor, the subsequent furor resulted in the resignation of both the pro-vice-chancellor and the vice-chancellor of the university.

78. "Hong Kong: Now the Hard Part," *Far Eastern Economic Review* (June 11, 1998), p. 13.

79. Michael C. Davis, "Constitutionalism and Hong Kong's Future," *Journal of Contemporary China* (July 1999), Vol. 8, No. 21, pp. 271.

80. *Ibid*., p. 269, 273.

81. Kai-Yin Lo, "A big awakening for Chinese rivals: Hong Kong and Shanghai look afar," *International Herald Tribune*, January 20, 2005

82. William Overholt, Testimony, "The Hong Kong Legislative Election of Sept 12, 2004: Assessment and Implications" (Testimony presented to the Congressional-Executive Commission on China on Sept. 23, 2004), (Santa Monica: RAND Corporation, 2004), p. 7.

Taiwan Map

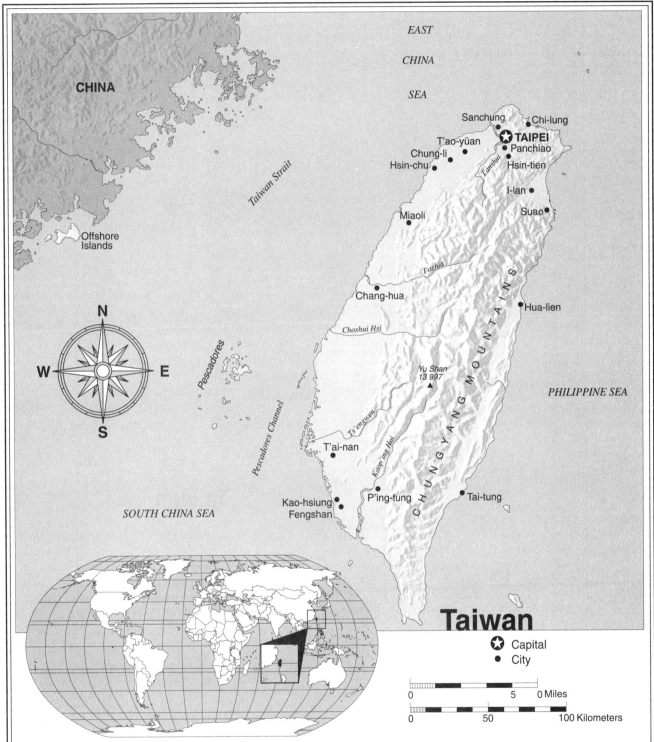

Taiwan has been considered the center of the government of the Republic of China (Nationalist China) since 1949. The province of Taiwan consists of the main island, 15 islands in the Offshore Islands group, and 64 islands in the Pescadores Archipelago. While the Pescadores are close to Taiwan, the Offshore Islands are only a few miles off the coast of mainland China.

Taiwan

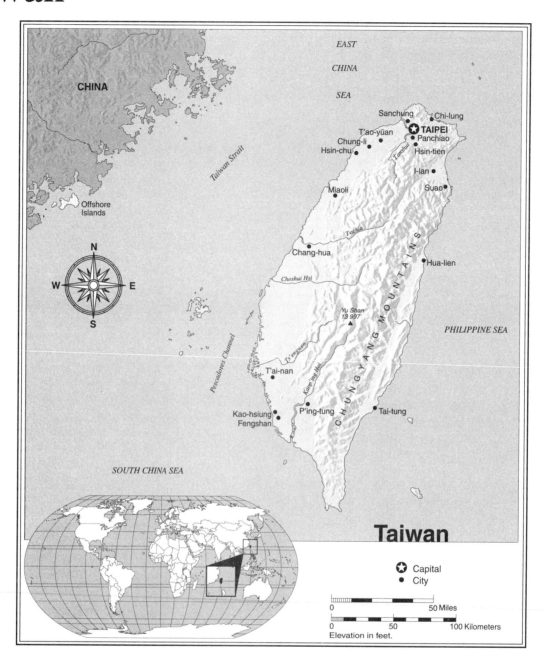

Taiwan Statistics

GEOGRAPHY

Area in Square Miles (Kilometers):

22,320 (36,002) (about the size of Maryland and Delaware combined)

Capital (Population): Taipei (2,722,600)

Environmental Concerns: water and air pollution; contamination of drinking water; radioactive waste; trade in endangered species

Geographical Features: mostly rugged mountains in east; flat to gently rolling plains in west

Climate: tropical; marine

PEOPLE

Population

Total: 22,749,838

Annual Growth Rate: 0.64%

Rural/Urban Population Ratio: 25/75

Major Languages: Mandarin Chinese; Taiwanese and Hakka dialects also used

Ethnic Makeup: 84% Taiwanese; 14% Mainlander Chinese; 2% aborigine

Religions: 93% mixture of Buddhism, Confucianism, and Taoism; 4.5% Christian; 2.5% others

Health

Life Expectancy at Birth: 74.3 years (male); 80.1 years (female)

Infant Mortality: 6.52/1,000 live births
Physicians Available: 71.7/10,000 people
HIV/AIDS Rate in Adults: na

Education

Adult Literacy Rate: 96.1%
Compulsory (Ages): 6–15; free

COMMUNICATION

Telephones: 13,355,000 main lines
Televisions: 8.8 million
Internet Users: 8,830,000

TRANSPORTATION

Highways in Miles (Kilometers): 20,940
  (35,931)
Railroads in Miles (Kilometers): 665
  (2,544)
Usable Airfields: 40
Motor Vehicles in Use: 5,170,000
  (passenger) 911,000 (commercial)

GOVERNMENT

Type: multiparty democratic regime
Head of State/Government: President
  Chen Shui-bian; Premier Yu Shyi-kun
Political Parties: Nationalist Party
  (Kuomintang); Democratic Progressive
  Party; Labor Party; People's First Party;
  others
Suffrage: universal at 20

MILITARY

Military Expenditures (% of GDP): 2.7%
Current Disputes: disputes with various
  countries over islands; long-term
  concern that P.R.C. might attack

ECONOMY

Currency ($ U.S. Equivalent): 34.42 new
  Taiwan dollars = $1
Per Capita Income/GDP: $23,400/$396
  billion
GDP Growth Rate: 3.2%

Inflation Rate: -0.3%
Unemployment Rate: 5%
Labor Force by Occupation: 57%
  services; 35% industry; 7.5% agriculture
Population Below Poverty Line: 1%
Natural Resources: coal; natural gas;
  limestone; marble; asbestos
Agriculture: rice; tea; fruit; vegetables;
  corn; livestock; fish
Industry: steel; iron; chemicals;
  electronics; cement; textiles; food
  processing; petroleum refining
Exports: $143 billion (primary partners
  China, United States, Japan)
Imports: $119.6 billion (primary partners
  China, South Korea)

SUGGESTED WEB SITES

http://www.taipeitimes.com/news/
http://www.cia.gov/cia/
  publications/factbook/geos/
  tw.html
http://www.gio.gov.tw/

Taiwan Report

Taiwan, today a powerful economic center in Asia, was once an obscure island off the coast of China, just 90 miles away. (Taiwan has also been known as Formosa, Free China, the Republic of China, and Nationalist China. Today, the government in Taiwan is called the "Republic of China on Taiwan.") It was originally inhabited by aborigines from Southeast Asia. By the seventh century A.D., Chinese settlers had begun to arrive. The island was subsequently "discovered" by the Portuguese in 1590, and Dutch as well as Spanish settlers followed. Today, the aborigines' descendants, who have been pushed into the remote mountain areas by the Chinese settlers, number under 400,000, a small fraction of the nearly 23 million people now living in Taiwan. Most of the population is descended from people who emigrated from the Chinese mainland's southern provinces before 1885, when Taiwan officially became a province of China. Although their ancestors originally came from China, they are known as *Taiwanese,* as distinct from the Chinese who fled the mainland from 1947 to 1949. The latter are called *Mainlanders* and represent less than 20 percent of the island's population. After 1949, the Mainlanders dominated Taiwan's political elite; but the "Taiwanization" of the political realm that began after Chiang Kai-shek's death in 1975 and the political liberalization since 1988 have al-

lowed the native Taiwanese to take up their rightful place within the elite.

The Manchus, "barbarians" who came from the north, overthrew the Chinese rulers on the mainland in 1644 and established the Qing Dynasty. In 1683, they conquered Taiwan; but because Taiwan was an island 90 miles distant from the mainland and China's seafaring abilities were limited, the Manchus paid little attention to it and exercised minimal sovereignty over the Taiwanese people. With China's defeat in the Sino–Japanese War (1894–1895), the Qing was forced to cede Taiwan to the Japanese. The Taiwanese people refused to accept Japanese rule, however, and proclaimed Taiwan a republic. As a result, the Japanese had to use military force to gain actual control over Taiwan.

For the next 50 years, Taiwan remained under Japan's colonial administration. Taiwan's economy flourished under Japanese rule. Japan helped to develop Taiwan's agricultural sector, a modern transportation network, and an economic structure favorable to later industrial development. Furthermore, by creating an advanced educational system, the Japanese developed an educated workforce, which proved critical to Taiwan's economic growth.

With Japan's defeat at the end of World War II (in 1945), Taiwan reverted to China's sovereignty. In the meantime, the Chinese had overthrown the Manchu Dy-

nasty (1911) and established a republican form of government on the mainland. Beginning in 1912, China was known as the Republic of China (R.O.C.). Thus, in 1945 it was Chiang Kai-shek who, as head of the R.O.C. government, accepted the return of the island province of Taiwan to R.O.C. rule. When some of Chiang Kai-shek's forces were dispatched to shore up control in Taiwan in 1947, tensions quickly arose between them and the native Taiwanese. The ragtag, undisciplined military forces of the KMT (the Kuomintang, or Nationalist Party) met with hatred and contempt from the local people. They had grown accustomed to the orderliness and professionalism of the Japanese occupation forces and were angered by the incompetence and corruption of KMT military and political officials. Demonstrations against rule by Mainlanders occurred in February 1947. Relations were badly scarred when KMT troops killed thousands of Taiwanese opposed to mainland rule. Among those murdered were many members of the island's political and intellectual elite.

Meanwhile, the KMT's focus remained on the mainland, where, under the leadership of General Chiang Kai-shek, it continued to fight the Chinese Communists in a civil war that had ended their fragile truce during World War II. Civil war raged from 1945 to 1949 and diverted the KMT's attention away from Taiwan. As a

McCARTHYISM: ISOLATING AND CONTAINING COMMUNISM

New York Public Library

The McCarthy period in the United States was an era of rabid anticommunism. McCarthyism was based in part on the belief that the United States was responsible for losing China to the Communists in 1949 and that the reason for this loss was the infiltration of the U.S. government by Communists. As a result, Senator Joseph McCarthy (pictured here) spearheaded a "witch-hunt" to ferret out those who allegedly were selling out American interests to the Communists. McCarthyism took advantage of the national mood in the Cold War era that had begun in 1947, in which the world was seen as being divided into two opposing camps: Communists and capitalists.

The major strategy of the Cold War, as outlined by President Harry Truman in 1947, was the "containment" of communism. This strategy was based on the belief that if the United States attempted—as it had done with Adolf Hitler's aggression against Czechoslovakia (the first step toward World War II)—to appease communism, it would spread beyond its borders and threaten other free countries.

The purpose of the Cold War strategy, then, was to contain the Communists within their national boundaries and to isolate them by hindering their participation in the international economic system and in international organizations. Hence, in the case of China, there was an American-led boycott against all Chinese goods, and the United States refused to recognize the People's Republic of China as the legitimate representative of the Chinese people within international organizations.

result, Taiwan continued, as it had under the Qing Dynasty's rule, to function quite independently of Beijing. In 1949, when it became clear that the Chinese Communists would defeat the KMT, General Chiang and some 2 million members of his loyal military, political, and commercial elite fled to Taiwan to establish what they claimed to be the true government of the Republic of China. This declaration reflected Chiang's determination to regain control over the mainland, and his conviction that the more than 600 million people living on the mainland would welcome the return of the KMT to power.

During the McCarthy period of the "red scare" in the 1950s (during which Americans believed to be Communists or Communist sympathizers—"reds"—were persecuted by the U.S. government), the United States supported Chiang Kai-shek. In response to the Chinese Communists' entry into the Korean War in 1950, the United States applied its Cold War policies of support for any Asian government that was anti-Communist, regardless of how dictatorial and ruthless it might be, in order to "isolate and contain" the Chinese Communists. It was within this context that in 1950 the United States committed itself to the military defense of Taiwan and the offshore islands in the Taiwan Strait, by ordering the U.S. Seventh Fleet to the Strait and by giving large amounts of military and

economic aid to Taiwan. General Chiang Kai-shek continued to lead the government of the Republic of China on Taiwan until his death in 1975, at which time his son, Chiang Ching-kuo, succeeded him.

Two Governments, One China

Taiwan's position in the international community and its relationship to the government in Beijing have been determined by perceptions and values as much as by actions. In 1949, when the R.O.C. government fled to Taiwan, the Chinese Communists renamed China the "People's Republic of China" (P.R.C.), and they proclaimed the R.O.C. government illegitimate. Later, Mao Zedong, the P.R.C.'s preeminent leader until his death in 1976, was to say that adopting the new name instead of keeping the name "Republic of China" was the biggest mistake he ever made, for it laid the groundwork for future claims of "two Chinas." Beijing claimed that the P.R.C. was the legitimate government of all of China, including Taiwan. Beijing's attempt to regain de facto control over Taiwan was, however, forestalled first by the outbreak of the Korean War and thereafter by the presence of the U.S. Seventh Fleet in the Taiwan Strait.

Beijing has always insisted that Taiwan is an "internal" Chinese affair, that international law is therefore irrelevant, and that other countries have no right to interfere. For its part, until 1995, the gov-

ernment of Taiwan agreed that there was only one China and that Taiwan was a province of China. But by 1995–1996, the political parties had begun debating the possibility of Taiwan declaring itself an independent state.

Although the Chinese Communists' control over the mainland was long evident to the world, the United States managed to keep the R.O.C. in the China seat at the United Nations by insisting that the issue of China's representation in the United Nations was an "important question." This meant that a two-thirds affirmative vote of the UN General Assembly, rather than a simple majority, was required. With support from its allies, the United States was able to block the P.R.C. from winning this two-thirds vote until 1971. Once Secretary of State Henry Kissinger announced his secret trip to Beijing in the summer of 1971 and that President Richard Nixon would be going to China in 1972, the writing was on the wall. Allies of the United States knew that U.S. recognition of Beijing as China's legitimate government would eventually occur, so there would no longer be pressure to block the P.R.C. from membership in the United Nations. They quickly jumped ship and voted for Beijing's representation.

At this critical moment, when the R.O.C.'s right to represent "China" in the United Nations as withdrawn, the R.O.C. could have put forward the claim that Tai-

CHIANG KAI-SHEK: DETERMINED TO RETAKE THE MAINLAND

Until his dying day, Chiang Kai-shek, pictured here with his wife, maintained that the military, led by the KMT (Kuomintang, or Nationalist Party), would one day invade the mainland and, with the support of the Chinese people living there, defeat the Communist government. During the years of Chiang's presidency, banner headlines proclaimed daily that the Communist "bandits" would soon be turned out by internal rebellion and that the KMT would return to control on the mainland. In the last years of Chiang Kai-shek's life, when he was generally confined to his residence and incapable of directing the government, his son, Chiang Ching-kuo, always had two editions of the newspaper made that proclaimed such unlikely feats, so that his father would continue to believe these were the primary goals of the KMT government in Taiwan. In fact, a realistic appraisal of the situation had been made long before Chiang's death, and most of the members of the KMT only pretended to believe that an invasion of the mainland was imminent.

Chiang Ching-kuo, although continuing to strengthen Taiwan's defenses, turned his efforts to building Taiwan into an economic showcase in Asia. Taiwan's remarkable growth and a certain degree of political liberalization were the hallmarks of his leadership. A man of the people, he shunned many of the elitist practices of his father and the KMT ruling elite, and he helped to bring about the Taiwanization of both the KMT party and the government. The "Chiang dynasty" in Taiwan came to an end with Chiang Ching-kuo's death in 1988. It was, in fact, Chiang Ching-kuo who made certain of this, by barring his own sons from succeeding him and by grooming his own successor, a native Taiwanese.

wan had the right to be recognized as an independent state, or at least to be granted observer status. Instead, the R.O.C. steadfastly maintained that there was but one China and that Taiwan was merely a province of China. As a result, today Taiwan has no representation in any international organization under the name of "Republic of China;" and it has representation only as "Chinese Taipei" in organizations (such as the Olympics and the Asian Pacific Economic Cooperation, or APEC, forum) in which the P.R.C. is a member—if the P.R.C. allows it any representation at all. With few exceptions, however, Beijing has been adamant about not permitting Taiwan's representation regardless of what it is called.[1]

It is important to understand that at the time Taiwan lost its seat in the United Nations, it was still ruled by the KMT mainlanders. The "Taiwanization" of the ruling KMT party-state did not begin until the mid 1970s. As a result, the near 85 percent of the population that was native Taiwanese lacked the right to express their preference for rule by native Taiwanese rather than KMT mainlanders, much less their desire for a declaration of independent statehood. Under martial law, those who had dared to demand independence were imprisoned, forcing the Taiwan independence move-

ment to locate outside of Taiwan. In short, most of the decisions that have shaped Taiwan's international legal standing through treaties and diplomatic relations were made when the KMT mainlanders held political power. Only when the Taiwanese gained their voice in the mid-1990s, did they start to assert a different view of Taiwan's future. By then, many would argue, it was too late.

INTERNATIONAL ACCEPTANCE OF THE P.R.C., AT TAIWAN'S EXPENSE

The seating of the P.R.C. in the United Nations in 1971 thus led to the collapse of Taiwan's independent political standing in international affairs. Not wanting to anger China, which has a huge and growing economy and significant military power, the state members of international organizations have given in to Beijing's unrelenting pressure to exclude Taiwan. Furthermore, Beijing insists that in bilateral state-to-state relations, any state wishing to maintain diplomatic relations with it must accept China's "principled stand" on Taiwan—notably, that Taiwan is a province of China and that the People's Republic of China is the sole representative of the Chinese people.

Commercial ventures, foreign investment in Taiwan, and Taiwan's investments abroad have suffered little as a result of

countries' severing their *diplomatic* relations with Taipei. After being forced to close all but a handful of its embassies as one state after another switched recognition from the R.O.C. to the P.R.C., Taipei has simply substituted offices that function as if they are embassies. They handle all commercial, cultural, and official business, including the issuance of visas to those traveling to Taiwan. Similarly, the states that severed relations with Taipei have closed down their embassies there and reopened them as business and cultural offices. The American Institute in Taiwan is a typical example of these efforts to retain ties without formal diplomatic recognition.

Still, Taiwan's government feels the sting of being almost completely shut out of the official world of international affairs. Its increasing frustration and sense of humiliation came to a head in early 1996. Under intense pressure to respond to demands from its people that Taiwan get the international recognition that it deserved for its remarkable accomplishments, Taiwan's president, Lee Teng-hui, engaged in a series of maneuvers to get the international community to confer de facto recognition of its statehood. Not the least of these bold forays was President Lee's offer of U.S. $1 billion to the United Nations in return for a seat for Taiwan—an offer rejected by the United Nations' secretary-general.

Lee's campaign for election in the spring of 1996 proved to be the final straw for Beijing. Lee had as one of his central themes the demand for greater international recognition of Taiwan as an independent state. Beijing responded with a military buildup of some 200,000 troops in Fujian Province across the Taiwan Strait, and the "testing" of missiles in the waters around Taiwan. Under pressure from the United States not to provoke a war with the mainland, and a refusal on the part of the United States to say exactly what it would do if a war occurred, President Lee toned down his campaign rhetoric. A military conflict was averted, but it was not until late 1998 that Taipei and Beijing agreed to move forward with their temporarily shelved plans to futher link Taiwan with the mainland. Since then, these plans have stalled repeatedly because of objections coming from one side or the other. Since 1996, however, the pattern of Taiwan aggressively discussing independence, followed by China threatening the use of military force and the U.S. telling Taiwan to refrain from rhetoric about independence, has become well-established.

THE OFFSHORE ISLANDS

Crises of serious dimensions erupted between China and the United States in 1954–1955, 1958, 1960, and 1962 over the blockading of supplies to the Taiwan-controlled Offshore Islands in the Taiwan Strait. Thus, the perceived importance of these tiny islands grew out of all proportion to their intrinsic worth. The two major island groups, Quemoy (about two miles from the Chinese mainland) and Matsu (about eight miles from the mainland) are located about 90 miles from Taiwan. As a consequence, Taiwan's control of them made them strategically valuable for pursuing the government's professed goal of retaking the mainland—and valuable for psychologically linking Taiwan to the mainland.

In the first years after their victory on the mainland, the Chinese Communists shelled the Offshore Islands at regular intervals. When there was not a crisis, their shells were filled with pro-Communist propaganda materials, which littered the islands. When the Chinese Communists wanted to test the American commitment to the Nationalists in Taiwan and the Soviet commitment to their own objectives, they shelled the islands heavily and intercepted supplies to the islands. In the end, China always backed down; but in 1958 and 1962, it did so only after going to the

brink of war with the United States. By 1979, most states had affirmed Beijing's claim that Taiwan was a province of China through diplomatic recognition of the P.R.C. Beijing's subsequent "peace initiatives" toward Taiwan moved the confrontation over the Offshore Islands to the level of an exchange of gifts by balloons and packages floated across the channel: As a 1986 commentary noted:

> The Nationalists load their balloons and seaborne packages with underwear, children's shoes, soap, toys, blankets, transistor radios and tape recorders, as well as cookies emblazoned with Chiang Ching-kuo's picture and audio tapes of Taiwan's top popular singer, Theresa Teng, a mainland favorite.
>
> The Communists send back beef jerky, tea, herbal medicines, mao-tai and cigarettes, as well as their own varieties of soap and toys.[2]

Because of the dramatic increase in contacts, tourism, trade, and smuggling since the early 1990s, however, such practices have ceased.

The lack of industry and manufacturing on the Offshore Islands led to a steady emigration of their natives to Southeast Asia for better jobs. The civilian population is about 50,000 (mostly farmers) in Quemoy and about 6,000 (mostly fishermen) in Matsu. Until the mid-1990s, the small civilian population in Quemoy was significantly augmented by an estimated 100,000 soldiers. Until recent years, the heavily fortified islands appeared to be somewhat deserted, since the soldiers lived mostly underground: hospitals, kitchens, sleeping quarters—everything was located in tunnels blasted out of granite, including two-lane highways that could accommodate trucks and tanks. Heavily camouflaged anti-artillery aircraft dotted the landscape, and all roads were reinforced to carry tanks.

Taiwan's military administration of Quemoy and Matsu ended in 1992.

Today, Taiwan's armed forces are streamlined, with only some 10,000 troops remaining on the Offshore Islands. Taiwan's Ministry of Defense believes, however, that these troops are better able than the former, larger forces to protect the Offshore Islands, because of improved weapons and technology. Many of the former military installations have become profit-making tourist attractions. Moreover, as of January 2001, the "three mini-links" policy was initiated, with both Quemoy and Matsu allowed to engage in direct trade, transportation, and postal exchange with the mainland. In truth, this change only le-

galized what had been going on for more than a decade—namely, a roaring trade in smuggled goods between the Offshore Islands and the mainland. Taiwan now seems far more interested in boosting the economy of these islands and increasing links with the mainland than it is concerned with the islands becoming hostages of the mainland. Indeed, Taiwan has moved forward with a plan to secure water for the islands from the mainland, to cope with an increased demand for water it assumes will be generated by growing tourism and business activities.[3] (The first PRC tourists went to Quemoy in Sept. 2004.)

Thus one fiction—that there was no direct trade between Taiwan's Offshore Islands and the mainland—has been abolished. Other fictions wait to be dismantled as Taiwan and the mainland move toward further integration. For example, there is still not supposed to be direct trade between the island of Taiwan and the mainland. All ships are supposed to transship their goods by way of another port, such as Hong Kong or the Ryuku Islands. But now, as soon as ships have left Taiwan, they simply process paperwork with port authorities in the Ryuku Islands by way of fax—that is, without actually ever going there—so they can go directly to the mainland. This eliminates the expensive formalities of transshipments via another port. In short, direct trade is already in existence.[4]

CULTURE AND SOCIETY

Taiwan is a bundle of contradictions: "great tradition, small island; conservative state, drastic change; cultural imperialism, committed nationalism; localist sentiment, cosmopolitan sophistication."[5] Over time, Taiwan's culture has been shaped by various cultural elements—Japanese, Chinese, and American culture; localism, nationalism, cosmopolitanism, materialism; and even Chinese mainland culture (in the form of "mainland mania"). At any one time, several of these forces have coexisted and battled for dominance. As Taiwan has become increasingly affected by globalization, the power of the central government to control cultural development has diminished. This has unleashed not just global cultural forces but also local *Taiwanese* culture.[6]

The Taiwanese people were originally immigrants from the Chinese mainland; but their culture, which developed in isolation from that of the mainland, is not the same as the "Chinese" culture of the defeated "Mainlanders" who arrived from 1947 to 1949. Although the Nationalists saw Taiwan largely in terms of security and as a bastion from which to fight against and defeat the Chinese Communist regime on the mainland, "it also cultivated Taiwan as the last outpost of traditional Chinese high culture. Taiwanese folk arts, in particular opera and festivals, did thrive, but as low culture."[7]

The Taiwanese continue to speak their own dialect of Chinese, distinct from the standard Chinese spoken by the Mainlanders, and almost all Taiwanese engage in local folk-religion practices. However, until the mid-1990s, the Mainlander-controlled central government dictated a cultural policy that emphasized Chinese cultural values in education and the mass media. As a result, the distinctions institutionalized in a political system that discriminated against the Taiwanese were culturally reinforced.

The Taiwanese grew increasingly resistant to efforts by the KMT Mainlanders to "Sinify" them—to force them to speak standard Chinese and to adopt the values of the dominant Chinese Mainlander elite. By the 1990s, state-controlled television offered many programs in the Taiwanese dialect. Today, with the presidency in the hands of a Taiwanese, and the vast majority of seats held by native Taiwanese (regardless of party affiliation) they appear to have won the battle to maintain their cultural identity. Taiwanese legislators are refusing to use Chinese during the Legislature's proceedings, so now Mainlanders in the Legislative Yuan must learn Taiwanese.[8] Indeed, the pendulum appears to have swung the other way, not just in language, but in terms of engineering an appropriate psycho-cultural milieu for an independent Taiwan. For example, Chen Shui-bian (at least, before his pro-independence allies were defeated in the December 2004 legislative elections) has planned to "de-Sinicise" Taiwan by ridding it of any linguistic connections with the mainland, such as by changing the names of government agencies and state-owned corporations that have "China" in them (including "China Airlines" and "China Steel Corporation").(Source: Frank Ching, "Perfect Time to Test the Water," *South China Morning Post* (December 15, 2004). (on line) But such changes are merely a change in window dressing, given the billions of dollars of investment flowing from Taiwan into the mainland. Investments are doing far more to link Taiwan with the mainland than any mere name change can counter.

Generally speaking, however, Taiwanese and Mainlander culture need not be viewed as two cultures in conflict, for they share many commonalities. As Taiwanese move into leadership positions in what used to be exclusively Mainlander institutions, and as intermarriage between the two groups becomes more common, an amalgamation of Taiwanese and traditional Chinese practices is becoming evident throughout the society. As is discussed later in this report, the real source of conflict is the Taiwanese insistence that their culture and political system not be controlled by Chinese from the mainland, whether they be Nationalists or Communists.

On the other hand, rampant materialism as well as the importation of foreign ideas and goods are eroding both Taiwanese *and* Chinese values. For example, there are more than 3,000 7-Eleven outlets in Taiwan, and customers in central Taipei rarely have to walk more than a few blocks to get to the next one. Starbucks coffee houses flourish. Most major American fast-food franchises, such as Kentucky Fried Chicken and McDonald's, are ubiquitous. They provide "snack" food for Taiwan's teenagers and children—before they settle down for "real" (Chinese) food. The Big Mac culture affects more than waistlines, for it brings with it many "modern" and even Western values (though at this point, these Western multinationals are no longer recognized as "Western.") Although the government has at various times engaged in campaigns to reassert traditional values, the point seems lost in its larger message, which asks all to contribute to making Taiwan an Asian showplace. The government's emphasis on hard work and economic prosperity has seemingly undercut its focus on traditional Chinese values of politeness, the sanctity of the family, and the teaching of culturally based ethics (such as filial piety) throughout the school system. Materialism and an individualism that focuses on personal needs and pleasure seeking are slowly undermining collectively oriented values.[9] The "I can do whatever I want" attitude is, in the view of many, leading to a breakdown in social order.[10] In the meantime, unlike their mainland cousins, who have had a 40-hour, five-day work week since 1996, the Taiwanese did not relinquish an exhausting 6-day, 48-hour work-week until 2001.While playing a part in Taiwan's economic boom of the past decade, the emphasis on materialism has contributed to a variety of problems, not the least of which are the alienation of youth, juvenile crime, the loosening of family ties, and the general decline of community values. The pervasive spread of illicit sexual activities through such phony fronts as dance halls, bars, saunas, "barber shops," and movies on-video/DVD and music-video/DVD establishments, as well as hotels and brothels, grew so scandalous and detrimental to social morals and social order that at one point the government even suggested cutting off their electricity.[11] Another major activity that goes virtually uncontrolled is gambling. Part of the problem in clamping down on either illicit sexual activities or gambling (both of which are often combined with drinking in clubs) is that organized crime is involved; and that in exchange for bribes, the police look the other way.[12]

THE ENVIRONMENT

The pursuit of individual material benefit without a concomitant concern for the public good has led to uncontrolled growth and a rapid deterioration in the quality of life, even as the people in Taiwan become richer. According to Taiwan's Environmental Protection Agency, from 1990 to 2000, the amount of garbage produced doubled. Part of the problem is that in addition to being a very small island, Taiwan is mostly mountainous, so there is virtually nowhere to construct new landfills. Old landfills are filled to capacity and are leaching toxins into the soil. The result is that it is not uncommon for garbage to simply be dumped into the river, or left to rot in gullies or wherever it can be dumped unobserved. Recycling is limited.[13] But new laws that require citizens to sort their garbage into three categories (regular waste, kitchen leftovers, and recyclable materials), and allow the government to fine citizens who do not sort properly, are intended to bring significant declines in the amount of garbage—above and beyond the 40 to 50 percent decline already realized under current recycling regulations.[14]

Taiwan must continue to battle against the polluting effects of rapidly increasing wealth; for with almost all Taiwanese families owning a refrigerator and air conditioner, and purchasing cars at the rate of 300,000 per year (without a commensurate expansion of the island's roads), air quality continues to deteriorate. Taipei's recent construction of a subway system has helped traffic flow and kept air pollution from worsening even more quickly. However, political roadblocks have confounded efforts to build a high-speed rail system, and led in 2000 to the temporary suspension of the building of Taiwan's fourth nuclear-power plant, a critical project for providing clean electrical power.[15] The antinuclear movement has become increasingly active, especially after an accident at one of Taiwan's nuclear-power plants and the discovery that there are more than 5,000 radioactive buildings in Taiwan, including more than 90 in the city of Taipei.[16]

As Taiwan struggles to catch up with its own success, the infrastructure has faltered. During the hot, humid summers in Taipei, both electricity and water are frequently shut off; roads are clogged from 7:00 A.M.

Religion is an integral part of everyday life in Taiwan. The inside of this busy Buddhist temple in Taipei shows Taiwanese making offerings of food and burning incense.

until 10:00 P.M.; and the city's air is so dense with pollution that eyes water, hair falls out, and many people suffer from respiratory illnesses. Inadequate recreational facilities leave urban residents with few options but to join the long parade of cars out of the city on weekends. Taiwan's citizens have begun forming interest groups to address such problems, and public protests about the government's neglect of quality-of-life issues have grown increasingly frequent. Environmental groups, addressing such issues as wildlife conservation, industrial pollution, and waste disposal, have burgeoned; but environmental campaigns and legislation have difficulty keeping pace with the rapid growth of Taiwan's material culture.

RELIGION

A remarkable mixture of religions thrives in Taiwan. The people feel comfortable with placing Buddhist, Taoist, and local deities—and even occasionally a Christian saint—in their family altars and local temples. Restaurants, motorcycle-repair shops, businesses small and large—almost all maintain altars. The major concern in prayers is for the good health and fortune of the family. The focus is on life in this world, not on an afterlife. People pray for prosperity, for luck in the stock market, and even more specifically for the winning lottery number. If the major deity in one temple fails to answer prayers, people will seek out another temple where other deities

have brought better luck. Alternatively, they will demote the head deity of a temple and promote others to his or her place. The gods are thought about and organized in much the same way as is the Chinese bureaucracy. In fact, they are often given official clerical titles to indicate their rank within the deified bureaucracy.

Numerous Taiwanese religious festivals honor the more than 100 local city gods and deities. Offerings of food and wine are made to commemorate each one of their birthdays and deaths, and to ensure that the gods answer one's prayers. It is equally important to appease one's deceased relatives. The annual Tomb-Sweeping Festival (*qing ming*) in April is when the whole family cleans up their ancestral grave sites, makes offerings of food, money, and flowers, and burns incense to honor their ancestors.[17]

If they are neglected or offered inadequate amounts of food, money, and respect, they will cause endless problems for their living descendants by coming back to haunt them as ghosts. Even those having trouble with their cars or getting their computer programs to work will drop by the temple to pray to the gods and ancestors— just in case their problems have arisen from giving them inadequate respect.

The seventh month of the lunar calendar is designated "Ghost Month." For the entire month, most do whatever is necessary "to live in harmony with the omnipotent spirits that emerge to roam the world of the living." This includes "preparing doorway altars full of meat, rice, fruit, flowers and

beverages as offerings to placate the anxious visitors. Temples [hang] out red lanterns to guide the way for the roving spirits.… Ghost money and miniature luxury items made of paper are burned ritualistically for ghosts to utilize along their desperate journey.…"

In addition, the people heed a long list of taboos that can have an adverse effect on business during Ghost Month.

The real estate industry is particularly hard hit, because people do not dare to move into new houses, out of fear that homeless ghosts might take up permanent residence. Many choose to wait until after Ghost Month to make major purchases, such as cars. Few choose to marry at this time as, according to folk belief, a man might discover that his new bride is actually a ghost! Pregnant women usually choose to undergo Caesarean sections rather than give birth during Ghost Month. And law suits decline because it is believed that ghosts dislike those who bring law suits.[18]

Finally, there continues to be a preference for seeking medical cures from local temple priests, over either traditional Chinese or modern Western medicine. The concern of local religion is, then, a concern with this life, not with salvation in the afterlife. The attention to deceased ancestors, spirits, and ghosts is quite different from attention to one's own fate in the afterlife.

What is unusual in the case of Taiwanese religious practices is that as the island has become increasingly modern and wealthy, it has not become less religious. Technological modernization has seemingly not brought secularization with it. In fact, aspiring capitalists often build temples in hopes of getting rich. People bring offerings of food; burn incense and bundles of paper money to honor the temple gods; and burn expensive paper reproductions of houses, cars, and whatever other material possessions they think their ancestors might like to have in their ethereal state. They also pay real money to the owner of their preferred temple to make sure that the gods are well taken care of. Since money goes directly to the temple owner, not to a religious organization, the owner of a temple whose constituents prosper will become wealthy. Given the rapid growth in per capita income in Taiwan since 1980, then, temples to local deities have proliferated, as a builder of a temple was almost guaranteed to get rich if its constituents' wealth grew steadily.

Christianity is part of the melange of religions. About 4 percent of the population is Christian; but it is subject to local adaptations, such as setting off firecrackers inside a church during a wedding ceremony to ward off ghosts, and the display of flashing neon lights around the Virgin Mary. The Presbyterian Church in Taiwan, established in Taiwan by missionaries in 1865, was frequently harassed by the KMT because of its activist stance on social and human rights issues and because it generally supported the Taiwan-independence viewpoint.[19] Now that the pro-independence parties are legal, the church vigorously continues its activist pro-independence role without restraint. There are more than 1,200 Presbyterian congregations in Taiwan; but in recent years, there has been a 10 percent drop in members—apparently due to the aging church leaders inflexibility in responding to modernization and ideas from the outside.[20] The Catholic Church in Taiwan is likewise witnessing a decline in membership; and it is also suffering from the aging of its priests, most of whom had emigrated in the 1940s from the China mainland and are now dying off. There is an ever smaller number of young priests in training to replace them. As for Confucianism, it is more a philosophy than a religion. Confucianism is about self-cultivation, proper relationships among people, ritual, and proper governance. Although Confucianism accepts ancestor worship as legitimate, it has never been concerned directly with gods, ghosts, or the afterlife. In imperial China, if drought brought famine, or if a woman gave birth to a cow, the problem was the lack of morality on the part of the emperor—not the lack of prayer—and required revolt.

When the Nationalists governed Taiwan, they tried to restore Chinese traditional values and to reinstitute the formal study of Confucianism in the schools. Students were apathetic, however, and would usually borrow Confucian texts only long enough to study for college-entrance exams. Unlike the system of getting ahead in imperial China through knowledge of the Confucian classics, students in present-day Taiwan need to excel in science and math. Yet though efforts to engage students in the formal study of Confucianism have fallen on deaf ears, Confucian values suffuse the culture. The names of streets, restaurants, corporations, and stores are inspired by major Confucian virtues; advertisements appeal to Confucian values of loyalty, friendship, and family to sell everything from toothpaste to computers; children's stories focus on Confucian sages in history; and the vocabulary that the government and party officials use to conceptualize issues is the vocabulary of Confucianism—moral government, proper relationships between officials and the people, loyalty, harmony, and obedience.

EDUCATION

The Japanese are credited with establishing a modern school system in Taiwan in the first part of the twentieth century. After 1949, Taiwan's educational system developed steadily. Today, Taiwan offers nine years of free, compulsory education. Almost all school-age children are enrolled in elementary schools, and most go on to junior high schools. More than 70 percent continue on to senior high school. Illiteracy has been reduced to about 6 percent and is still declining. Night schools that cater to those students anxious to test well and make the cut for the best senior high schools and colleges flourish. Such extra efforts attest to the great desire of Taiwan's students to get ahead through education.

Taiwan has one of the best-educated populations in the world, a major factor in its impressive economic development. Its educational system is, however, criticized for its insistence on uniformity through a unified national curriculum, a lecture format that does not allow for student participation, the grueling high school and university entrance examinations, tracking, rote memorization, heavy homework assignments, and humiliating treatment of students by teachers. Its critics say that the system inhibits creativity.[21] There is also a gender bias in education, which results in women majoring in the humanities and social sciences, while men choose science and math majors. Reforms in recent years have tried to modify some of these prac-

tices; and Taiwan's burgeoning information-technology and high-tech sectors have added to pressures to train more women in science and technology.

The number of colleges and universities has more than doubled since martial law was lifted. But the more than 120 colleges and universities now in existence cannot meet the demand for spaces for all qualified students. As a result, many students go abroad for study. From 1950 to 1978, only 12 percent of the some 50,000 students who studied abroad returned, a reflection of both the lack of opportunity in Taiwan and the oppressive nature of government in that period. Beginning in the late 1980s, as Taiwan grew more prosperous and the political system more open, this outward flood of human talent, or "brain drain," was stemmed. Nevertheless, the system of higher education has been unable to keep up with the demand for high-tech workers. This has led to a loosening of restrictions on importing high-tech workers from mainland China.[22]

As the economies of mainland China and Taiwan have become more intertwined, the options that Taiwanese have in the educational arena have increased as well. A growing number of Taiwan's high school graduates are choosing to go to a university on the mainland. This is in no small part because the vast majority of Taiwan's businesspeople are investing their money in the mainland economy. However, in spite of the fact that many students from Taiwan attend China's leading universities, which produce outstanding graduates whose degrees are readily accepted in the West, Taiwanese students find that their degrees are not properly honored once they return to Taiwan. Indeed, in a poll conducted in Taiwan, in 2000, 40.5 percent of respondents believed that academic credentials earned by Taiwan's students in the mainland should have a stricter standard applied than to students graduating in Taiwan; and 6.3 percent believed that they should not be recognized at all.[23] Students have to take a set of exams upon returning from the mainland to validate the legitimacy of their degrees.[24] It would appear, however, that the real reason is political—namely, to challenge the quality of any institutions under the control of the Chinese Communist Party.

HEALTH CARE AND SOCIAL SECURITY

Although Taiwan's citizens have received excellent health care for many years, the health-care system is now facing a crisis. Health care has been a major political issue with candidates for office promising to do more than the others if elected. But slower economic growth has produced lower governmental revenues for health care. The elderly, moreover, tend to visit their doctors on a weekly, if not a daily, schedule, not only because their visits are virtually free, but also because that's where all the other elderly people are! Lonely, with time on their hands and an obsession with longevity, the elderly often hang out at hospitals and health clinics. This contributes to the bankruptcy of the "medicare" system, for doctors are more than happy to see dozens of patients each hour, as they are paid by the number of patients they see. Indeed, citizens complain that their visits usually last less than a minute!

The Labor Standards Law requires that Taiwanese enterprises provide a pension plan for their employees; but the old labor standards regulations required that only employees who had worked in the same company for 25 years would be covered. Since "the average length of employment in most small- and medium-sized businesses is 13 years, ... 90 percent didn't qualify for a pension." Once the draft National Annuity Act is passed, everyone will be covered.[25] So far, however, the government's plan to implement a system of social security for retired persons has not been implemented. As a result, Taiwan's citizens continue to rely on their families for support in old age.

WOMEN

The societal position of women in Taiwan reflects an important ingredient of Confucianism. Traditionally, Chinese women "were expected to obey their fathers before marriage, their husbands after, and their sons when widowed. Furthermore, women were expected to cultivate the "Four Virtues": morality, skills in handicrafts, feminine appearance, and appropriate language."[26] In Taiwan, as elsewhere throughout the world, women have received lower wages than men and have rarely made it into the top ranks of government and business—this in spite of the fact that it was women who, from their homes, managed the tens of thousands of small businesses and industries that fueled Taiwan's economic boom.

HEALTH/WELFARE

Taiwan has one of the highest population densities in the world. Education is free and compulsory to age 15, and the country boasts more than 100 institutions of higher learning. Social programs, however, are less developed than those in Singapore, Japan, and some other Pacific Rim countries.

In the workplace outside the home, women are treated differently than men. For example, women are not allowed to serve in the armed forces; but until the 1990s, "all female civil servants, regardless of rank, [were] expected to spend half a day each month making pants for soldiers, or to pay a substitute to do this."[27] Women, who make up 40 percent of Taiwan's civil service, find themselves "walking on glue" when they try to move from the lower ranks to the middle and senior ranks of the civil service. By the beginning of 2000, only 12 percent of the total senior level civil service, and less than a third of intermediate ranks, were made up of women. Those figures had not changed at all by the end of 2004.[28] There have, however, been greater opportunities for women in the last decade. Women are more visible in the media and politics than before. In 2000, for the first time, a woman was elected as vice-president (and was reelected in 2004); but since 1949, not one of the five branches of government has been headed by a woman. On the other hand, there are 9 women in the Cabinet, 7 with a ministerial post, and 2 without portfolio; and in 2004, the president appointed the first female vice premier. About 18 percent of the legislators in the Legislative Yuan are women; and an even higher percentage of women are city or county councilors.[29] This training ground for leaders is, then, laying the basis for the advancement of women as leaders in the future. Because women may now receive the same education as men, and because employment in the civil service is now based on an examination system, women's social, political, and economic mobility has increased.[30] Better education of women has been both the cause and the result of greater advocacy by Taiwan's feminists of equal rights for women. It has also eroded the typical marriage pattern, in which a man is expected to marry a woman with an education inferior to his own.

THE ECONOMY

The rapid growth of Taiwan's economy and foreign trade has enriched Taiwan's population. A newly industrialized economy (NIE), Taiwan long ago shed its "Third World," underdeveloped image. With a gross domestic product per capita income that rose from U.S.$100 in 1951 to more than U.S.$23,400 by 2004, and a highly developed industrial infrastructure and service industry, Taiwan has entered the ranks of some of the most developed economies in the world. As with the leading industrial nations, however, the increasing labor costs for manufacturing and

industry have led to a steady decline in the size of those sectors as companies relocate to those countries with cheaper raw materials and lower wages. Taiwan's economic growth rate since 1953 has averaged a phenomenal 8-percent-plus annually; but in recent years, growth has slowed down significantly, and a growing percentage of even this growth is due to production on the mainland, where most of Taiwan's manufacturing base has moved.

The government elite initiated most of the reforms critical to the growth of Taiwan's economy, including land redistribution, currency controls, central banking, and the establishment of government corporations. Taiwan's strong growth and high per capita income does not, however, bring with them a lifestyle comparable to that in the most developed Western states. Taiwan's cities are crowded and badly polluted, and housing is too expensive for most urbanites to afford more than a small apartment. The overall infrastructure is inadequate to handle traffic, parking, sewage, electricity, and other services expected in a more economically advanced society.[31]

Taiwan's economic success thus far may also be attributed to a relatively open market economy in which businesspeople have developed international markets for their products and promoted an export-led economy. Taiwan's highly productive workers have tended to lack class consciousness, because they progress so rapidly from being members of the working-class "proletariat" to becoming capitalists and entrepreneurs. Even factory workers often become involved in small businesses.[32]

Now that Taiwan is privatizing those same government corporations that used to have complete control over many strategic materials as well as such sectors as transportation and telecommunications, workers are resisting the loss of their "iron rice bowl" of permanent employment in state enterprises and the civil service. Much like mainland China, the government is concerned that social instability may result if organized labor, instead of accepting the "global trend" toward privatization, resists it through street protests.[33] As an increasing number of industrial workers in Taiwan move into white-collar jobs and are replaced by relatively poorly organized immigrant laborers, who are often sent home when the jobs disappear, this problem seems to have been temporarily sidelined.

Sometimes called "Silicon Island," Taiwan has some 1.2 million small and medium-size enterprises, and only a handful of mega-giants. Most of these smaller enterprises are not internationally recognized names, but they provide the heart and even the backbone of technological products

(Photo courtesy of Suzanne Ogden)
Motorcycles are parked on public sidewalks due to the lack of adequate parking facilities in Taipei. The 7-Eleven logo is ubiquitous in Taipei, one indication that Taiwan's economy is tied into the "global village."

worldwide. They make components, or entire products (such as computer hardware), according to specifications set by other, often well-known, large firms, whose names go on the final product. Furthermore, because Taiwan's firms tend to be small, and flexible, they can respond quickly to changes in technology. This is particularly true in the computer industry. Thanks to the many students who have gone to the United States to study and then stayed to work in the computer industry's Silicon Valley, there are strong ties with, and dependency on, Taiwan's entrepreneurs.[34] As a result of the emphasis on the information-technology sector in Taiwan, it has become the leading Asian center for information technology (IT) and software.

A stable political environment facilitated Taiwan's rapid growth. So did Taiwan's protected market, which brought protests over Taiwan's unfair trade policies from those suffering from an imbalance in their trade with Taiwan. Slowly Taiwan has shed most of the regulations that have protected its industries from international trade competition. But the Asian financial crisis that began in 1997 made the government reconsider deregulation, as it was precisely because of some of the tight controls over investments, banks, credit, and currency flows that Taiwan survived the crisis better than most of its Asian brethren. Although Taiwan was not able to insulate itself from the effects of the crisis, its growth rate recovered significantly—be-

fore the presidential elections in 2000 led to the ouster of the Nationalists' government and a 50 percent decline in the value of Taiwan's stock market.

Since the 2000 elections, however, the "industrial policy" model, whereby the government, banks, and businesses essentially colluded for the purpose of economic growth, no longer works; for the new government is unable to get the banks to give loans to declining industries to keep them alive. Furthermore, with Taiwan's entry into the World Trade Organization (WTO) in 2002, the agricultural sector has had to readjust, because Taiwan is no longer able to use import regulations to protect itself from a flood of agricultural goods from mainland China, where virtually every agricultural product is produced more cheaply. In particular, rice and poultry imports have soared. But at the same time, agricultural exports have increased, a result attributed to the promotion of Taiwan products abroad, such as mango, tea and litchi, as well as other fruits, and flowers and seeds.[35]

For the economy as a whole, there has been a dramatic turnaround since the 1997-2000 Asian financial crisis. By 2004, GDP growth has moved from negative numbers to a positive 3.2 percent. Taiwan came out of its economic downturn largely by increasing commercial links with China, where Taiwanese businesspeople can get better returns on their investment. There are more than 50,000 Taiwanese factories already in China; 70 percent of China's IT products that are exported are made in Taiwanese factories on the mainland; and Taiwanese firms control a quarter of China's export licenses. Some 10,000 Taiwanese business people travel to the mainland each day. Taiwan's government has lifted the limits on investment in China by companies and enterprises from U.S.\$5 million to U.S.\$50 million; but in reality, there is no longer any real limit on the flow of investment funds into China. In short, Taiwan's growth is coming from its investments in China.[36]

Indeed, most of Taiwan's business people are concerned about the survival of their businesses, not about national security vis-à-vis the mainland. As a result, they have been sending delegations to China (without authorization from the government) to reassure its leaders that they will not let Taiwan declare independence and will continue to develop economic ties with the mainland. They believe (as does Beijing) that once the two economies are fully integrated, a declaration of independence of Taiwan will be highly unlikely.[37]

Taiwanese have invested well over U.S. \$100 billion in China. Because of restrictions on trade with the mainland, however, money must first move to Hong Kong or elsewhere, and only then to China. These sorts of maneuvers complicate business investments and irritate Taiwan's business community. They contend that Chen has also been slow in fulfilling his promises to liberalize current restrictions on travel to and from mainland China, which still require that they travel first, like their money, to a country other than China. As part of his 2004 election campaign, President Chen Shui-bian promised to open direct air passenger and cargo links, to ease restrictions on travel to and from the mainland, and to liberalize regulations prohibiting Taiwanese from raising capital for their China businesses on the Taiwanese stock ex-

change, the Taiex.[38] But, as of 2005, he had not done so. Many in Taiwan's business community are, in fact, distressed that Chen is not doing more to protect and promote business ties with the mainland (which may explain why the business community generally supported the KMT in the December 2004 elections).

DEVELOPMENT

Taiwan has vigorously promoted export-oriented production, particularly of electronic equipment. In the 1980s, manufacturing became a leading sector of the economy, employing more than one third of the workforce. By 2000, most manufacturing facilities had been relocated to the Chinese mainland and Southeast Asia. Taiwan now leads Asia in the area of computer software and technology. Taiwanese households are equipped with virtually every household appliance. In 2001, Taiwan became the 144th member of the World Trade Organization. Taiwan remains peaceful and prosperous.

Taiwanese enterprises run what amounts to a parallel economy on the mainland that is completely entangled with China's own. In fact, as of 2003, China had become Taiwan's number one trade partner. Taiwan has a substantial trade surplus with the mainland (U.S.\$735 million in 2003), and much of China's trade surplus with the United States is from Taiwanese enterprises doing business in China. It is estimated that up to one-third of exported consumer goods labeled 'Made in China' are actually made in Taiwanese-owned firms in China. "Analysts attribute more than 70 percent of the growth in America's trade deficit with China to the exports of Taiwanese firms."[39] In 2003, Taiwan exported more than three times as much to China as it imported; and, many of the imports were from their own factories on the mainland. The result is that Taiwan holds more than U.S.\$200 billion in its central reserve bank—the third largest holding of reserve currency in the world. U.S. pressure on Taiwan to purchase U.S.\$18 billion of military equipment from the U.S. in order to draw down some of these reserves has become a point of bitter contention between Taipei and Washington. The stalemate has continued because the "pan-blue alliance," led by the KMT, continues to control the legislature, and views any such purchases as a waste of resources and of little real use to Taiwan's defense. Indeed, they argue that it will simply lead to an acceleration of the arms race with China, and could even lead to war.

Finally, although Taipei insists that the government will not establish direct trade

links with China until Beijing meets certain conditions, it has passed legislation so that free-trade zones can be established near Taiwan's international airports and harbors. The purpose of these tax-free areas is to encourage foreign business people, including Chinese from the Mainland, to establish companies in Taiwan for the purpose of trade, processing, and manufacturing.[40] Clearly Taipei is trying both to stem the outward flow of Taiwan's investment moneys and to lure more investors to Taiwan by making its business conditions competitive with those of the mainland, Hong Kong, and Asian countries with free-trade zones.

Agriculture and Natural Resources

After arriving in Taiwan, the KMT government carried out a sweeping land-reform program: The government bought out the landlords and sold the land to their tenant farmers. The result was equalization of land distribution, an important step in promoting income equalization among the farmers of Taiwan. Today, farmers are so productive that Taiwan is almost self-sufficient in agriculture—an impressive performance for a small island where only 25 percent of the land is arable.

The land-reform program was premised upon one of Sun Yat-sen's famous Three Principles, the "people's livelihood." One of the corollaries of this principle was that any profits from the increase in land value attributable to factors not related to the real value of farmland—such as through urbanization, which makes nearby agricultural land more valuable—would be turned over to the government when the land was sold. As a result, although the price of land has skyrocketed around Taiwan's major cities, and although many farmers are being squeezed by low prices for their produce, they would get virtually none of the increased value for their land if they sold it to developers. Many farmers have thus felt trapped in agriculture. In the meantime, the membership of both China and Taiwan in the World Trade Organization means that cheaper mainland produce flows quite freely into Taiwan; but, while this has undercut the profits of farmers in such areas as rice, their profits in fruits and other products have benefited by the lifting of trade barriers.

Natural resources, including land, are quite limited in Taiwan. Taiwan's rapid industrialization and urbanization have put a strain on what few resources exist. Taiwan's energy resources, such as coal, gas, and oil, are particularly limited. The result is that the government has had to invest in the building of a number of nuclear-power plants to provide sufficient energy to fuel Taiwan's rapidly modernizing society and economy. Popular protest against further nuclear-power plants, however, brought an energy crisis in the 1990s. Hopes are that it will be abated by the completion of Taiwan's fourth nuclear plant. Thus far, however, Taiwan has been able to postpone its energy and resource crisis by investing in the development of mainland China's vast natural resources. Taiwan's businesspeople have also moved their industries to other countries where resources, energy, land, and labor are cheaper. They now control a vast network of manufacturing and distribution facilities throughout the world.[41]

Internationalization of its economy is also part of Taiwan's strategy to thwart China's efforts to cut off its relationships with the rest of the world. With Taiwan's important role in the international economy, it is virtually impossible for its trade, commercial, and financial partners to ignore it. This saves Taiwan from even greater international diplomatic isolation than it already faces in light of its current "non-state" status. In the meantime, its economy is becoming increasingly integrated with that of the Chinese mainland, to the mutual benefit of both economies.

Taiwan as a Model of Economic Development

Taiwan is often cited as a model for other developing countries seeking to lift themselves out of poverty. They could learn some useful lessons from certain aspects of Taiwan's experience, such as the encouragement of private investment and labor productivity, an emphasis on basic health care and welfare needs, and policies to limit gross extremes of inequality. But Taiwan's special advantages during its development have made it hard to emulate. These advantages include its small size, the benefits of improvements to the island's economic infrastructure and educational system made under the Japanese occupation, massive American financial and technical assistance, and a highly favorable international economic environment during Taiwan's early stages of growth.

What has made Taiwan extraordinary among the rapidly developing economies of the world is the government's ability—and commitment—to achieve and maintain a relatively high level of income equality. Although there are homeless people in Taiwan, their numbers are small. Government programs to help the disabled and an economy with moderate unemployment (about 5 percent in 2004) certainly helps, as does a tight-knit family system that supports family members in difficult times. The government's commitment to Sun Yat-

sen's principle of the "people's livelihood," or what in the West might be called a "welfare state," is still an important consideration in policy formation.

The growth of Taiwan's stock market—a market often floating on the thin air of gossip and rumor—has created (and destroyed) substantial wealth almost overnight. Nevertheless, Taiwan's economic wealth is still fairly evenly distributed, contributing to a strongly cohesive social system.

Taiwan's economy is not without growing pains. The government has so far been unable to solve many of the problems arising from its breathtakingly fast modernization: massive air and water pollution (only 3 percent of Taipei's sewage is treated, and diseases related to pollution have skyrocketed);[42] an inadequate urban infrastructure for housing, transportation, electricity, and water; and rampant corruption as everyone tries to get ahead in a now relatively open economy. In the cities, the rapid acquisition of air conditioners and automobiles has made the environment unbearable and transportation a nightmare. In spite of—and in some cases because of—Taiwan's astounding economic growth, the quality of life has deteriorated greatly. Complaints of oily rain, ignitable tap water, stunted crops due to polluted air and land, and increased cancer rates abound. "Garbage wars" over the "not-in-my-back-yard" issue of sanitary landfill placement have led to huge quantities of uncollected garbage.[43] Numerous public-interest groups have emerged to pressure the government to take action. Antinuclear activists even tried to use the recall vote to remove legislators who favored building Taiwan's fourth nuclear-power plant.[44]

ACHIEVEMENTS

From a largely agrarian economic base, Taiwan has been able to transform its economy into an export-based dynamo with international influence. Today, only about 10 percent of the population works in agriculture, and Taiwan ranks among the top 20 exporters in the world.

In spite of labor's demand for higher wages, inflation is still under control. To keep Taiwan's labor-intensive products from being priced out of the international market, however, Taiwan's own entrepreneurs have had to set up shop outside of Taiwan, notably in Thailand, the Philippines, and mainland China, where labor is far cheaper. Taiwan has also become Asia's leader in information technology and software, where Taiwan's educated la-

bor force is at an advantage, and low wage labor is not yet an issue. At the same time, Taiwan's rapid economic growth rate, combined with a relatively low birth rate, has led businesses to import foreign laborers to do unskilled jobs that Taiwan's better-paid residents refuse to do. These foreign workers, largely from Thailand, the Philippines, and Indonesia, work at wages too low, with hours too long and conditions too dangerous for Taiwan's own citizens. By the end of 2000, there were more than 300,000 foreigner workers in Taiwan.[45]

Foreign workers (now 1.5 percent of Taiwan's population) introduce their own invisible "costs" to Taiwan: lonely and isolated within a society where they are considered socially inferior, and where they rarely speak the language, they tend to engage in heavy drinking, gambling, and other socially dysfunctional behaviors.[46] Taiwan's security concerns result in its importing few mainland Chinese workers, who, because of their common ethnicity and language, could easily melt into Taiwan's population. They tend to be offered jobs that isolate them from the island, such as the fishing industry.[47]

THE POLITICAL SYSTEM

From 1949 to 1988, the KMT justified the unusual nature of Taiwan's political system with three extraordinary propositions. First, the government of the Republic of China, formerly located on the mainland of China, was merely relocated temporarily on China's island province of Taiwan. Second, the KMT was the legitimate government not just for Taiwan but also for the hundreds of millions of people living on the Chinese mainland under the control of the Chinese Communist Party. Third, the people living under the control of the Communist "bandits" would rush to support the KMT if it invaded the mainland to overthrow the Chinese Communist Party regime. Taiwan's political and legal institutions flowed from these three unrealistic propositions, which reflected hopes, not reality. Underlying all of them was the KMT's acceptance, in common with the Chinese Communist Party, that there was only one China and that Taiwan was a province of that one China. Indeed, until the early 1990s, it was a *crime* in Taiwan to advocate independence.[48]

The Constitution

In 1946, while the KMT was still the ruling party on the mainland, it promulgated a "Constitution for the Republic of China." This Constitution took as its foundation the same political philosophy as the newly founded Republic of China adopted in 1911 when it overthrew China's Manchu rulers

SUN YAT-SEN: THE FATHER OF THE CHINESE REVOLUTION

New York Public Library

Sun Yat-sen (1866–1925) was a charismatic Chinese nationalist who, in the declining years of the foreign-ruled Manchu Dynasty, played upon Chinese-nationalist hostility to both foreign colonial powers and to the Manchu rulers themselves.

Sun (pictured at the left) drew his inspiration from a variety of sources, usually Western, and combined them to provide an appealing program for the Chinese. This program was called the Three People's Principles, which translates the American tenet "of the people, by the people, and for the people" into "nationalism," "democracy," and the "people's livelihood."

This last principle, the "people's livelihood," is the source of dispute between the Chinese Communists and the Chinese Nationalists, both of whom claim Sun Yat-sen as their own. The Chinese Communists believe that the term means socialism, while the Nationalists in Taiwan prefer to interpret it to mean the people's welfare in a broader sense.

Sun Yat-sen is, in any event, considered by all Chinese to be the father of the Chinese Revolution of 1911, which overthrew the feeble Manchus. He thereupon declared China to be a republic and named himself president. However, he had to relinquish control immediately to the warlord Yuan Shih-K'ai, who was the only person in China powerful enough to maintain control over all other contending military warlords in China.

When Sun died, in 1925, Chiang Kai-shek assumed the mantle of leadership of the Kuomintang, the Chinese Nationalist Party. After the defeat of the KMT in 1949, Sun's widow chose to remain in the People's Republic of China and held high honorary positions until her death in 1982.

on the mainland: Sun Yat-sen's "Three People's Principles" ("nationalism," "democracy," and the "people's livelihood"). Democracy was, however, to be instituted only after an initial period of "party tutelage." During this period, the KMT would exercise virtually dictatorial control while in theory preparing China's population for democratic political participation.

The Constitution provided for the election of a National Assembly (a sort of "electoral college"); a Legislative Yuan ("branch") to pass new laws, decide on budgetary matters, declare war, and conclude treaties; a Judicial Yuan to interpret the Constitution and all other laws and to settle law suits; an Executive Yuan to run the economy and manage the country generally; a Control Yuan, the highest supervisory organ, to supervise officials through its powers of censure, impeachment, and auditing; and an Examination Yuan (a sort of personnel office) to conduct civil-service examinations. The Examination Yuan and Control Yuan were holdovers from Chinese imperial traditions dating back thousands of years.

Because this Constitution went into effect in 1947 while the KMT still held power on the mainland, it was meant to be applicable to all of China, including Taiwan. The KMT government called nationwide elections to select delegates for the

National Assembly. Then, in 1948, it held elections for representatives to the Legislative Yuan and indirect elections for members of the Control Yuan. Later in 1948, as the Civil War between Communists and Nationalists on the mainland raged on, the KMT government amended the Constitution to allow for the declaration of martial law and a suspension of regular elections; for by that time, the Communists were taking control of vast geographical areas of China. Soon afterward, the Nationalist government under Chiang Kai-shek fled to Taiwan. With emergency powers in hand, it was able to suspend elections and all other democratic rights afforded by the Constitution.

By October 1949, the Communists had taken control of the Chinese mainland. As a result, the KMT, living in what it thought was only temporary exile in Taiwan, could not hold truly "national" elections for the National Assembly or for the Legislative and Control Yuans as mandated by the 1946 Constitution. But to foster its claim to be the legitimate government of all of China, the KMT retained the 1946 Constitution and governmental structure, as if the KMT alone could indeed represent all of China. With "national" elections suspended, those individuals elected in 1947 from all of China's mainland provinces (534 out of a total 760 elected had fled with

General Chiang to Taiwan) continued to hold their seats in the National Assembly, the Legislative Yuan, and the Control Yuan—usually until death—without standing for reelection. Thus began some 40 years of a charade in which the "National" Assembly and Legislative Yuan in Taiwan pretended to represent all of China. In turn, the government of the island of Taiwan pretended to be a mere provincial government under the "national" government run by the KMT. At no time did the KMT government suggest that it would like Taiwan to declare independence as a state.

Although the commitment to retaking the China mainland was quietly abandoned by the KMT government even before Chiang Kai-shek's death, the 1946 Constitution and governmental structure remained in force. This was in spite of the fact that over those many years, numerous members of the three elected bodies died. Special elections were held just to fill their vacant seats. The continuation of this atavistic system raised serious questions about the government's legitimacy. The Taiwanese, who comprised more than 80 percent of the population but were not allowed to run for election to the Legislative Yuan or National Assembly, accused the KMT Mainlanders of keeping a stranglehold on the political system and pressured them for greater representation. Because the holdovers from the pre-1949 period were of advanced age and often too feeble to attend meetings of the Legislative Yuan (and some of them no longer even lived in Taiwan), it was virtually impossible to muster a quorum. Thus, in 1982, the KMT was forced to "reinterpret" parliamentary rules in order to allow the Legislative Yuan to get on with its work.

Chiang Ching-kuo, Chiang Kai-shek's son, decided to concede reality and began bringing Taiwanese into the KMT. By the time a Taiwanese, Lee Teng-hui, succeeded Chiang Ching-kuo in 1988 as the new Nationalist Party leader and president of the "Republic of China," 70 percent of the party's membership were Taiwanese. Pressures therefore built for party and governmental reforms that would diminish the power of the old KMT Mainlanders. In July 1988, behind the scenes at the 13th Nationalist Party Congress, the leadership requested the "voluntary" resignation of the remaining pre-1949 holdovers. Allegedly as much as U.S. $1 million was offered to certain hangers-on if they would resign, but few accepted. Finally, the Supreme Court forced all those Chinese mainland legislators who had gained their seats in the 1948 elections to resign by the end of 1991.

Under the Constitution, the Legislative, Judicial, Control, and Examination branches hold certain specific powers. Theoretically, this should result in a separation of powers, preventing any one person or institution from the arbitrary abuse of power. In fact, however, until after the first completely democratic legislative elections of December 1992, none of these branches of government exercised much, if any, power independent of the Nationalist Party or the president who was chosen by the Nationalist Party instead of by democratic election until 1996. In short, the KMT and the government were merged, in much the same way as the CCP was merged with the government on the mainland. Indeed, property and state enterprises owned by the government were claimed by the KMT as their own property when they lost control of the presidency and the KMT was no longer the majority party.

Thanks to changes made by President Lee, however, Taiwan now has a far greater separation of powers as well as a two-headed government: The president is primarily responsible for Taiwan's security, and the prime minister (premier) is responsible for the economy, local government, and other broad policy matters. But problems remain because the president may appoint the prime minister without approval by the Legislature—yet the Legislature has the power to cast a no confidence vote and

require new elections if it finds the government's actions unacceptable.

A final consequence of the three propositions upon which political institutions in Taiwan were created was that, after the KMT arrived in 1947, Taiwan maintained two levels of government. One was the so called national government of the "Republic of China," which ruled Taiwan as just one province of all of China. The other was the actual provincial government of Taiwan, which essentially duplicated the functions of the "national" government, and reported to the "national" government, which for so many years pretended to represent all of China. In this provincial-level government, however, native Taiwanese always had considerable control over the actual functioning of the province of Taiwan in all matters not directly related to the Republic of China's relationship with Beijing. Taiwan's provincial government thus became the training ground for native Taiwanese to ascend the political ladder once the KMT reformed the political system after 1988. This expensive and unnecessary redundancy of governments has at long last been eradicated.

Martial Law

The imposition of martial law in Taiwan from 1948 to 1987 is critical to understanding the dynamics of Taiwan's politics. Concerned with the security of Taiwan against subversion or an invasion by the Chinese Communists, the KMT government had imposed martial law on Taiwan. This allowed the government to suspend civil liberties and limit political activity, such as organizing political parties or mass demonstrations. Thus it was a convenient weapon for the KMT Mainlanders to control potential Taiwanese resistance and to quash any efforts to organize a "Taiwan independence" movement. Police powers were widely abused, press freedoms were sharply restricted, and dissidents were jailed. As a result, the Taiwan Independence Movement was forced to organize abroad, mostly in Japan and the United States. Taiwan was run as a one-party dictatorship supported by the secret police.

Non-KMT candidates were eventually permitted to run for office under the informal banner of *tangwai* (literally, "outside the party"); but they could not advocate independence for Taiwan; and they had to run as individuals, not as members of opposition political parties, which were forbidden until 1989. The combination of international pressures for democratization, the growing confidence of the KMT, a more stable situation on the China mainland, and diminished threats from Beijing led the KMT to lift martial law in July

1987. Thus ended the state of "Emergency" under which almost any governmental use of coercion against Taiwan's citizens had been justified.

Civil Rights

Until the late 1980s, the rights of citizens in Taiwan did not receive much more protection than they did on the Chinese mainland. The R.O.C. Constitution has a "bill of rights," but most of these civil rights never really existed until martial law was lifted in 1987. Civil rights were repeatedly suspended when their invocation by the citizenry challenged KMT power or policies; and opposition political parties were not allowed to organize. Because the "Emergency" regulations provided the rationale for the restriction of civil liberties, the KMT used military courts (which do not accord basic protection of defendants' civil rights)[49] to try what were actually civil cases,[49] arrested political dissidents, and used police repression, such as in the brutal confrontation during the 1980 Kaohsiung Incident.[50]

FREEDOM

For nearly 4 decades, Taiwan was under martial law. Opposition parties were not tolerated, and individual liberties were limited. A liberalization of this pattern began in 1987 when martial law was lifted. In 1991, 5,574 prisoners, including many political prisoners, were released in a general amnesty. In the 2000 presidential elections, a member of the opposition party was elected for the first time. Today, the media flourish and legislative races are hotly contested.

Today, it is highly unlikely that individuals would be imprisoned for political crimes. Since Chen Shui-bien won the presidency in 2000 and the Democratic Progressive Party (DPP) became the largest party (although it only controlled a minority of seats) in the Legislative Yuan in 2001there has been a far greater focus on civil and human rights. This reflects the sensitivity of the DPP to rights issues, since so many of its members were themselves deprived of civil and human rights because of their advocacy of Taiwan independence. The 2002 Human Rights Index report, based on a survey by the Taipei-based Chinese Association for Human Rights, indicates that the protection of judicial rights had improved slightly, to 2.85 points on a scale of five points. This reflected some improvement in the interrogation of suspects, but the report concluded that the protection of suspects in police custody remained inadequate. Police were urged to follow proper procedures in taking affidavits, and to allow suspects access to lawyers while being interrogated. In late 2002, the Executive Yuan approved a draft bill to limit police authority in law enforcement. If the bill becomes law, search warrants and police raids of businesses suspected of illegal activities would require "more stringent criteria for probable cause." Furthermore, if police have failed to follow due process and therefore infringed on a suspect's rights, the suspect can immediately file an administrative appeal.[51]

The 2002 survey also indicated that political rights (civil rights and freedom, equality, political efficiency, democratic consolidation, and media independence and objectivity) had shown modest progress over the year. In the specific category of "civil rights and freedom," the score was 3.21 points out of five and was the only category in which Taiwan scored over three points. Protection of women's rights showed a modest improvement, from 2.45 to 2.77, but was still a low score on the five-point scale. The report found that the improvement was due to better implementation of the Domestic Violence Prevention Law, and the promulgation of the Gender Equality Labor Law in March 2002. The latter is aimed at eradicating sexual harassment and discrimination, requires employers to offer up to two years' paid maternity leave, and embodies principles of equal pay for equal work and equal rights in places of employment. Taiwan also made progress in rights for children, elderly, and the handicapped.[52]

Political Reform in Taiwan

The Kuomintang maintained its political dominance until the turn of the century in part by opening up its membership to a broader segment of the population. It thereby allowed social diversity and political pluralism to be expressed *within* the KMT. The "Taiwanization" of both the KMT and of governmental institutions after Chiang Kai-shek's death in 1975 actually permitted the KMT to ignore the demands of the Taiwanese for an independent opposition party until the late 1980s. The KMT also hijacked the most appealing issues in the platform of the Democratic Progressive Party—notably, the DPP's demand for more flexibility in relations with the P.R.C., environmental issues, and greater freedom of the press. By the time of the 1996 elections, the dominant wing of the KMT had even co-opted the DPP's platform for a more independent Taiwan. In short, as Taiwan became more socially and politically diverse, the KMT relinquished many of its authoritarian methods and adopted persuasion, conciliation, and

open debate as the best means to maintain control.[53]

External pressures played a significant role in the democratization of Taiwan's institutions. American aid from the 1950s to the 1970s was accompanied by considerable American pressure for liberalizing Taiwan's economy, but the United States did little to force a change in Taiwan's political institutions during the Cold War, when its primary concern was to maintain a defense alliance with Taiwan against the Chinese Communists. Taiwan's efforts to bolster its integration into the international economy has allowed it to reap the benefits of internationalization. The government has also, after much prodding, responded positively to demands from its citizens for greater economic and cultural contact with China,[54] and for reform of the party and government. As a result, the KMT was able to claim responsibility for Taiwan's prosperity as well as for its eventual political liberalization; but as noted elsewhere, it must also take responsibility for the many problems it left on the platter for the incoming DPP administration in 2000.

The KMT's success in laying claim to key elements of the most popular opposition-party policies forced the opposition to struggle to provide a clear alternative to the KMT. Apart from the issue of Taiwan independence, when the DPP was acting as the opposition party it demanded more rapid political reforms and harshly criticized the KMT's corrupt practices. The DPP's exposure of the KMT's corruption, and of the infiltration into the political system by criminal organizations, brought public outrage. Over the decades of its rule, the KMT used money and other resources to help tie local politicians and factions to itself. The KMT's power at the grassroots level was lubricated through patronage, vote buying, and providing services to constituents.[55] The public has demanded that the interweaving of political corruption, gangsterism, and business (referred to as "black and gold politics" in Taiwan) be brought under control, and that the KMT divest itself of corporate holdings that involve conflicts of interest and permit it to engage in corrupt money politics. (When it also controlled the government, the KMT as a political party possessed an estimated U.S.$2.6 billion of business holdings, a sizable percentage of corporate wealth in Taiwan.) According to statistics published in the *Taipei Times*, "two-thirds of gangs in Taiwan have lawmakers running on their behalf in the legislature, while one-quarter of elected public representatives have criminal records… [T]hey are invulnerable because, as the KMT's legislative majorities have slowly declined, the ruling party needs the support

of independent lawmakers—including those with organized crime backgrounds—to pass its legislative agenda.[56]

By the time of the first democratic elections for the president of Taiwan in 1996, much had changed in the platforms of both the KMT and the DPP. The KMT, which had by then developed a powerful faction within it demanding greater international recognition of Taiwan as an independent state, adopted what amounted to an "independent Taiwan" position. Although President Lee stated this was a "misinterpretation" by Beijing and the international community, and that Taiwan merely wanted more international "breathing space," his offer of U.S.$1 billion to the United Nations if it would give Taiwan a seat was hardly open to interpretation.

Angered by Lee's efforts to gain greater recognition for Taiwan as an independent state, China began missile "tests" in the waters close to Taiwan in the weeks leading up to the 1996 elections. Fortunately, none of the missiles accidentally hit Taiwan. As a result of Beijing's saber-rattling and pressures from the United States, the KMT had, by the time of the elections, retreated from its efforts to gain greater recognition as an independent state. Many members of the KMT regretted Lee's pushing for an independent Taiwan. By the 2000 election, however, many of those who did favor independence left the KMT to form a new party. In the December 2004 legislative elections, the KMT ran on a platform rejecting independent statehood and supporting a positive and stable relationship with China to protect their business interests. It managed to keep the DPP and other parties sympathetic to declaring independence from winning a majority in the Legislative Yuan.

Taiwan's large middle class, with its diverse and complex social and economic interests arising from ownership and concern for private property, has proven to be a catalyst for political liberalization. Moreover, Taiwan has not suffered from vast economic disparities that breed economic and social discontent. The government's success in developing the economy meant at the least that economic issues did not provide fuel for political grievances. Thus, with the end of martial law in 1987, the KMT could undertake political reform with a certain amount of confidence. Its gradual introduction of democratic processes and values undercut much of its former authoritarian style of rule without its losing political power for more than a decade. Reform generated considerable tensions, but by the 1990s the KMT realized that street demonstrations would not bring down the government and that it was unnecessary to use harsh measures to suppress the opposition. This being the case, the KMT liberalized the political realm still further. Today, Taiwan's political system functions in most respects as a democracy; but as the KMT is the first to admit, it was so busy democratizing Taiwan that it neglected to democratize itself. Once it lost the presidency in the 2000 elections, and only narrowly remained in control of the Legislative Yuan, it rethought its political platform and tried to rid itself of the serious corruption and elitism that has alienated Taiwan's voters.[57] This may partially explain why in the 2004 legislative elections, the KMT, together with its political allies (the "Pan-Blue" alliance), was able to retain a narrow margin of control in the legislature—114 seats, as opposed to the 101 seats for the "Pan-Green" alliance.

Political Parties

Only in 1989 did the KMT pass new laws legalizing opposition political parties. The Democratic Progressive Party, a largely Taiwanese-based opposition party, was officially recognized as a legal party. As with other political reforms, this decision was made in the context of a growing resistance to the KMT's continued restriction of democratic rights. Even after 1989, however, the KMT continued to regulate political parties strictly, in the name of maintaining political and social stability.

Angry factional disputes within both the DPP and the KMT have marred their ability to project a unified electoral strategy. The KMT was particularly hurt by vitriolic disputes between pro-unification and pro-independence factions, and between progressive reformers who pushed for further liberalization of the economic and political systems, and conservative elements who resisted reform. These internal conflicts explain the KMT's inability to move forward on reform and its defeat in the presidential elections of 2000.

In their first years in the Legislature, DPP legislators, no doubt frustrated by their role as a minority party that could not get through any of its own policies, sometimes engaged in physical brawls on the floor of the Legislature, ripped out microphones, and threw furniture. As the DPP steadily gained more power and influence over the legislative agenda, its behavior became more subdued, but its effectiveness has continued to be undermined by internal factionalism. As a result, policy gridlock has continued to plague the legislature.[58]

Infighting became so serious in both parties that it led to the formation of three new breakaway parties, which competed for the first times in the 1998 elections.[59]

The Taiwan Independence Party—the more radical faction within the DPP that broke away in 1998—refused to be intimidated by Beijing's possible military response to a public declaration of independence. Today, it has been replaced by (renamed) the Taiwan Solidarity Union. Within the KMT, one faction that grew angry with the KMT leadership for not moving more actively to bring about reunification with the mainland of China and was generally opposed to the liberalizing reforms of the KMT, broke off to form the New Party, which first ran candidates in the 1994 elections. Its bitter disputes with former colleagues in the KMT continue to make consensus difficult. By the 2000 presidential elections, angry debates over who should lead the KMT led to the expulsion of James Soong, who then formed the People's First Party. Soong's effective campaign, which split the KMT vote, contributed to the victory of the DPP candidate, with Soong himself placing a close second and the KMT candidate a distant third.[60] Today, the KMT, the New Party, and the People's First Party are in the "Pan-Blue" alliance in the Legislature, but are deeply divided on many issues.

Since the 1996 election, the most divisive issue has been whether or not to press for an independent Taiwan. Polls have indicated that although the Taiwanese people would like Taiwan to be independent, only a small minority has been willing to incur the risks of Beijing using force against Taiwan if the government were to endorse independence as a stated policy goal. The preference is for candidates who have promised a continuation of the status quo—namely, not openly challenging Beijing's stance that Taiwan is a province of China, but continuing to act as if Taiwan is in fact an independent sovereign state.

The preconditions for reunification that the KMT set when it was in power are not likely to be met easily: democratization of the mainland to a (unspecified) level acceptable to Taiwan, and an (unspecified) level of economic development that would move the mainland closer to Taiwan's level of development. Ironically, although President Chen Shui-bian pushes aggressively to lay the groundwork for declaring independence, he has at the same time permitted, and even encouraged investment in China (though not as much as Taiwan's business people would like). He accepts the argument that Taiwan's economy cannot continue to grow, or perhaps even survive, without commercial ties to the mainland. Symbolically, the most important step Chen has made in advancing "the three links" (trade, postal, and transportation links) between China and Taiwan is al-

lowing direct flights to accommodate citizens at both ends to visit the other over the Chinese New Year's period (about 5 weeks in January and February 2005.) Although direct flights were permitted in 2003, there has been a huge leap in the details: now commercial airliners from both sides (not just Taiwan) may take passengers across the Taiwan Strait; passengers may board in multiple cities on both sides; and China's planes only have to fly through the air space, not land in, either Hong Kong or Macao on their way to Taiwan, saving considerable time. Already President Chen is being condemned by extremists as selling out Taiwan's interests and taking further steps toward unification.

Money politics, vote-buying practices, and general dishonesty have plagued all of Taiwan's parties. In fact, they have burgeoned over the years, in part because of the growing importance of the legislature;[61] for now that the Legislative Yuan is no longer a body of officials who fled the mainland in 1949 but a genuinely elected legislature with real power to affect Taiwan's policies, who wins really matters. As a result, candidates throw lavish feasts, make deals with businesspeople, and spread money around in order to get the vote. Equally disturbing is the growing amount of election-related violence, and the influence of the "underworld" on the elections.[62] By the time of the 2004 legislative elections, the positions of the KMT and DPP had moved apart on the issue of Taiwan independence. Still, there is little real difference between the two major parties on social and economic issues. Both are also committed to democracy, advocate capitalism, and support an equitable distribution of wealth. In these respects, the DPP does not significantly threaten policies that the KMT so carefully laid out during its many decades in power.

Finally, although Mainlanders are almost all within the KMT or the New Party, the KMT has become so thoroughly "Taiwanized" that the earlier clear divide between the DPP and KMT based on Mainlander or Taiwanese identity has eroded considerably. Even those who strongly favor reunification with the mainland identify themselves not as Mainlanders but as "new Taiwanese." They are, in short, "born-again Taiwanese."[63]

Interest Groups

As Taiwan has become more socially, economically, and politically complex, alternative sources of power have developed that are independent of the government. Economic-interest groups comprised largely of Taiwanese, whose power is based on wealth, are the most important; but there are

THE U.S. SEVENTH FLEET HALTS INVASION

In 1950, in response to China's involvement in the Korean War, the United States sent its Seventh Fleet to the Taiwan Strait to protect Taiwan and the Offshore Islands of Quemoy and Matsu from an invasion by China. Because of improved Sino–American relations in the 1970s, the enhanced Chinese Nationalist capabilities to defend Taiwan and the Offshore Islands, and problems in the Middle East, the Seventh Fleet was eventually moved out of the area. The aircraft carrier *Enterprise,* a part of the Seventh Fleet, is shown at left. In 1996, however, part of the Seventh Fleet briefly returned to the Taiwan Strait when China threatened to use military force against Taiwan if its leaders sought independent statehood.

also public interest groups that challenge the government's policies in areas such as civil rights, the environment, women's rights, consumer protection, agricultural policy, aborigine rights, and nuclear power. Even before the lifting of martial law in 1987, these and other groups were organizing hundreds of demonstrations each year to protest government policy.

On average, every adult in Taiwan today belongs to at least one of the thousands of interest groups. They have been spawned by political liberalization and economic growth, and in turn add to the social pluralism in Taiwan. They are also important instruments for democratic change. Taiwan's government, then, has successfully harnessed dissent since the 1990s, in part by allowing an outlet for dissent through the formation of interest groups—and opposition parties.

Mass Media

With the official end to martial law in July 1987, the police powers of the state were radically curtailed. The media abandoned former taboos and grew more willing to openly address social and political problems, including the "abuse of power and corruption." A free press, strongly critical of the government and *all* the political parties, now flourishes. Taiwan, with under 23 million people, boasts close to 4,000 magazines; about 100 newspapers, with a total daily circulation of 5 million; 150 news agencies; three domestic television channels; and more than 30 radio stations,

which now include foreign broadcasts such as CNN, NHK (from Japan), and the BBC. Although television and radio are still controlled by the government, they have become far more independent since 1988; and 1.2 million households receive 20 to 30 channels through a satellite dish.[64] Television stations show programs from all over the world, exposing people to alternative ideas, values, and lifestyles, and contributing to social pluralism. Political magazines, which are privately financed and therefore not constrained by governmental financial controls, have played an important role in undercutting state censorship of the media and developing alternative perspectives on issues of public concern. New technology that defies national boundaries (including satellite broadcasts from Japan and mainland China), cable television, the Internet, and VCRs are diminishing the relevance of the state monopoly of television.[65]

THE TAIWAN–P.R.C.–U.S. TRIANGLE

From 1949 until the 1960s, Taiwan received significant economic, political, and military support from the United States. Even after it became abundantly clear that the Communists effectively controlled the China mainland and had legitimacy in the eyes of the people, the United States never wavered in its support of President Chiang Kai-shek's position that the R.O.C. was the legitimate government of all of China. U.S. secretary of state Henry Kissinger's secret

trip to China in 1971, followed by President Richard M. Nixon's historic visit in 1972, led to an abrupt change in the American position and to the final collapse of the R.O.C.'s diplomatic status as the government representing "China."

Allies of the United States, most of whom had loyally supported its diplomatic stance on China, soon severed diplomatic ties with Taipei, a necessary step before they could, in turn, establish diplomatic relations with Beijing. Only one government could claim to represent the Chinese people; and with the KMT in complete agreement with the Chinese Communist regime that there was no such thing as "two Chinas" or "one Taiwan and one China," the diplomatic community had to make a choice between the two contending governments. Given the reality of the Chinese Communist Party's control over 1 billion people and the vast continent of China, and, more cynically, given the desire of the business community throughout the world to have ties to China, Taipei has found itself increasingly isolated in the world of diplomacy. It should be noted, however, that even when the R.O.C. had represented "China" in the United Nations; and at the height of the diplomatic isolation of the People's Republic of China from 1950 to 1971, Taipei was never officially recognized by any of its neighbors in Southeast Asia (unless they were in a defense alliance with the United States)—even though they distrusted China. That this was the case even at the height of China's unpopularity in the region was a bad omen for Taipei's dream of obtaining international legitimacy."[66]

Eventually the United States made the painful decision to desert its loyal Cold War ally, a bastion against communism in Asia, if not exactly an oasis of democracy. The United States had, moreover, heavily invested in Taiwan's economy. But on January 1, 1979, President Jimmy Carter announced the severing of diplomatic relations with Taipei and the establishment of full diplomatic relations with Beijing.

Taiwan's disappointment and anger at the time cannot be overstated, in spite of the fact that an officially "unofficial" relationship took its place. American interests in Taiwan are overseen by a huge, quasi-official "American Institute in Taiwan"; while Taiwan is represented in the United States by multiple branches of the "Taipei Economic and Cultural Office." In fact, the staffs in these offices continue to be treated in most respects as if they are diplomatic personnel. Except for the handful of countries that officially recognize the R.O.C., Taiwan's commercial, cultural, and political interests are represented abroad by these unofficial offices. The only states that recognize the R.O.C. as an independent state are those who receive large amounts of aid from Taiwan.

The United States' acceptance of the Chinese Communists' "principled stand" that Taiwan is a province of China and that the People's Republic of China is the sole legal government of all of China, made it impossible to continue to maintain a military alliance with one of China's provinces. Recognition of Beijing, therefore, required the United States to terminate its mutual-defense treaty with the R.O.C. In the Taiwan Relations Act of 1979, the United States stated its concern for the island's future security, its hope for a peaceful resolution of the conflict between Taiwan's government and Beijing, and its decision to put a moratorium on the sale of new weapons to Taiwan.

Renewal of Arms Sales

The Taiwan Relations Act was, however, largely ignored by the newly ensconced Ronald Reagan administration, which almost immediately upon taking office, announced its intention to resume U.S. arms sales to Taiwan. The administration argued that Taiwan needed its weapons upgraded in order to defend itself. Irate, Beijing demanded that, in accordance with American agreements and implicit promises to China, the United States phase out the sale of military arms over a specified period. Thus far, this has not happened. As a result, the issues of U.S. arms sales to Taiwan and congressional appropriations for Taiwan's

defense continue to plague relations with China.[67] Similarly, the U.S. Congress's proposed Taiwan Security Enhancement Act in 2000—which, had it passed, would have amounted to a military alliance with Taiwan—and the possibility that Congress may authorize a theater missile defense for the island, have been major irritants to the U.S.–China relationship.

When it took office in January 2001, the George W. Bush administration immediately stated its intention to go forward with deepening military ties with Taiwan. Tensions with Beijing generated by the U.S. sales of military equipment to Taiwan are aggravated by China's own sales of military equipment, such as medium-range missiles to Saudi Arabia, Silkworm missiles to Iran (used against American ships), nuclear technology to Pakistan,[68] and massive sales of semiautomatic assault weapons to the United States (one of which was used to attack the White House in 1994). These sales have undercut the P.R.C.'s standing on the moral high ground to protest American sales of military equipment to Taiwan. But views even within Taiwan concerning the purchase of U.S. military weapons and equipment are complicated. The DPP has argued that they are necessary to protect Taiwan against an attack by China. The KMT is, however, firmly opposed to further purchases, considering them a waste of money; for in the event of an attack, its defense would depend almost entirely on the United States—assuming that the Americans decided to come to its defense. Even worse, the KMT argues, further arming Taiwan would accelerate the arms race with China, and further destabilize the Taiwan Straits. The KMT has even accused President Chen of wanting greater military power as a further effort to lay the groundwork for declaring Taiwan's independence.

This is a bizarre twist to the issue of Taiwan's defense; for in the past, the United States and other countries have repeatedly backed out of proposed arms deals in the face of Beijing's threatened punitive measures. Now it is Taiwan that is backing out. With or without arms purchases from the U.S., Taiwan knows that its own defenses would be overwhelmed by a military onslaught from the mainland.

The United States wants Beijing to agree to the "peaceful resolution of the Taiwan issue," but "on principle," China refuses to make any such a commitment. China insists that Taiwan is an "internal" affair, not an international matter over which other states might have some authority. From China's perspective, then, it has the right as a sovereign state to choose to use force to settle the Taiwan issue. There is general recognition, however, that the

purpose of China's military buildup across from Taiwan is not as much to attack Taiwan as to prevent it from declaring independence as a sovereign state.[69] Indeed, apart from a mild statement from Japan protesting China's "testing" of missiles over Taiwan in 1996, no Asian country questioned China's right to display force when President Lee pressed for independent statehood for Taiwan. Still, Beijing knows that American involvement may be critical to getting Taiwan to agree to unification with the mainland. One thing is certain: the day Beijing no longer threatens to use force against Taiwan if it declares independence, Taipei will declare independence. So we can expect Beijing to continue its bluster about using military force until Taiwan is reunified with the mainland.

Minimally, Taiwan wants the United States to insist that any solution to the China unification issue be *acceptable to the people of Taiwan.* One possible solution would be a confederation of Taiwan with the mainland: Taiwan would keep its "sovereignty" (that is to say, govern itself and formulate its own foreign policy), but China could say there was only one China. An interim solution might be that Taiwan would promise not to declare independence for 15 years, and Beijing would promise not to use force against Taiwan for 15 years.[70]

CHINA'S "PEACE OFFENSIVE"

Since the early 1980s, Beijing has combined its threats and warnings about Taiwan's seeking independence with a "peace offensive." This strategy aims to draw Taiwan's leaders into negotiations about the future reunification of Taiwan with the mainland. Beijing has invited the people of Taiwan to visit their friends and relatives on the mainland and to witness the progress made under Communist rule. The vast majority of Taiwan's adult population has traveled to the mainland. In turn, less than 0.1 percent of Chinese from the mainland has been permitted by Taipei to visit Taiwan (although more have entered through a third area, such as Hong Kong or the United States).[71] However, the government has arranged for mainland Chinese students studying abroad to come for "study tours" of Taiwan. They have treated them as if they were visiting dignitaries, and the students usually returned to their universities full of praise for Taiwan.

China's "peace offensive" is based on a nine-point proposal originally made in 1981. Its major points include Beijing's willingness to negotiate a mutually agreeable reintegration of Taiwan under the mainland's government; encouragement of trade, cultural exchanges, travel, and communications between Taiwan and the mainland; the offer to give Taiwan "a high degree of autonomy as a special administrative region" within China after reunification (the same status it offered to Hong Kong when it came under Beijing's rule in 1997); and promises that Taiwan could keep its own armed forces, continue its socioeconomic systems, maintain political control over local affairs, and allow its leaders to participate in the national leadership of a unified China. This far exceeds what China offered to Hong Kong.

The KMT's original official response to the Beijing "peace offensive" was negative. The KMT's bitter history of war with the Chinese Communists, and what the KMT saw as a pattern of Communist duplicity, explained much of the government's hesitation. So did Beijing's refusal to treat Taiwan as an equal in negotiations. Nevertheless, since 1992, Taiwan has engaged in "unofficial" or "track 2" and "track 3" discussions on topics of mutual interest, such as the protection of Taiwan's investments in the mainland, tourism, cross-Strait communication and transportation links, and the dumping of Taiwan's nuclear waste on the mainland. Even when the KMT has been out of power, KMT businesspeople still traveled regularly to China to discuss such issues, albeit without authorization from the government.

Taipei remains sensitive, however, to the Taiwanese people's concern about the unification of Taiwan with the mainland. The Taiwanese have asserted that they will never accede to rule by yet another mainland Chinese government, especially a Communist one. When the DPP is in power, Taiwanese have fewer fears that the leadership will strike a deal with Beijing at their expense; for, as the long-time advocate of the interests of the Taiwanese people and an independent Taiwan, the Taiwanese tend to believe that the DPP can be trusted not to sell out their interests. To speak of the "Taiwanese" as a united whole is, however, misleading; for it must be remembered that the overwhelming majority of the KMT membership are Taiwanese, and that Taiwanese who are doing business with the mainland are eager to see integration—if not necessarily political unification at this point in history—progress much more rapidly. In any event, the U.S. position as to whether it would come to Taiwan's aid if it were attacked by China is complicated by the war in Iraq. With more than 150,000 troops in Iraq, and others tied down in Afghanistan, and with no end in sight, the U.S. simply cannot participate in a conflict with China.

With only a handful of countries recognizing the R.O.C., and with Beijing blocking membership for the R.O.C. in most international organizations, Taipei is under pressure to achieve some positive results in its evolving relationship with Beijing. While officially the "three mini-links" (trade, postal exchange, and transportation) are unacceptable to Taiwan, they are in fact developing rapidly. With the introduction of direct commercial flights from Taiwan to Shanghai over the 2003 Chinese New Year, for example, the first steps toward direct air links were initiated;[72] but there has been no progress in this regard since that time. On the other hand, "indirect" trade between China and Taiwan by way of Hong Kong has continued to soar, although the opening of Taiwan's Offshore Islands (Quemoy and Matsu) to direct trade with the mainland in 2001 has somewhat undercut the need for the Hong Kong connection. China has, in fact, become Taiwan's largest export market. It is also the single largest recipient of investment from Taiwan.

Nevertheless, Taiwan's government still hesitates on many issues that would further bind Taiwan with the mainland. Its policy on mainland spouses is particularly stringent and clearly indicates a perception that China's citizens are potential enemies, not compatriots. Relatively few Taiwanese residents who marry individuals from the mainland are permitted to live with them in Taiwan. Since 1992 several hundred have each year been allowed to join their spouses in Taiwan, but thousands are still prohibited from doing so. This is in startling contrast to Beijing's policy, which welcomes Taiwan spouses to come live on the mainland. Taiwan's government argues that the mainland spouses could be spies and that internal security forces are inadequate to follow them around. As Mainlanders are smuggled into Taiwan in ever-larger numbers (primarily to satisfy the needs of entrepreneurs in Taiwan for cheap labor), the issue of surveillance has become a growing concern. At the same time, Taiwan spying on the mainland has grown steadily as its contacts with the mainland have increased. The Military Intelligence Bureau recruits from its hundreds of thousands of citizens working, living, and touring on the mainland. They in turn form Taiwan spy networks that recruit local Chinese with sex, money, and "democratic justice" (an appeal to their sense of injustice at the hands of the Chinese government).[73]

In the meantime, China continues to deepen and widen harbors to receive ships from Taiwan; wine and dine influential Taiwanese; give preferential treatment to Taiwan's entrepreneurs in trade and invest-

ment on the mainland; open direct telephone links between Taiwan and the mainland; rebuild some of the most important temples to local deities in Fujian Province, favorite places for Taiwanese to visit; establish special tourist organizations to care solely for people from Taiwan; and refurbish the birthplace of Chiang K'ai-shek, the greatest enemy of the Chinese Communists in their history.

Taiwan's business people and scholars are eager for Taiwan's relationship with the mainland to move forward. They seek direct trade and personal contacts, and try to separate political concerns from economic interests and international scientific exchanges. The manufacturing and software sectors are particularly concerned with penetrating and, if possible, controlling the China market. Otherwise, they argue, businesspeople from other countries will do so.

The business community, faced with Taiwan's ever-higher labor cost and Taiwan's lack of cheap raw materials, has flocked to China. It has moved Taiwan's now outdated labor-intensive factories and machinery to the mainland, where cheap labor allows these same factories to remain profitable. Even industries investing in cutting-edge technology benefit from China's cheap, hard-working, well-educated labor force.

Others are concerned, however, that, with more than 15 percent of its total foreign investment in the mainland, Taiwan could become "hostage" to Beijing. That is, if China were to refuse to release Taiwan's assets or to reimburse investors for their assets on the mainland in case of a political or military conflict between Taipei and Beijing, Taiwan's enterprises would form a pressure point that would give the advantage to Beijing. Furthermore, without diplomatic recognition in China, Taiwan's businesses on the mainland are vulnerable in case of a conflict with local businesses or the government. High-level members of the government have even denounced those who invest in the mainland as "traitors." And members of the pro-independence press are attempting to whip up fear among its citizens that if Beijing passes its newly proposed anti-secession law, Taiwanese visiting or living in China will be vulnerable to "shakedown artists." "You can easily imagine the kind of scenario—'give us a stake in your enterprise or we will denounce you and you'll end up in a labor camp in Qinghai.' [A]fter the law passes, almost no Taiwanese will be safe in China."[74] So far, affairs have turned out quite the opposite: China has actually favored Taiwan's businesses over all others; and Taiwan's investors have tended to turn a quick profit, and to construct many safeguards, so that any seizure of assets would result in negligible losses.

PROSPECTS FOR THE REUNIFICATION OF TAIWAN WITH THE MAINLAND

Although the December 2004 legislative elections kept an anti-independence majority in the legislature, most Taiwanese remain opposed to reunification. Most agree that, given China's threats to use force if Taiwan were to declare independence, it would be foolish to do so; and that the government's policy toward China should be progressive, assertive, and forward-looking. Lacking a long-term plan and simply reacting to Beijing's initiatives puts the real power to determine the future relationship in Beijing's hands.

For the last 15 years, reform-minded individuals in Taiwan's political parties have insisted that the government needs to actively structure how the cross-Straits relationship evolves. One of the most widely discussed policies favored by those who opposed a declaration of independence has been a "one country, two regions" model that would approximate the "one country two systems" model China has with Hong Kong. The problem is determining who would govern in that "one China" after reunification with the mainland. To call that country the People's Republic of China would probably never find acceptance in Taiwan—a point understood by Beijing. As a result, it now frequently drops the "people's republic" part of the name in its pronouncements. Symbolically, this eliminates the issue of two different governments claiming to represent China, and whether that China would be called a "people's republic" or a "republic." It would be called neither. But, would Beijing be in charge of the new unified government? Would the government of the "Republic of China," even were free and democratic elections for all of the mainland and Taiwan to be held, be the winner? There is certainly no evidence to suggest that most people on the mainland would welcome being ruled by Taiwan's leaders.

In spite of the negative rhetoric, changes in Taipei's policies toward the P.R.C. have been critical to improving cross-Strait ties. For example, Taiwan's government ended its 40-year old policy of stamping "communist bandit" on all printed materials from China and prohibiting ordinary people from reading them; Taiwan's citizens, even if they are government officials, are now permitted to visit China; scholars from Taiwan may now attend international conferences in the P.R.C., and Taipei now permits a few P.R.C. scholars to attend

conferences in Taiwan; and KMT retired veterans, who fought against the Communists and retreated to Taiwan in 1949, are actually encouraged to return to the mainland to live out their lives because their limited KMT government pensions would buy them a better life there! Certainly some members of Taiwan's upper class are acting as if the relationship will eventually be a harmonious one when they buy apartments for their mistresses and purchase large mansions in the former international sector of Shanghai and elsewhere. (There are an estimated 500,000 individuals from Taiwan living in China, primarily in the Shanghai area). A 2000 survey indicated, in fact, that after the United States and Canada, mainland China was the preferred place to emigrate for Taiwan's citizens![75] And, from the perspective of the Democratic Progressive Party , things are getting worse: as many as 1 million Taiwanese are relocating themselves to the mainland, and they are taking with them venture capital estimated at U.S.$100 billion—money that might otherwise be pumped into Taiwan's economy. Some Taiwanese share with Beijing a common interest in establishing a Chinese trading zone in East Asia. A "Chinese common market" would incorporate China, Taiwan, Hong Kong, Macau, and perhaps other places with large ethnic-Chinese communities such as Singapore and Malaysia. Economically integrating these Chinese areas through common policies on taxes, trade, and currencies would strengthen them vis-à-vis the Japanese economic powerhouse, which remains larger than all the other Far Eastern economies put together.

Yet, with an ever-smaller number of first-generation Mainlanders in top positions in the KMT and Taiwan's government, few are keen to push for reunification. Indeed, the majority of people in Taiwan still oppose reunification under current conditions. They are particularly concerned about two issues. The first is the gap in living standards. Taiwan is fully aware of the high price that West Germany paid to reunify with East Germany. Obviously the price tag to close the wealth gap with mammoth China would be prohibitive for tiny Taiwan; and any plan that would allow Beijing to heavily tax Taiwan's citizens would be unacceptable. Yet China is developing so quickly, at least along the densely populated eastern coast that faces Taiwan, that this issue should disappear. Indeed, at this point it might be predicted that Shanghai will easily match Taipei's living standard within this generation's lifetime.

The second issue is democracy. Fears that they might lose certain political freedoms and control over their own institu-

tions, have made the Taiwanese wary of reunification.

Finally, whether or not "one country, two systems" succeeds in protecting Hong Kong from Beijing's intervention, Taiwanese reject a parallel being drawn between Hong Kong and Taiwan. Hong Kong was, they argue, a British colony, whereas Taiwan, even if only in the last ten years, is a fledgling democracy. When the KMT was in power, it stated that once China had attained a certain (unspecified) level of development and democracy, China and Taiwan would be reunified; but the DPP has made no such statement. In any event, with so much left open to its own arbitrary interpretation as to what is sufficient development and democracy, even the KMT seems to have made no real commitment to reunification. For its part, the DPP-led government supported growing interdependency, yet tried desperately to move toward the declaration of an independent state, all to no avail thus far.

China has done a number of things that should make reunification more palatable to Taiwan, including its significant progress in economic development and increasing the rights of the Chinese people in the last 25 years. Greater contacts and exchanges between the two sides should in themselves help lay the basis for mutual trust and understanding.

Many members of Taiwan's political and intellectual elite, including the Mainland Affairs Council, the DPP and KMT party leaders, think tanks, and the Ministry of Foreign Affairs, seem to spend the better part of each day pondering the meaning of "one China." In general, they would like Beijing to return to the agreement that they reached with Beijing in 1992: namely, that each side would keep its own interpretation of the "one China" concept. Beijing would continue to see "one China" as the government of the P.R.C. and would continue to deny that the R.O.C. government existed. Taipei would continue to hold that Taiwan and the mainland are separate but politically equivalent parts of one China. Unfortunately for Taipei, Beijing no longer embraces this view. Instead, it states that no further negotiations can occur until Taipei accepts that it is not a state, and hence not the equal of Beijing at the negotiating table.

Today, China has 496 (non-nuclear) missiles in positions facing Taiwan; but unless Taiwan again pushes seriously for recognition as an independent state, an attack is unlikely; for the mainland continues to benefit from Taiwan's trade and investments, remittances and tourism; and it is committed to rapid economic development, which could be seriously jeopardized by even a brief war—a war that it might not win if the United States were to intervene on behalf of Taiwan. In spite of the fact that the U.S. has repeatedly made it clear to President Chen Shui-bian that it is angered by his efforts to advance the cause of independence, Beijing could well worry that the U.S. might suddenly be persuaded to come to Taiwan's support.

It is in Taiwan's best interests for the relationship with China to develop in a careful and controlled manner, and to avoid public statements on the issue of reunification versus an independent Taiwan, even as that issue haunts every hour of the day in Taiwan. It is also in Taiwan's interest to wait and see how well China integrates Hong Kong under its formula of "one country, two systems." In the meantime, Taiwan's international strategy—agreeing that it is a mere province of China, while acting like an independent state and conducting business and diplomacy with other states as usual—has proved remarkably successful. It has allowed Taiwan to get on with its own economic development without the diversion of a crippling amount of revenue to military security. A continuation of the status quo is clearly the preferred alternative for Taiwan,[76] the United States, and mainland China; for it avoids any possibility of a military conflict, which none would welcome; and it does not interrupt the preferred strategy of both the DPP and the KMT—"closer economic ties, lower tensions, and more communication with the mainland."[77]

At the same time, this strategy allows time for the mainland to become increasingly democratic and developed, in a way that one day might make reunification palatable to Taiwan. Meanwhile, Beijing's leadership knows that Taiwan acts as a de facto independent state, but it is willing to turn a blind eye as long as Taipei does not push too openly for recognition. Thus far, the reality has mattered less to Beijing than recognition of the symbolism of Taiwan being a province of China. Beijing's leadership is far more interested in putting resources into China's economic development than fighting a war with no known outcome. In part for this reason, and in part because of continuing sales of American armaments to Taiwan, Beijing has made it clear to Washington that China's own buildup of missiles targeted at Taiwan would be linked to American sales: As long as those sales continue, so will their buildup; and if they stop, so will the increasing deployment of missiles facing Taiwan. Because China's sovereignty over Taiwan has been an emotional, patriotic, even nationalistic, issue for the Chinese people, however, Beijing does not make a

"rational" cost-benefit analysis of the use of force against a rebellious Taiwan. Taiwan does not like Beijing's militant rhetoric, but some mainland Chinese analysts believe that China's leadership is forced to sound more militant than it feels, Taiwan does not like Beijing's militant rhetoric, but some mainland Chinese analysts believe that China's leadership is forced to sound more militant than it feels, thanks to the militant nationalism of ordinary Chinese people and the Chinese military. Indeed, some go so far as to say that, were China an electoral democracy, the people would have voted out the CCP leadership because it has done little to regain sovereignty over Taiwan.

As Taiwan's relationship with China deepens and broadens, it is possible that more arrangements could be made for the representation of both Taiwan and China in international organizations, without Beijing putting up countless roadblocks. Indeed, Beijing welcomed Taiwan's membership in the WTO, because it allows it to pry open Taiwan's market. Further, Taiwan's WTO membership has led to even more investments in the mainland and further economic integration.

Taiwan eagerly embraced membership in the WTO, not because it would necessarily benefit from WTO trade rules, but so that it could become a player in a major international organization. But, the very trade practices that led to Taiwan's multibillion-dollar trade surplus will have to be abolished to gain compliance with WTO regulations. Still, had Taiwan not joined the WTO, it would have eventually lost its competitiveness in agriculture and the automobile industry anyway.[78]

To conclude, the concrete results of Taiwan diplomatic relations with any country that officially recognizes the Republic of China as the legitimate government of China. Those countries that do recognize the R.O.C. as a sovereign state also cannot trade with the P.R.C. Given the size of the China market, this is an unacceptable price for most countries to pay. Beijing would no doubt use this trump card to punish those that would dare to recognize Taiwan as an independent and sovereign state, just as it does now.

Taiwan is looking for a place for itself in the international system, and it can't seem to find it. But its government realizes that the island is a small place, and that if Taiwan ever were to stop demanding international status and attention, it might well discover that it had suddenly become, de facto, a province of China while the international community looked the other way. If only for this reason, it is in Taiwan's interest to continue to press its case for

greater international recognition, and to continue to engage in pragmatic unofficial diplomacy and trade with states throughout the world. It may not buy Taiwan statehood, but it may well buy the government's continued independence of Beijing.

Timeline: PAST

A.D. 1544
Portuguese sailors are the first Europeans to visit Taiwan

1700s
Taiwan becomes part of the Chinese Empire

1895
The Sino-Japanese War ends; China cedes Taiwan to Japan

1945
Japan is forced to return the colony of taiwan to China when Japanese forces are defeated in World War II

1947–49
Nationalists, under Chiang Kai-shek, retreat to Taiwan

1950s
A de facto separation of Taiwan from China; The Chinese Communist Party is unable to bring its civil war with the KMT to an end because the U.S. interposes its Seventh Fleet between the mainland and Taiwan in the Taiwan Straits

1971
The People's Republic of China replaces the Republic of China (Taiwan) in the United Nations as the legitimate government of "China"

1975
Chiang Kai-shek dies and is succeeded by his son, Chiang Ching-Kuo

1980s
38 years of martial law end and opposition parties are permitted to campaign for office

1990s
Relations with China improve; the United States sells F-16 jets to Taiwan; China conducts military exercises to intimidate Taiwanese voters

PRESENT

2000s
Trade and communication with China continue to expand

The opposition Democratic Progressive Party wins the presidency and controls the Legislature for several years

NOTES

1. Before it won its bid to host the 2008 Olympics, Beijing had said that it would not allow Taipei to cohost them unless it first accepted the "one China" principle. Now that Beijing has won its bid, it is go-

ing ahead without Taipei, which has not accepted this principle.

2. John F. Burns, "Quemoy (Remember?) Bristles with Readiness," *The New York Times* (April 5, 1986), p. 2.

3. Mainland Affairs Council, "Report on the Preliminary Impact Study of the 'Three Mini-links' Between the Two Sides of the Taiwan Strait" (October 2, 2000); and discussions at The National Security Council and Ministry of National Defense in Taiwan (January 2001).

4. Meetings at the National Security Council and Ministry of National Defense in Taiwan (January 2001).

5. Edwin A. Winckler, "Cultural Policy on Postwar Taiwan," in Steven Harrell and Chun-chieh Huang, eds., *Cultural Change in Postwar Taiwan* (Boulder, CO: Westview Press, 1994), p. 22.

6. *Ibid.*, p. 29.

7. Thomas B. Gold, "Civil Society and Taiwan's Quest for Identity," in Harrell and Huang, p. 60.

8. Parris Chang, "The Impact of the 1996 Missile Crisis on US–China–Taiwan Relations," seminar at Harvard University (November 13, 1998). Chang is a DPP legislator in the Legislative Yuan.

9. Thomas A. Shaw, "Are the Taiwanese Becoming More Individualistic as They Become More Modern?" Taiwan Studies Workshop, *Fairbank Center Working Papers,* No. 7 (August 1994), pp. 1–25.

10. David Chen, "From Presidential Hopeful, Frank Words on Democracy," *The Free China Journal* (September 9, 1994), p. 6.

11. "Premier Hau Bristling about Crime in Taiwan," *The Free China Journal* (September 13, 1990), p. 1.

12. Winckler, in Harrell and Huang, p. 41.

13. Paul Li, "Trash Transfigurations," *Taipei Review* (October 2000), pp. 46–53.

14. Central News Agency, Taipei, "Nearly 90 percent support new garbage classification policy, survey finds," *Taiwan News Online* (December 27, 2004). http://www.etaiwannews.com/ Taiwan/Politics/2004/12/27/ 1104112881.htm

15. The public concerns about building nuclear-power plants are understandable. Taiwan sits astride an active earthquake zone, and the potential damage to the environment from a nuclear reactor hit by an earthquake is incalculable. Recent major earthquakes have measured as high as 6.7 on the Richter scale and caused significant damage to the island.

16. A dormitory for employees of Tai Power is one of these buildings. Ninety percent of Taiwan's nuclear waste is stored on Orchid Island, where one of Taiwan's 9 aboriginal tribes lives. Christian Aspalter, *Understanding Modern Taiwan: Essays in Economics, Politics, and Social Policy* (Burlington, Vt.: Ashgate Publishers, 2001), pp. 103–107.

17. http://www.settlement.org/ cp/english/taiwan/holidays. html

18. Lee Fan-fang, "Ghosts' Arrival Bad for Business," *The Free China Journal* (August 7, 1992), p. 4. On the mainland of China, the Chinese Communist Party's emphasis on science and the eradication of su-

perstition means that it is far less common than in Taiwan for people to worry in such a systematic way about propitiating ghosts.

19. Marc J. Cohen, *Taiwan at the Crossroads* (Washington, D.C.: Asian Resource Center, 1988), pp. 186–190. For further detail, see his chapter on "Religion and Religious Freedom," pp. 185–215. Also, see Gold, in Harrell and Huang, p. 53.

20. "Taiwan Presbyterian Church says NO to the 'Three no's' Taipei, August 28, 1998 http://www.taiwandc.org/ nws-9843.htm. "Presbyterian Church in Taiwan Calls for Reform," February 18, 2004. Posted on http://www. christiantoday.com/news/ asip/42.htm

21. See *Free China Review*, Vol. 44, No. 9 (September 1994), which ran a series of articles on educational reform, pp. 1–37.

22. Brian Cheng, "Foreign Workers Seen as a Mixed Blessing," *Taipei Journal* (October 20, 2000), p. 7.

23. Election Study Center, National Chengchi University, *Taipei: Face-to-Face Surveys* (February 2000). Funded by the Mainland Affairs Council.

24. Mainland Affairs Council, discussions in Taiwan in January 2001.

25. Building a High-Quality Environment for an Aging Society: An Interview with CEPD Chairman Hu Sheng-cheng. Interview by Teng Sue-feng/tr. by Paul Frank. *Sinorama*, 2004. On line: http://www.sinorama. com.tw/en/current_issue/show_ issue.php3?id=2004119311018E. TXT&page=1

26. Cohen, p. 107.

27. *Ibid.*, p. 108. For more on women, see the chapter on "Women and Indigenous People," pp. 106–126.

28. "Taiwan's Civil Service Markes Headway on Gender Equality," *Taiwan Update*," Vol. 5, No. 3 (March 30, 2004), p. 7. Jim Hwang, "The Civil Service: Walking on Glue," *Taipei Review* (October 2000), pp. 22–29.

29. Cher-jean Lee, "Political Participation by Women of Taiwan," *Taiwan Journal* (August 20, 2004), p. 7.

30. Ibid.

31. James A. Robinson, "The Value of Taiwan's Experience," *The Free China Journal* (November 6, 1992), p. 7.

32. Taiwan's workers could not get higher wages through strikes, which were forbidden under martial law. The alternative was to try starting up one's own business. Gold, in Harrell and Huang, pp. 50, 53.

33. Kelly Her, "Not-So-Iron Rice Bowl," *Free China Review* (October 1998), pp. 28–35.

34. "Taiwan: In Praise of Paranoia," *The Economist* (November 7, 1998), pp. 8–15.

35. William C. Pao, "COA (Council of Agriculture) to help agriculture sector upgrade its competitiveness," *The China Post* (Taipei), 2003. On line http://www. gio.gov.tw/taiwan-website/ 5-gp/apec/ap3_12.htm

36. Ho Mei-yueh, vice-chair, Council for Economic Planning and Development, Executive Yuan, discussions in Taipei (January 2001).

37. Discussions, Taipei (January 8–12, 2001).

38. Peter Morris, "Taiwan business in China supports opposition," *Asia Times Online*

(Feb. 4, 2004). http://www.atimes. com/atimes/China/FB04Ad04.html

39. For 2004, Taiwan had an overall trade surplus (with all countries) of well over US$7 billion, somewhat lower than the 2003 surplus of U.S.$8.84 billion, "Taiwan: In Praise of Paranoia," p. 17.

40. Francis Li, "Taiwan to Set Up Free-Trade Zones," *Taipei Journal* (October 11, 2002), p. 3.

41. "Taiwan: In Praise of Paranoia," p. 16.

42. Lynn T. Whyte III, "Taiwan and Globalization," in Samuel S. Kim, ed., *East Asia and Globalization* (New York: Rowman & Littlefield, 2000), p. 163.

43. Robert P. Weller, "Environmental Protest in Taiwan: A Preliminary Sketch," Taiwan Studies Workshop, *Fairbank Center Working Papers*, No. 2 (1993), pp. 1, 4.

44. Susan Yu, "Legislature Acts to Protect Lawmakers from Recall Movements," *The Free China Journal* (October 14, 1994), p. 2.

45. Cheng, p. 7.

46. Dianna Lin, "Alien Workers Face Problems in New Life," *The Free China Journal* (September 9, 1994), pp. 7–8.

47. Cheng, p. 7.

48. The maps of China in Taiwan's schools until the mid-1990s included not just Taiwan and all of China proper but also Tibet, Inner Mongolia, and even Outer Mongolia, which is an independent state.

49. From 1950 to 1986, military courts tried more than 10,000 cases involving civilians. These were in violation of the Constitution's provision (Article 9) that prohibited civilians from being tried in a military court. Hung-mao Tien, *The Great Transition: Political and Social Change in the Republic of China* (Palo Alto, CA: Hoover Institution, Stanford University, 1989), p. 111.

50. The Kaohsiung rally, which was followed by street confrontations between the demonstrators and the police, is an instance of KMT repression of *dangwai* activities, activities that were seen as a challenge to the KMT's absolute power. The KMT interpreted the Kaohsiung Incident as "an illegal challenge to public security." For this reason, those arrested were given only semi-open hearings in a military, not civil, tribunal; and torture may have been used to extract confessions from the defendants. Tien, p. 97.

51. Lin Fang-yan, "Rights Group Reports on ROC Progress," *Taipei Journal* (January 3, 2003), p. 1.

52. Ibid.

53. Tien, p. 72.

54. A February 2000 poll conducted in Taiwan indicated that only 6.1 percent of the respondents opposed conditional or unconditional opening up of direct transportation links with the mainland. Election Study Center, National Chengchi University, Taipei. Face-to-face surveys. Funded by the Mainland Affairs Council, Executive Yuan (February 2000).

55. Shelley Rigger, "Taiwan: Finding Opportunity in Crisis," *Current History* (September 1999), p. 290.

56. Shelley Rigger, "Taiwan Rides the Democratic Dragon," *The Washington Quarterly* (Spring 2000), pp. 112–113. Reference is to the editor, *Taipei Times* (January 4, 2000).

57. Discussions with SHAW Yu-ming, deputy secretary-general, KMT, at KMT headquarters (January 2001).

58. Cal Clark, "Taiwan in the 1990s: Moving Ahead or Back to the Future?" in William Joseph, *China Briefing: The Contradictions of Change* (Armonk, NY: M.E. Sharpe, 1997), p. 206.

59. Myra Lu and Frank Chang, "Election Trends Indicate Future of Taiwan Politics," *Free China Journal* (November 27, 1998), p. 7. The breakaway parties include the Taiwan Independence Party and the New Nation Alliance, which split off from the DPP, and the Green Party.

60. According to Shaw Yu-ming, the deputy secretary general of the KMT, the KMT was defeated not because of its policies but because it was the KMT leadership that chose the candidates to run in the election. What the KMT needs is a party primary, in which the membership will choose the candidates. Discussions at KMT Headquarters in Taipei (January 2001).

61. Clark, in Joseph, p. 206.

62. Myra Lu, "Crackdown on Vote-Buying Continues," *Free China Journal* (November 27, 1998), p. 2.

63. Lee Changkuei, "High-Speed Social Dynamics," *Free China Review* (October 1998), p. 6.

64. Yu-ming Shaw, "Problems and Prospects of the Democratization of the Republic of China on Taiwan," Taiwan Studies Workshop, *Fairbank Center Working Papers,* Harvard University, No. 2 (October 1993), pp. 1–2.

65. Chin-chuan Lee, "Sparking a Fire: The Press and the Ferment of Democratic Change in Taiwan," in Chin-chuan Lee, ed., *China's Media, Media China* (Boulder, CO: Westview Press, 1994), pp. 188–192.

66. Chen Jie, *Foreign Policy of the New Taiwan: Pragmatic Diplomacy in Southeast Asia* (Northampton, MA: Edward Elgar Publishing, 2002), pp. 63–64.

67. "The Omnibus Appropriation Act and the 1999 Fiscal Year Department of Defence Authorization Act serve to interfere in China's internal affairs by supporting Tibetan separatist elements, putting Taiwan into theatre missile defence (TMD) system and allowing the sale of arms to Taiwan...." Ma Chenguang, "Resentment Expressed on Anti-China US Bills," *China Daily* (October 30, 1998), p. 1.

68. In 2000, however, China agreed to stop selling nuclear and missile technology to Pakistan.

69. Thomas J. Christensen, "The Contemporary Security Dilemma: Deterring a Taiwan Conflict," *The Washington Quarterly* (Autumn 2002), pp. 7–21.

70. Kau Ying-mao, National Security Council, discussions in Taipei (January 2001). The latter proposal was put forth by Joseph Nye and Kenneth Lieberthal, both on the NSC during the Clinton administration.

71. As excerpted in *China News Daily*, on the Internet (May 5, 1996). A different perspective is offered by these figures: 12 million visits by 3 million people from Taiwan to the mainland since 1988, but only 10,000 per year from China to Taiwan since 1988. Ninety percent of the visits are

family visits, not pure tourism. From "The Koo-Wang Talks: Constructive Dialogue," seminar at Harvard University (October 30, 1998).

72. Taiwan's airplanes must still touch down in Hong Kong or Macau before entering the mainland, making what would be a one-hour direct flight to Shanghai into a four-hour trip. Nevertheless, the flight time has been cut by several hours, as passengers do not have to change planes, since Taiwan's own airlines are now permitted to continue on to Shanghai.

73. Wendell Minnick, "The Men in Black: How Taiwan Spies on China," Asia Times Online Co., http://www.atimes.com), 2004.

74. Editorial, "The Chinese gulag beckons," *Taipei Times,* (January 10, 2005). http://www.taipeitimes.com/News/edit/archives/2005/01/10/2003218825

75. Chien-min Chao, "Introduction: The DPP in Power," *Journal of Contemporary China,* vol. 11, no. 33 (November 2002), p. 606.

76. In October 1999, 84 percent of those polled in Taiwan indicated a preference for the "status quo" over "immediate independence" or "immediate unification." Taiwan's Mainland Affairs Council Web site (January 6, 2000). www.mac.gov.tw/english/POS/. Cited by Shelly Rigger, p. 114.

77. "Taiwan Stands Up," *The Economist* (March 25, 2000), p. 24.

78. Ho Mei-yueh, Vice Chair, Council for Economic Planning and Development, Executive Yuan, discussions in Taipei (January 2001).

Images of Dynasty

China's golden age of archaeology

By Albert E. Dien

Only after the establishment of the People's Republic in 1949 did large-scale excavations begin in China, research that has proved so fruitful that the second half of the twentieth century may be remembered by Chinese archaeologists as a golden age. This unprecedented efflorescence has been brought about largely by the enormous scale of industrial development, leading to the discovery of ancient sites. Also crucial is state support, forthcoming because discoveries are seen as confirming the Marxist interpretation of history, in which societies defined by class struggle are viewed as evolving toward communism. Such discoveries also foster a sense of nationalism and cultural pride.

Traditional Confucian principles stress the importance of precedent and universal truths articulated by sages of the past, so a deep interest in antiquity has always pervaded China. But this past was seen to consist of heroes and wise kings such as the mythic Yellow Emperor, who led the Chinese from savagery to civilization. As early as the first and second centuries A.D., one finds accounts of a historical transition from a stone age to a jade, bronze, and iron age, but this sequence was speculation based on legend, not historical investigation. Antiquarian interest emerged in the Song Dynasty (A.D. 960–1279) but was largely limited to collecting and publishing ancient bronze vessels and ink rubbings of inscriptions on bronze and stone, reflecting an interest in epigraphy that has continued to the present.

Westerners introduced modern techniques of prehistoric archaeology to China. One early example was J.G. Andersson, a Swede working for the Chinese Geological Survey in the 1920s, who led the team that discovered Peking Man *(Homo erectus)* at Zhoukoudian, southwest of Beijing, and the Neolithic site at Yangshao, in Shanxi. Western-trained Chinese soon took up the task. A series of fortuitous discoveries of writing on bone, used in divinations, led to large-scale excavations at Anyang, a capital of the Shang Dynasty, sponsored by Academia Sinica, a national research institute in Taipei, Taiwan, and directed by the Harvard-trained Li Chi and others between 1928 and 1937. Wartime conditions led to a hiatus in archaeological activity, which was not resumed in a systematic way until 1949.

China has a vast written record extending far into the past. It has been estimated that a translation only of official histories, compiled from the second century B.C. on, would fill 400 volumes of 500 pages each. There is in addition an enormous body of classical literature, nonstandard historical works, philosophy, and so forth, which sheds light on still earlier periods. But while the written materials have provided tremendous insight, little was known about how the ancients lived, and almost nothing was known about the prehistoric period beyond legend. Recent discoveries allow us for the first time to place people in their own temporal contexts, to know how a person of the Han Dynasty differed from one of the Tang.

Despite the limitations of the Marxist interpretation of history, archaeological discoveries of the past 50 years have had a significant impact on the way in which the past is viewed in China. After a long-standing insistence that the Yellow River basin was the birthplace of Chinese civilization, the archaeological establishment is recognizing, rather grudgingly, that ancient remains found on the peripheries of China may represent independent evolved cultures.

THE NEOLITHIC
(6000–2200 B.C.)

Finds from the Neolithic are revealing ancient cultures far removed from the presumed cradle of Chinese civilization. A case in point is the discovery by Japanese archaeologists in 1935 of the Hongshan culture, a people who lived between 4000 and 3000 B.C. in an area bounded by the eastern Mongolian Plateau and including all of modern Liaoning Province in northeast China. The discovery by Chinese archaeologists of a Hongshan cult center at Niuheliang, Liaoning, published in 1986, has aroused great interest and speculation about the nature of Neolithic worship.

Dating to ca. 3500 B.C., the complex included two buildings roughly aligned along a north-south axis; the larger one to the north was 60 feet long and had several rooms, its walls originally decorated with murals. Clay fragments of hands, shoulders, and arms scattered nearby belonged to female figures of varying sizes, which had been modeled in a seated position with their backs attached to the walls. There were also fragments of a statue of a pig.

Further south were a series of low-lying stone cairns. One of the cairns was a stepped platform, very probably an altar, while the others were set atop stone cist graves, in one case with only one burial but in another with some 15 accompanying cist graves around a central one. The grave sites were quite elaborate, with stone boundary walls along which hundreds of pottery tubes of unknown use were placed in rows. The graves contained numerous carved jade pieces, the most distinctive being pig-dragon pendants, so-called because of their rounded, dragon-like bodies and pig-like heads. No settlements have been found in the vicinity, additional evidence that the site was used only for ritual purposes.

	CHINA TIMELINE
CA. 6000	NEOLITHIC
CA. 2200	XIA
CA. 1700	SHANG
CA. 1100	WESTERN ZHOU
771	EASTERN ZHOU
221	QIN
206 B.C.	HAN
0	HAN
A.D. 220	THREE KINGDOMS
265	WESTERN JIN
317	SIX DYNASTIES, or
386	NORTHERN & SOUTHERN DYNASTIES
591	SUI
618	TANG
906	FIVE DYNASTIES
960	SONG
1279	YUAN
1368	MING
1644	QING
1912	REPUBLIC OF CHINA
1949	PEOPLE'S REPUBLIC OF CHINA

Some scholars have claimed that the Hongshan culture represents one of the cradles of Chinese civilization and view the cult complex, consisting of temples and altars, as presaging such later Chinese architectural devices as the outdoor altar at the fifteenth-century Temple of Heaven in Beijing. The early reports on this site thus emphasized possible ties to the cultures of the heartland, but more recently Hongshan has been viewed by Chinese scholars as an independent culture of unknown origin. The presence of the female figurines and of the pig-dragon pendants has led some to speculate that the temple complex was dedicated to a cult of a fertility goddess with the pig-dragons as fertility symbols, an iconography unlike any other known in China.

There never was any doubt of a foreign origin for the mummies found in modern Xinjiang, which have attracted so much attention recently (see ARCHAEOLOGY, March/April 1995, pp. 28–35). Their existence has been known since European explorers found them in the early years of this century, but the announcement by Chinese archaeologists in 1981 of the discovery of a Neolithic blonde mummy made headlines in the Western press. They are mummified Europoids, with blonde or light brown hair and high noses, dating from as early as ca. 2000 B.C., whose bodies have been preserved by the dryness of the Taklamakan Desert. It remains to be seen whether these people, probably members of the Afanasievo or later Andronova cultures of the Russian steppes, played any role in the development of Chinese civilization. Archaeological efforts in the desert have uncovered many early sites, including the buried cities of Niya, Dandan-Oiliq, and Miran, which date from the third century A.D. A joint French-Chinese project of some years' duration has pushed even further into the desert to explore previously unknown sites, and promises to lead to a better understanding of the history of this pivotal area in the contact between East and West.

THE SHANG DYNASTY
(CA. 1700–1100 B.C.)

The Shang Dynasty saw the development of a bronze technology, the chariot, and a writing system, each probably due to foreign influences, but each with its own distinctively Chinese elements. Excavations at the dynasty's capital, known as Yinxu, near Anyang, began before World War II, and by 1976 archaeologists had explored 11 royal tombs and more than 1,000 graves to the northeast of the city and located the foundations of palaces, temples, and shrines erected on platforms of rammed earth within the city. But it was the discovery under one of these shrines of the tomb of Fu Hao, the consort of the Shang ruler Wu Ding, that proved the most spectacular. The only undisturbed Shang royal sepulcher ever excavated at Yinxu, the tomb was in the form of a vertical pit, 18 by 13 feet at the mouth and 25 feet deep, extending four feet below the water table. Various ledges and niches contained 16 sacrifices of men, women, and children, and six dogs. Only fragments remained of a wooden chamber that once housed Fu Hao's lacquered coffin. Below the coffin was a small pit containing a dog and a human sacrifice. The tomb was modest in size compared to royal burials, yet more than 1,900 objects were recovered, including some 200 bronze ritual vessels; 250 bronze bells, knives, mirrors, weapons, and other objects; nearly 750 jades; 560 bone objects; 6,900 cowrie shells, the currency of the time; and stone sculptures and ivory carvings.

The richness of the royal tombs at Anyang is difficult to assess because all of them had been looted before they could be properly excavated. The undisturbed tomb of Fu Hao gives some idea of the incredible quantities of bronze, jade, and other goods that the kings' tombs must originally have held. The bronzes in her tomb had been used in ancestor worship and had diverse origins; many were perhaps taken as booty or tribute after victories in battle. A clue to their origin comes from an oracle bone which records that Fu Hao led 13,000 men to attack the Qiang, one of the many enemies that bordered Shang territory. As Oxford University's Jessica Rawson points out, the diversity of decorative motifs on these bronzes indicates the influence of or manufacture by neighboring, contemporary societies of some sophistication. The discovery made it clear that the Shang by no means had a monopoly on culture or industrial production, as claimed in later accounts.

Further evidence that Bronze Age China was a much more complex system of competing civilizations than suggested by passages from contemporary oracle bones emerged in 1986 at Sanxingdui, Sichuan Province. Workmen digging for clay at a brickyard reported finding some jade pieces; subsequent excavation by archaeologists uncovered two pits. The first, five feet deep, was dated to ca. 1300–1200 B.C., while the second was the same depth and slightly later, ca. 1200–1000 B.C. Laid out in layers were objects of gold, bronze, jade, stone, and pottery; cowrie shells; more than 80 elephant tusks; and, most surprising, bronze masks, heads, and figures, one life-size, of humans with grotesque features. The large standing figure and the individual heads have slanting eyes and grim expressions, while the masks have protruding eyeballs. The heads and masks may once have been attached to bodies fashioned from perishable materials. All the objects revealed evidence of having been burned, after which they were ritually buried, the pits then being covered with earth. The pits were located inside what had been a large city, whose earthen walls enclosed an area of more than one square mile. Probably occupied between 1700 and 1000 B.C., with closely packed buildings and a high population density, Sanxingdui was obviously a major city of the Shang period. It was in contact with the cultures on the Yangzi, as well as to the north. The artifacts found here reveal a culture very different from the Shang; there are hardly any parallels that might explain their meaning.

WESTERN AND EASTERN ZHOU
(1100–221 B.C.)

With ongoing discoveries in the Wei River Valley, upriver from Xi'an, we can now trace the emergence and growth of the Zhou, once a feudatory and eventually the successors to the Shang. What we knew previously of the Zhou period came from Confucian classics such as the *Book of History*, the *Spring and Autumn Annals*, and its attached text, the *Zuo Commentary*, as well as other writings from the Confucian school. In the past, undocumented Zhou and Shang ritual bronze vessels would periodically appear on the market, but controlled excavations of Zhou tombs and hoards in the Wei River area over the last few decades have enabled archaeologists to understand the role of these bronzes in ritual and burials, and to clarify their use as status and hierarchical symbols. The lengthy inscriptions contained on some bronze vessels have also corrected a long-held view that the Zhou were philistines who sharpened their military skills by fighting off frontier tribes.

Surely one of the most spectacular finds to come to light in recent years is the tomb of Marquis Yi of Zeng in modern Suixian, Hubei Province (see ARCHAEOLOGY, January/February 1994, pp. 42–51). Archaeologists have recovered a staggering number of artifacts here, more than 7,000 in all, including bronze ritual vessels, utilitarian objects, weapons, armor, gold, jade, and lacquered wood. The bronze items weigh nearly 11 tons. The marquis, ruler of a state absent from documents of that period, was encased in a double lacquered coffin in tomb that included 21 women. Eight women, perhaps His consorts, shared the marquis' burial chamber, the easternmost of three chambers and thought to represent his living quarters. The other 13 women, in the west chamber, may have been musicians. The central chamber, modeled after a ceremonial hall, stored musical instruments including zithers, mouth-organs, panpipes, transverse flutes, drums of various types, a lithophone of 32 stones, and what has attracted the most attention, 65 bronze bells, still mounted in three layers on an immense L-shaped wooden frame. The timbers are covered with black lacquer, and among their supports are six bronze figures of warriors, each weighing 660 pounds. The bells still retain their original sounds (approximating the Western key of C major), with a range of 5 1/2 octaves in 12 semitones, Each has a gold-inlaid inscription that records the name of that bell's particular tone, providing insights into the music of fifth-century B.C. China.

THE QIN EMPIRE
(221–206 B.C.)

The Qin state marked the beginning of centralized imperial rule in China. Like the Zhou, this state had its beginnings in the Wei River Valley, and by dint of arms imposed its rule over all of China then known. In fact, the name China is probably derived from the dynasty's name. For the first time figures such as Shihuangdi ("The First Theocrat"), founder of the dynasty, begin to emerge from the mists of tradition and myth to be seen as more fully developed personalities in histories compiled from the second century B.C. on. They recount his building a forerunner of the Great Wall even further north than the present one, constructing roads to all parts of the new empire, and building hundreds of palaces.

The discovery in 1974 of an army of more than 7,000 terracotta soldiers revealed just how completely state resources were controlled by Shihuangdi. These life-size warriors are distributed among three pits east of the emperor's mausoleum, as if to stand guard against invaders who might threaten his seat of power. The terracotta army may have symbolized the might of this emperor and served his spirit in the other world so that he could take his rightful place next to Shangdi, the supreme deity, where he was to intercede on behalf of his descendants and look after their interests. Today, Shihuangdi's tomb itself remains unexcavated. Archaeologists are hard-pressed to keep up with necessary salvage work stemming from the enormous modern expansion of the economy, and available resources are simply not adequate to undertake such a huge project.

THE ALL-POWERFUL STATE

Fostering pride in the past while validating Marxist dogma

W<small>HY IS ARCHAEOLOGY SO IMPORTANT TO THE PEOPLE'S</small> republic? As Cambridge University scholar Victor Purcell has pointed out, it seems paradoxical that a government dedicated to the destruction of the vestiges of its "feudal" past, and which has as its goal building a new society, should allocate such significant financial and human resources to the study of that discredited past.

The archaeological apparatus in China is truly impressive. The State Bureau of Cultural Relics administers numerous provincial, county, and local archaeological units and museums. In 1985 there were 1,980 of these units, distributed in every province and autonomous region. Also at the national level is the Institute of Archaeology, which has various field stations, its own publications, and the authority to pick its own sites for investigation.

When we look at the reasons for the state's interest in archaeology, we see that the science does indeed meet a number of important needs in this new society. Archaeology fosters a pride in the past and a sense of self-identification, while demonstrating the indigenous nature of Chinese culture. It provides grounds for celebrating what is said to be the genius of the common people, the anonymous creators of the artifacts archaeologists uncover. The discoveries, finally, serve to validate Marxist doctrines of historical development, which describe a shift from early matrilineal communal clans to patrilineal, property-owning families with social stratification, and thence by stages through a slave-owning society, feudalism, and capitalism. In time, Marxist theory predicts the arrival of socialism and a return to a communist society without the class struggle that had driven the previous stages of development.

In China, archaeological reports, especially those on prehistory, take the Marxist framework for granted, and there have been almost no attempts to revise these tenets. It is not necessary to prove Marxist theories; rather, Chinese archaeologists concern themselves with fitting the archaeological evidence into its proper temporal sequence. If there is any controversy, it is one of periodization: When did the slave-owning society give way to feudalism? and so forth. The difficulty of the Marxist framework is that archaeological proof of a matrilineal communal clan society, an early stage in the Marxist model, is difficult to come by, and the demonstration that these early stages existed in China is far from convincing. Further, as Harvard's K.C. Chang has suggested, by being content uncritically to abide by the Marxist theories of anthropologist Lewis Henry Morgan (1818–1881) and socialist Friedrich Engels (1820–1895), Chinese archaeologists have generally failed to broach a number of significant questions for which their abundant data would be most useful.

Another dogma that vitiates scholarship derived from archaeological discovery, and one certainly not limited to China, is nationalism. In China, this takes the form of a model in which the central Yellow River Valley is seen as the source of all of Chinese civilization.

More recently, this concept has been expanded to encompass the various Neolithic and early Bronze Age cultures as contributing to the emergence of a unified civilization. However, as Lothar von Falkenhausen of the University of California, Los Angeles, has observed, the increasing importance of regional archaeology does not signify a broadening of intellectual perspectives; one dogma is simply being replaced by another.

There is thus a price to pay for government support of archaeology in China. One notes little initiative or exploration of new ideas. There is an enormous output of publications and reports with much description and little interpretation. What little interpretation there is is confined to typological classification of artifacts, stratigraphical dating, seriation, and so forth, and the results are incorporated into prevailing dogmas.

—A.D.

HAN DYNASTY
(206 B.C. TO A.D. 220)

The Qin state was succeeded by the Han, which ruled for the next four centuries, a time of economic growth. The imperial mausolea of the Han Dynasty, marked by huge burial mounds clearly visible north of Xi'an, Shaanxi Province, have also not been opened due to limited resources. An ancillary pit of the tomb of Emperor Jing (ruled 156–141 B.C.) was excavated in 1991 because it lay in the path of a highway project, and an army was found, in this case of pottery figures about two feet tall, largely unclothed because their silk garments and lamellar wooden armor had not survived. Why genitalia were included on the figures, when such details would not be visible once clothed, is a matter of some interest. This may well indicate that the figures were not seen just as inanimate symbols, and so had to be crafted in as much detail as possible.

Such martial displays were not limited to imperial burials. A relatively modest, unidentified tomb also in the Xi'an area, probably of a high military officer, contained an army of some 2,000 figurines of similar size, with clothing and armor modeled in clay. Others with arrays of figurines have been found elsewhere in China as well. The tombs of imperial princes who served as governors did not yield such armies. The undisturbed

96

tomb of the prince of Zhongshan, at Mancheng, Hebei, contained only a couple of small stone figures, one of jade, and another pair of bronze, hardly enough to have served the prince in the other world. Perhaps even in death it was thought too dangerous to give princes, potential rivals to the throne, access to military forces. But what this prince and his consort did have, as did other members of the royal family, were suits made up of thousands of jade plates joined together by gold and silver wire, intended to preserve the bodies.

While the Mancheng bodies disintegrated, a tomb built for the marquise of Dai (died ca. 168 B.C.) at Mawangdui, Changsha, was more successful: her body was so perfectly preserved that her skin and muscles still retained a certain degree of elasticity, and it was possible to determine what she had eaten just before death. Layers of clay and charcoal insulation, as well as mercurial compounds, were responsible for the excellent preservation of the body and tomb contents. Large quantities of clothing made of plain silks, brocades, damasks, and gauze reveal much about Han weaving techniques, indicating that rather elaborate looms were already in use. Lacquerware, musical instruments, articles of bamboo and wood, herbs, foodstuffs, and important texts written on bamboo strips and on silk were also found, all of which have added much to our knowledge of the material and intellectual life of the Han.

Another source about life during Han times is the pottery models of buildings, household goods, and farm equipment that were placed in tombs, a custom that also continued in subsequent centuries. The structure of tombs also changed during the Han from the pit type to brick chambers built into the side of a hill or cave dug at the end of a sloping passageway. These chambers provided adequate space for grave goods alongside the coffin, a boon for archaeologists.

SUI AND TANG DYNASTIES
(A.D. 589–618 AND 618–906)

Following a period of disunion, the Sui and Tang dynasties ushered in another period of relative prosperity. The brick chamber tomb by then had become almost universal in China, the style differing by period and region. The size, of course, depended upon the status of the deceased, and the largest tombs were reserved for the imperial family. None of the imperial mausolea of the Tang emperors has been opened, but those of some of the royal family have been excavated. These had been rather thoroughly looted in the past, leaving behind only ceramic figurines disdained by tomb robbers.

What these tombs did yield were murals and paintings in lacquer, a trove of information about Sui-Tang life. Examples of early painting in China had survived primarily in later copies, and there was always a question as to how well these reflected the originals. The painted bricks of third- to fourth-century tombs in Gansu Province depict scenes of the hunt, farming, and household tasks. Northern Qi (A.D. 550–577) tomb murals depict the Xianbei rulers of north China riding in a splendid procession, and in the Tang royal tombs are scenes of life in the palace, the reception of foreign envoys, and the polo field. The best artists probably were not called upon to descend underground to decorate tomb walls, but still these wall paintings are generally of high quality and have filled in a missing chapter in the history of Chinese art.

BUDDHISM

It may seem strange that early tombs contain almost nothing of an overt religious nature, especially pertaining to Buddhism, which flourished after it was introduced into China in the late Han Dynasty. However, numerous statues of bronze and stone, stone stelae, and cave temples along the Silk Road, especially at Dunhuang, provide an enormous amount of information concerning the development of Buddhist teachings, iconography, and art (see ARCHAEOLOGY, November/December 1997, pp. 60–63). In 1996, hundreds of sculptures of Buddhas and bodhisattvas from the fourth to the twelfth centuries, most carved from limestone but some from marble and other materials, were found at the site of a former temple at Qingzhou, Shandong. While a persecution of Buddhism in the sixth century has been cited as the reason for other, deliberately buried caches of Buddhist material, in this case the reason for the burial is still under investigation. An additional source of information about Buddhism lies in sealed crypts beneath pagodas, some of which have come to light during reconstruction or excavation of these structures.

One of the most momentous discoveries of this sort was at the Famen Temple, some 75 miles west of Xi'an. When much of the brick pagoda collapsed during heavy rains in 1981, local authorities decided to rebuild it, and during its reconstruction, which included a complete disassembly of its 13 stories and strengthening of its foundation, a crypt, sealed in 873, was discovered. This was an elaborate affair, with a passageway and three chambers, filled with hundreds of gold, silver, porcelain, and glass objects, and precious woven materials from the emperor and his court. What tradition holds are the finger bones of the Buddha were encased in elaborate reliquaries, one in each chamber, and a fourth, perhaps thought at the time to be the only genuine one, in a secret room under the last chamber.

Another vital source of information about classical works and law, medicine, and religion of this period have been ancient texts on bamboo and wood slips, on silk, and even on jade tablets. Most of these textual materials have been found in graves, but in 1997, a cache of tens of thousands of bamboo and wooden slips comprising a third-century A.D. administrative archive was found at the bottom of an ancient well, where ground water had preserved it (see ARCHAEOLOGY, May/June 1997, p. 26).

LOOKING TO THE FUTURE

This is only a small sampling of the incredible finds by Chinese archaeologists in recent decades. The scope of their activities can be measured in the area of Echang, Hubei, where some 400 tombs dating to A.D. 220–589 have been excavated since 1949, but only 40 published. For all of China, some 2,000 tombs of the same period were published, indicating, if the same ratio holds, that 20,000 tombs were opened. Excitement over these discoveries has been tempered by a lack of funds that has created a publication backlog. Increasingly, a low-paying career in this memoir

THE SEARCH FOR SHANG

WE HAD TO DECIDE WHICH TEXTS TO TRUST
AND HOW MANY OF THEM WERE WORTHY OF EVALUATION

By K.C. Chang

In 1988, I was invited to Beijing as a representative of Harvard's Peabody Museum to select an archaeological site to excavate with colleagues from the Institute of Archaeology of the Chinese Academy of Social Sciences. I was interested in searching for the holy city of Shang, which historical sources say is the site where Shang Dynasty rulers were buried and a royal house of worship was erected. I remember speaking with the institute's director at that time, Xu Pingfang, a very learned man. "I fully agree with you on the importance of the city, and its discovery will be interesting," Xu said with a twinkle in his eyes. "But I think it is the most difficult subject you could choose to pursue in our first collaborative effort." I smiled, and we laughed together, shaking hands. That's how the project began. The Tiananmen Square uprising delayed the formal signing of the agreement until 1992, and field research didn't begin until 1994.

Where was Shang? In his collection of articles known as the *Guantang Jilin,* Wang Guowei (1877–1927), citing the Chinese history *Zuozhuan* and many other sources, located Shang in the walled town of Shangqiu, in Henen Province. Since the 1930s, a good number of archaeologists have investigated Shangqiu, but Neolithic mound sites are all they have found. Finding the real Shang posed many difficulties. To begin with, Shangqiu (literally the Mound of Shang) is the name of the small town surrounded by a mile-long walled enclosure built in the Ming Dynasty (1368–1644). But it is also the name of the second-largest city of Henan Province, now a rapidly developing transportation hub, and of a Henan Province district as well, which incorporates seven *xian,* or counties, of which Shangqiu *xian* is one, including the city of Shangqiu. There are no obvious geographical features that would identify where the ancient city of Shang was located. Its ruins might have been anywhere within or without these modern administrative units. Furthermore, we correctly predicted that loess, the windblown soil typical of the northern Chinese landscape, would cover the site.

Here, then, is where historical texts came in handy. We had to decide which texts to trust and how many of them were worthy of evaluation. Sometimes scholars are fortunate enough to have texts that not only help them find ancient sites, but also help in interpreting them. Such was our case. In oracle-bone inscriptions from Yin, near Anyang, the last capital city of the Shang Dynasty, Shang is widely used as the name of a major town connected with religious activities of Shang royalty. In these inscriptions, the king is often described as "performing a telling ritual," which means informing his ancestors about some event. Of special interest in the phrase *"gao yu Da yi Shang,"* or "performed the telling ritual at Da Yi Shang," written in the final period of the Shang Dynasty, during the military campaigns against the Ren Fang and the Yu Fang, neighboring states to the east. Dong Zuobin, a scholar of oracle-bone inscriptions at Taipei's Academia Sinica in the 1940s, explained that "Da Yi Shang, or Great City Shang, was the site of the [Shang] people's ancient capital; here lay the lineage temples of their earlier kings."

CLUES TO SHANG'S LOCATION

If we can rely on the ancient records—and *Zuozhuan* is the most reliable of the Chinese classics—then it is entirely reasonable to accept Wang Guowei's hypothesis that the ruling Shang Dynasty acquired its name from the first seat of their power, a town named Shang, where at least the first three of their dynastic rulers, Xie, Zaoming, and Xiangtu, had their political capital. The subsequent names of the predynastic kings and their capital cities are not clear, but from oracle-bone inscriptions it seems probable that although the royal capital later moved from Shang and was located at several places in present-day Henan, Shandong, and Hebei provinces, Shang remained a royal cult center throughout the Shang Dynasty, the site where later kings would return to perform certain rites, including the telling rituals.

Zuozhuan's accounts mention Ebo, the marquis of E, a small state that had occupied Shangqiu before the arrival of the Shang. In these annals Ebo is described as making sacrifices to Dahuo, the fire star, and living in Shangqiu among the ancestors of the Shang. So we felt passages about Ebo offered clues to the city's location. *Zuozhuan* also identifies a city named Ebo, placing it in one of the two modern Shangqius. Throughout Chinese history, folktales about the city of Ebo all relate that it was located where there is now a temple near the walled town of Shangqiu honoring Ebo, the historical figure. In the *Bamboo Books,* historical accounts on bamboo found ca. A.D. 280, a passage describes how the duke of Zhou, after his conquest of the Shang ca. 1100 B.C., moved the dynasty's crown prince and the rest of his court to the city of Shang so that they could continue worshiping their ancestors. A temple honoring the founders of the Xia, Shang, and Zhou dynasties may have been built in the same place as the modern Ebo temple near the walled town of Shangqiu. We know that people in Shangqiu have been worshiping at the Ebo temple since remote antiquity. Toward the end of the Song Dynasty, Shangqiu was neglected, but in the Ming, new city walls were built that presently delineate the town as well as a new Ebo temple.

So we thought the location of the present Ebo temple could guide us to the ruins of the city of Shang, and we concentrated our archaeological efforts around it. After three years of boring holes in the ground, our geophysicists knew a great deal about the Holocene (ca. 10,000 years ago) stratigraphy of the Shangqiu area, but did not find the compacted layers of earth known as *hangtu* that would have indicated ancient occupation.

In 1994 it was suggested that we should move on and try a location several miles away from where we were working, in an area to the southwest of the present city wall. I declined. That the city of Shang should be near the Ebo temple is supported by the best of the ancient historians. My confidence in these historians was not misplaced. *Hangtu* were discovered in the spring of 1997 by geophysicists within Shangqiu county. The *hangtu* came from the remains of an Eastern Zhou town wall that led us to a city two miles wide and over two miles long. There is no question in our minds that this is the remnant of the Song-state city, the successor town of the city of Shang. Now it is just a matter of time before Shang sees light.

K.C. Chang, born in China, received his Ph.D. from Harvard in 1960, and taught at Yale University from 1961 to 1977, then at Harvard University from 1977 until his retirement in 1996.

field is less appealing to university students. And as social controls have relaxed and ideological commitment weakened since 1987, there is an increasing amount of tomb pillaging and the smuggling of archaeological artifacts into foreign markets. The bulging inventories of antique stores in Hong Kong, Taipei, and elsewhere are testimony to this devastation. A display of material confiscated by the authorities at the National History Museum in Beijing a few years ago filled hall after hall. In 1994, Scotland yard confiscated more than 3,000 artifacts illegally smuggled into England and found in warehouses in London. After a long and complicated trial, they were returned to China early in 1998 and put on display, also in the National History Museum, as a demonstration of the government's resolve to fight the smuggling of archaeological artifacts. Civil engineering projects, too, are taking their toll on archaeological sites. The Three Gorges Dam, now being built on the middle reaches of the Yangzi River, poses a grave threat to countless Neolithic sites, whose ceramics are more varied in shape, color, and decoration than anything produced by contemporary Yellow River Valley civilizations (see ARCHAEOLOGY, November/December 1996, pp. 38–45).

As financial support by the government has decreased, a number of binational expeditions, largely foreign-financed, have taken place, facilitated by changes in the law which now permit foreign archaeologists to take part in excavations. This can only be a positive development, and it may cause Chinese archaeologists to move beyond constricting ideologies that now hinder the full appreciation of their country's archaeology.

Albert E. Dien is emeritus professor of Asian languages at Stanford University. He is completing a book, Six Dynasties Civilization, *to be published by Yale University Press as part of their Early Chinese Civilizations series.*

Reprinted with permission of *Archaeology Magazine*, Volume 52, Number 2 (Copyright the Archaeological Institute of America, 1999).

The Mao® Industry

"One can indeed say that political reform has been visited upon China. It came, however, not in the form of an institutional transformation of the state-based political system … but in a far subtler yet profoundly life-transforming manner. No longer are people enthralled by the political, or even intimidated by it."

MICHAEL DUTTON

In 1984, I was a student at Beijing University when China's official de-Maoification program was at its height. Throughout the city, statues of Chairman Mao were being torn down. It seemed as if the reassessment of the chairman's thought (70 percent correct, 30 percent incorrect) advanced by the Communist Party's new, reform-era central committee had been transformed into a plan for urban renewal. The cult of Mao that the new central committee wanted to dismantle had left China awash with images of the chairman. A Mao statue could be found in the heart of every town square, another at every work unit entrance and, once inside these compounds, one in every office, as well as in every home. Invariably the statues stood in places formally occupied by ancestral shrines. It was not just Mao's thought that needed moderating, but also his omnipresence.

It was around this time that a Slovenian student friend discovered something that she knew would be of interest to me. At the back of the university campus, outside an old 1950s-style administration building, she found a huge pile of smashed white plaster statues. On closer inspection, the pile turned out to be the university's own contribution to the de-Maoification program. On the ground beneath our feet lay the splintered remains of hundreds of little Mao statues that had been taken down from their exalted positions in administrative offices and unceremoniously dumped into piles awaiting collection. In effect, what my friend had discovered was, if not the burial ground of Chairman Mao, then at the very least, the beginnings of a graveyard of his auric presence. As foreigners, we felt no qualms scavenging about in the pile looking for body parts to souvenir. Soon, we were joined by a couple of inquisitive primary school children. They asked us what we were up to and when we told them, they joined in. Before long, we were happily trading body parts: "I'll give you an arm and a leg of the chairman for his nose," said one child. "No, I want his ear," came my reply. Such irreverence didn't last long.

A Communist Party cadre came screaming out of the administrative building: "You are not allowed to touch the chairman's parts, it's disrespectful." The kids scampered but we held our ground. "We're not being disrespectful," we responded. "That honor goes to the cadre who had the chairman's statues removed from the offices, smashed to pieces, then shoveled into this pile." Our response did little to appease the cadre on what was clearly a sensitive issue. He ordered us off the little Mao mountain with the warning that security guards were on their way. Gathering up what booty we could, we loaded it onto our bikes and tried to make a run for it, but a detachment of security officers pursued and caught us. So we were ourselves "de-Mao-ified," sent back to our respective departments and criticized by our teachers and fellow students.

MYSTICAL MAO, SENTIMENTAL MAO

That was China in 1984. Less than 10 years later it was a different story. In 1993 one did not have to dig to find the chairman's image on the campuses of Beijing. He was, once again, just about everywhere. He re-emerged when a growing wave of popular nostalgia was bolstered by official party promotions to mark the one-hundredth anniversary of his birth. With the party sanctioning a variety of activities that were centered around Mao, the popular response was to join in with alacrity. A decade later, in 2003, he would return once more, but this time accompanied by marketing strategists and advertising executives. Where 1993 had the air of spontaneity and the hint of trouble, 2003 had only the hallmarks of a well-oiled advertising campaign.

There is something odd in this constant reappearance of Mao during the long march toward consumerism. Perversely, the chairman's return is more a sign of his fall from grace than of any lingering sense of revolutionary devotion.

No one really knows how the current Mao craze started. Most commentators agree that it probably began after a multicar pileup on a Guangdong highway in 1989. The crash was fatal for many, and the only person to walk away unscathed was the driver of a car with a Mao talisman. With this story, the mystical Mao was born. Transformed from revolutionary leader into a god of good fortune, Mao became a soothsayer for troubled times. From then on other "miracles" confirmed his beatification. His shadowy apparition appeared on the surface of a pebble drawn from the waters of the Yangtze River; it reappeared as a ghostly image on the surface of a peasant's household wall where his portrait once hung; it even resurfaced as nature's own handiwork, etched onto the rock face of a mountainside in Hainan Island that has become known as Mao Mountain (Maogongshan).

This spirit figure, however, was not the only guise in which the chairman re-emerged. Indeed, a wide range of different Maos made their debut in this early phase of the Mao craze. Popular movies and books started to appear. These tell of a very different Mao from the mystic and suggest, as well, a different origin.

Like the mystic Mao, this other, more sentimental Mao came to prominence in 1989. In that year a new "post-scar" literary trend described by Geremie Barmé in *Shades of Mao* as a "search for Mao" began to emerge.[1] This literary movement broke the mold of Mao writing, for rather than adding to the chairman's mystique as both scar literature and party hagiography had done in their different ways, it offered a radical dénouement.

Greatness (hagiography) and great excess (scar literature) were now replaced by a focus on the everydayness of Mao. A *Woman's Day* voyeurism combined with nostalgia for simpler times to produce this new and highly popular genre. Perversely, this nostalgia also had a political face, which reached its apogee in 1989 in the protests at Tiananmen Square.

While students fought party corruption with calls for democracy and liberty, the pragmatic worker participants fell back on nostalgia, resurrecting the image of the chairman to remind the party of their now forgotten obligations to the working class. For years, Mao's image had been the flag under which labor disputes in the reform era had been used to dramatize the loss of worker security and welfare provisions. With the Tiananmen Square protests, this well-worn weapon of worker protest reappeared in the heart of the capital. From such political nostalgia, a more general and genteel form of longing for past security and lost youth fed into a growing romanticism already seduced by movies, magazines, and novels.

By the early 1990s, these aberrant and contradictory trends started to congeal. What tied them together had less to do with the fact that they were all about Mao, and more to do with what they produced: sales. In this respect, origins are unimportant because what is central to this craze is the entrepreneurial discovery that Mao sells. Slowly, the hallowed images of Mao and the revolution were being transformed. They were now well on their way to becoming pure commodities.

MONEY IN THE BANK

This process of commodification began innocently enough. In the spring of 1990, a number of publishing houses discovered that the republication of Mao's official portrait had a greater market appeal than they had initially imagined. As many as 3.5 million copies of Mao's official poster portrait were sold in that first year of republication and, in the three years that followed, more than 11 million copies were sold. Seizing the opportunity, other entrepreneurs began to produce their own tokens of Mao.

Indeed, almost anything revolutionary now seemed to have a market. Even the songs of that era began to be sung again. Naturally, they were songs with a very different beat from those of the past. A 1992 remake of revolutionary tunes put to a disco beat was an overnight karaoke sensation, selling 70,000 to 80,000 copies in its first week of release, and over 1 million copies in its first month. In addition, long-neglected revolutionary sites started to become tourist attractions. By 1990, Shaoshan, the place of Mao's birth, boasted around 2,500 visitors per day, reaching a peak of nearly 3,000 per day the following year.

As this "love of Mao" grew, so, too, did the fears and suspicions of party cadres, who were still hostage to the belief that any revival of his image must mean a revival of his politics. Initially, the Communist Party read the various heterodox trends that fed into the Mao craze not as fashion or as fad, but as potentially troublesome and political. Party leaders grew concerned that they could well be looking into the image of a backlash against their liberal economic reform policies. Mao's image, they feared, might become the clarion call that would rally the masses to revolution once more.

They were not the first ruling class to be worried about mass entertainment providing a venue for popular unrest and revolution: from the very first moment business began promoting mass entertainment and the mass tourist industry, political rulers had worried about its effects. The first world trade exhibition at London's Crystal Palace in 1852 provoked anxiety within the political elite. Fearful of gathering together so many proletarians for leisure rather than work, the English ruling class at first tried to limit worker entry.

Soon, however, it discovered that far from fomenting revolution, mass entertainment actually quelled it. Once the mesmeric effects of the culture industry were fully appreciated by political leaders, they did a complete about-face. Instead of limiting participation, the English ruling classes started giving away tickets! Such generosity did not last long. Realizing that workers would pay for their enchantment, Thomas Cook began organizing train-travel tours to the exhibition site and, in so doing, initiated the modern tourist industry.

In China, the Communist Party simply skipped a few stages. It never gave away free tickets and instead moved directly into hiring marketing consultants to manage the growth of this new industry. From this, the so-called Red industry was born. It was brought to consumers, not by the bottlers of Coca-Cola, but by those who would bottle the revolution: the Communist Party.

With an official endorsement of Shaoshan as a national tourist site in 1993, the party fully embraced the Mao craze. By that stage, however, almost anything that made money seemed to be permissible. Even the most heterodox and mystical of Mao sites seemed able to gain official endorsement at this time. In November 1992, the State Environmental Protection Bureau even put a protection notice on Hainan's Mao Mountain and upgraded it from a county- and provincial-level site to one of national significance. By this stage, Mao was becoming irrelevant. Anything that promoted local tourism seemed to be acceptable, and Mao Mountain was certainly doing that. Indeed, Hainan's Mao Mountain became something of a pacesetter with at least three other rival Mao Mountain bids emerging in other parts of China.

While the party showed an initial reluctance to officiate over the spread of the burgeoning Mao business, Chinese entrepreneurs exhibited no such constraint. For them, the growth of Mao iconography and paraphernalia was money in the bank. One of the earliest entrepreneurs to cash in on the late chairman's earning potential was "Old Lady Tang" of Shaoshan, who renamed her capitalist enterprise "Maoist." Beginning as a humble greenbean noodle-soup stall owner, she started to repackage her business in 1978 as a Mao restaurant selling his favorite dishes. It proved a huge commercial success. Having conquered Mao's hometown of Shaoshan, she then moved on to bigger and better things. Currently she operates a franchise of more than 40 Mao-themed restaurants throughout China.

THE GREAT LOGO

During the Mao era, the chairman sang a siren song, enchanting the masses with his politics and poetry. Later, he would orchestrate the marshal music that was the Cultural Revolution. Millions would join his choir, march in step, and fight and die in his name. The re-emergence of his face on billboards and posters thus carries special significance—not because it threatens a new wave of radicalism, but because it does the opposite. Through the years of reform, Mao's siren song of revolution has been transformed into simply another soothing love song. He has become, in effect, a mere commodity-shell through which the very idea of being for sale is offered for sale. This is the chairman's afterlife in the era of reform, the foundation for the Mao Industry.

> *No longer is every event the basis of a political question. Now every event is a question of market opportunity.*

As economic reform developed into a consumer revolution, Mao's image was transformed from Great Helmsman to great logo. He became the fashionable embodiment of that mass-produced art form of the common and the everyday: kitsch. The production of kitsch plays on his shadow-image afterlife, and its import lies in the fact that it is not just born on the gravesite of revolutionary siren politics, but actually constitutes one of its most determined gravediggers.

Like the chairman himself, this siren form of politics—a form in which millions expressed an undying faith in their political cause and a willingness to die for it—has been reduced to an artless art form that merely entertains. What we are witnessing, the German sociologist Ulrich Beck would argue, is a very political form of depoliticization. In this sense, one can indeed say that political reform has been visited upon China. It came, however, not in the form of an institutional transforma-

tion of the statebased political system that swept away the Communist Party, but in a far subtler yet profoundly life-transforming manner. No longer are people enthralled by the political, or even intimidated by it.

TURNING REVOLUTION INTO MONEY

There is no better example of this political depoliticization of the masses than what has happened to the once taboo subject of the Cultural Revolution. In 1986, the Chinese author and critic Ba Jin controversially called for a Cultural Revolution museum to be built along the lines of the Holocaust Museum under construction in Washington, D.C. History, he argued, must never be repeated; a site where "real objects" and "striking scenes" would bring forth feelings of discomfort offered the best guarantee that the Chinese would never forget their own past atrocities. "Masks would fall, each will search his or her conscience, the true face of each one will be revealed, large and small debts from the past will be paid," Ba Jin wrote, and "the flowers that bloom in blood" that appear "bright and beautiful" but that are, ultimately, poisonous will never be planted again. For years, Ba fought official opposition to his idea. The Communist Party did not want to be reminded of its own past failings. Silence reigned while the party struggled with the fear that any revival of that memory spelled trouble. By 2004, all such fears were gone—swept from the party's thoughts by the profit motive.

In April 2004, entrepreneurs in Dayi county, Sichuan Province, turned the earth on a site that became China's first Cultural Revolution museum complex. Consisting of a restaurant, a hotel, and a teahouse, this "museum" was suitably outfitted so that it could take its place as part of a broader theme-park experiment to create the experience of revolutionary extremism without the excess. Promoting nostalgia rather than revolution, the museum was designed to appeal to tourists in need of a break from the capitalist rat race. With suitably attired staff and the accoutrements of revolution adorning its walls, the museum complex's restaurant was a perfect simulacrum of the "worker-peasant-soldier large canteen" (*gong nong bing dashitang*) and the complex's hotel, a Red Guard "reception center" (*Hong-weibing jiedaizhan*); the teahouse was renamed Chunlai (Spring Cometh) after one made famous in a Jiang Qing (Madam Mao) model opera, Shajiabang. "This venture definitely isn't just about social service, but about remembering the past, and, most importantly, about the management of a museum business," said one of the investors bankrolling the project. As a cynical reporter from the Qianlong web news service noted when he covered the opening of the "museum," "It was more than social service all right, it was about profit!"

Perversely, it is the very profit orientation and theme-park quality of the museum that makes it so politically powerful—but always in an antipolitical way. What might at first appear as little other than a parody of Ba's solemn vow to remember the victims offers quite literally the best antidote available to the infectious allure of the revolutionary sirens' song.

Ba Jin's solemnity was little different from the anger displayed by the responsible cadre who pulled me off the Mao rubbish pile at Beijing University. Both remain caught in the aura of politics in a way that the tourists of the museum, the karaoke crooners of Mao songs, or the diners at Old Lady Tang's Mao restaurant chain are not. While the former still pick at the scars of the sirens' return, the latter relax at the crossroads of distraction and nostalgia. Fear is replaced by fetishization. Such forms do not merely attract paying customers—they help produce a consumer mentality. In the process of producing this commodity mentality, they bring forth a far more effective and life-changing antidote to the Cultural Revolution than solemnity or outrage could ever evince. In speaking to an everyday anti-politics of fashion and fad, seriousness is caught off-guard. Yet the cost of this anti-political subversion is the reification of the commodity form.

THE POLITICS OF DISTRACTION

It is in this ability to transform even the greatest expressions of political commitment and political terror into items of consumption and enjoyment that one can see the power of the culture industry. This power to diminish our capacity to be thrown by the sirens' song of politics lies precisely in the ability it has to make even the most horrendous political event seem, in the eyes of the moderns, unthinkable. The commodified simulacrum of the Cultural Revolution makes the past unthinkable as a horrendous event by remembering it nostalgically, and by offering it to modern consumers as a light-hearted form of distraction.

Politically, this proves far more effective in halting the re-emergence as horror than any truth or revelation. Truths can always be contradicted and countered in ways that commodities cannot. After all, how does one challenge a theme park? How does one reverse the trend that makes remnants of siren politics fashionable or "cool?"

In transforming events like the Cultural Revolution and figures such as Mao Zedong into forms of consumer distraction and nostalgia, the commodity process actually robs them of their original transformative ability. In making everything into a commodity, that which is commodified is itself transformed. No longer is every event the basis of a political question. Now every event is a question of market opportunity. Our ears become plugged to the music of the political precisely because we are deafened by the hip-hop or disco versions of the revolutionary song. Increasingly, even when we desire change, we become tied to the very logic that stops it from taking effect. It is this political reform that the Mao Industry has helped solidify, and it is this that constitutes the long-term political legacy of economic reform.

In 1992, Deng Xiaoping undertook his now famous southern tour of China, effectively launching the consumer revolution in China. With this in mind, the dates of the Mao craze suddenly take on added significance. That craze constitutes the first postliberation consumer frenzy witnessed in China. The leader who once made the siren call to politics and then directed politics against capitalism now becomes the image form that leads the counter-revolution. In this new consumer counter-revolution, forms of enchantment remain. Nevertheless, like rootless migrant workers, the forms have left the tightly knit village of commitment politics and moved to the polymorphous consumer capitals of the eastern coastal cities. These are the places of seduction—seductive because they offer forms of enchantment that do not confront the political. Instead, they morph politics into a safer form of dreaming.

They do this by bleaching away the unnatural power of the political and replacing it with the artificial power of the simulacra. The aura remains but the passion is drained away. In each and every domain, and irrespective of their particular localized distinctions, the practices once undertaken out of passionate commitment to a political program are forgotten. Instead of political distinction, we now have economic distractions; instead of being uplifted, we are now stupefied. Caught between the Scylla of passion and excess and the Charybdis of alienation and order, China heads for the latter.

This dilemma is a feature of all liberal democracies. It is now beginning to emerge within the Chinese state. It is the Mao Industry that puts it into focus, for it is the Mao Industry that has led the charge. In this new, passionless world of the commodity, the life-affirming exhilaration of revolution gives way to the faux excitement of manufactured desire. The excesses of the past have gone, but so too have the life-affirming victories of revolutionary zeal. Perhaps this is the cost of any ethic of limit. Life-threatening intensity can only be limited by limiting life itself.

Note

[1]Revolutionary realism dominated art and literature during the Cultural Revolution. When it came to an end in 1976, a new form of realism emerged. Called "scar," it described the psychological traumas suffered by intellectuals, rusticated youth, and the common people during the Cultural Revolution. The name for this genre came from a short story by Lu Xinhua called "Scar" that was published in the *Wenhui Daily* on August 11, 1978.

MICHAEL DUTTON *is an associate professor in political science at the University of Melbourne. He is the author most recently of* Streetlife China *(Cambridge University Press, 1998) and* Policing Chinese Politics: A History *(Duke University Press, forthcoming, 2005). He is also the coeditor of the journal* Post-colonial Studies.

From *Current History*, September 2004, pp. 268-272. Copyright © 2004 by Current History, Inc. Reprinted by permission.

INSIDE THE NEW CHINA

Part communist, part capitalist—and full speed ahead

By Clay Chandler

WHEN THE NOON BELL RINGS ON THE SHENZHEN CAMPUS OF Chinese telecommunications equipment giant Huawei Technologies, an army of engineers falls out from the gleaming office towers. Like terra-cotta warriors sprung to life, they march past manicured lawns and soaring palms to the newly built cafeteria. The menu is distinctly local: pigs' feet with peanuts, dried bean curd with chives. But forget the Mao-era stereotypes about work units and "iron rice bowls." At Huawei employees pay for their own meals. Even management of the cafeteria is put up for bidding from outside vendors. In today's China, jokes Huawei executive vice president William Xu, "nobody gets a free lunch."

Back at their workstations, Huawei's troops are hatching plans to eat the lunch of global competitors, among them Cisco, Lucent Technologies, Siemens, and Alcatel. Founded in 1988 by a former People's Liberation Army sergeant, Huawei has come a long way from its start as a distributor of switchboards made in Hong Kong. Today the company employs 22,000 people, dominates China's market for high-speed switches and routers, and is muscling in overseas. Last year Huawei edged out half-a-dozen rivals, including Alcatel and Siemens, for rights to build a third-generation wireless-phone network in the United Arab Emirates, and it has signaled its U.S. ambitions by striking an alliance with 3Com, a Cisco rival. Huawei's sales this year are expected to top $5 billion, twice what they were in 2002, with $2 billion coming from outside China. Within three or four years, says Xu, foreign sales will account for more than 70% of Huawei's revenue.

Huawei's global prowess is just one of many signs that the world's most populous nation is coming into its own. After the communists seized power in 1949, Mao Zedong declared that China had "stood up." But as Chinese today like to point out, it wasn't until the diminutive, chain-smoking Deng Xiaoping opened the country to the forces of market competition after Mao's death that China was able to run. In the quarter-century since that second revolution, China has emerged as one of the most powerful—and unpredictable—forces in the global economy. It has sucked in hundreds of billions of dollars from multinationals eager

to tap its vast pool of cheap labor and secure positions in its burgeoning market. It has become the world's manufacturing hub (half of the planet's clothes and a third of its mobile phones are made in China), driving down the prices of toys sold at Wal-Mart and laptops shipped by Dell. It has surpassed Japan as the country that has the largest trade surplus with the U.S. and trails only Japan as the largest holder of U.S. Treasury bills.

China is pulling in imports from Asian neighbors—machinery from Japan, steel from South Korea, palm oil from Thailand—and its appetite for raw materials of all sorts, from peanuts to pig iron, has sparked an unprecedented boom in world commodity markets. It has become the world's largest consumer of copper, aluminum, and cement, and last year overtook Japan as the world's second-largest importer of oil. China is the world's No. 1 market for mobile phones and the No. 2 market for personal computers; many analysts believe it will become the world's second-largest auto market by the end of the decade. No wonder consultants at McKinsey's Shanghai office report visits from CEOs of more than 40 major multinationals so far this year. For FORTUNE 500 executives, says Ian Davis, McKinsey's worldwide managing director, China is "absolutely center stage right now."

Analysts disagree about whether China's economy can sustain this torrid pace. Last year's GDP growth rate, 9.8%, fanned fears that China was overheating. In May, Beijing moved to cool things down, ordering local governments to curb public-works spending and banks to rein in lending. Other threats to expansion loom. By Western accounting standards, not one major Chinese bank is solvent. The country's stock exchanges are dens of thieves, planning is incoherent and often contradictory, and corruption is rampant. Once the most egalitarian of societies, China now has one of the world's widest gaps between rich and poor.

CLSA Asia economist Jim Walker urges the long view. China, he predicts, will grow faster this year than last, despite banking woes and government attempts to hit the brakes, and may well keep barreling forward at its current speed for years. "Capitalism in China is only ten years old,"

Walker says. "We're at the early stages of one of the greatest industrial revolutions in world history." Analysts at Goldman Sachs have projected that China's economy (2003 GDP: $1.4 trillion) will overtake Japan's by 2015 and America's by 2039. Perhaps. But econometric models can't compute all the variables in the China equation. It is, after all, an experiment without precedent in economic history. No other communist nation has managed the transition to the global marketplace with greater aplomb. Then again, no country has ever attempted to do it in quite this way.

HUAWEI'S HOMETOWN of Shenzhen is the cradle of the new China. It was little more than a fishing village when Deng emerged from exile in the late 1970s. But its location—at the mouth of the Pearl River on China's southern coast, just across the border from Hong Kong—made it the obvious choice for the first of Deng's special economic zones, where foreign trade and investment could flourish. Three decades of isolation, central planning, and political upheaval had laid waste to China's economy. It was stagnating, while Japan and Asia's "tiger economies" were booming. Deng's solution: revive Chinese communism by injecting it with limited doses of capitalism. The remedy proved more potent than anyone dreamed. Investment poured in from Hong Kong. Trucks filled with toys, textiles, and consumer electronics rumbled back the other way. Within a decade Shenzhen was a boomtown with a forest of shiny office towers and its first McDonald's.

But the experiment was nearly scuttled. The 1989 crackdown in Tiananmen Square put Deng and his allies on the defensive. Conservative rivals, decrying the collapse of communist regimes in Eastern Europe, clamored for retreat from economic reform. Deng made no significant public appearances for more than a year, fueling speculation about his health and his grip on power. Then, in January 1992, he popped up in Shenzhen. He visited a botanical garden and rubbed the smooth bark of a jade tree said to bring fortune to those who touched it. He admired Shenzhen's metropolitan sprawl from the revolving restaurant atop the new trade center. And he toured factory after factory, uttering aphorisms like "Development is the hard truth" and hailing freewheeling Shenzhen as a model for the rest of the nation. As political theater goes, it was tame stuff—a far cry from Mao's heroic swims in the Yangtze. But Deng's message got through. After a few weeks of silence, the state media fell in step and other government leaders were parroting Deng's assertion that socialism's only salvation lay in markets and material gain. In short order Beijing announced plans for a dozen new special economic zones. Foreign investment flooded in. Xu, then in his first year at Huawei, remembers the shift in sentiment. "Deng's statements set a tone," he recalls. "He made it clear to everyone that there would be further opening. For years after, we had annual growth of more than 100%."

IN HINDSIGHT, Deng's southern tour was an event comparable to the collapse of the Berlin Wall—the moment that one-fifth of humanity made its great and irrevocable leap into the global economy. By Deng's death in 1997, China's course was set. "For Deng's successors, undoing the economic reforms or ending the interdependence with the outside world is not a real option," China scholars David Goodman and Gerald Segal wrote that same year. "The clock of reform cannot be turned back."

Today Shenzhen pays tribute to Deng with an imposing bronze statue. The city's population has ballooned to eight million, outstripping Hong Kong's, and its economy is growing by double digits. Average annual income, about $3,000, is the highest of any mainland city. Nearly everyone in Shenzhen seems to come from someplace else—like 25-year-old Chen Chao, an electrical engineer who arrived three years ago from Hebei province. "The village I grew up in," Chen recalls, "didn't even have electricity until I was 7."

The reforms Deng set in motion have buoyed living standards far beyond Shenzhen, helping lift 250 million people out of poverty since 1978, according to the World Bank. "This is modern China's great triumph," says World Bank economist Deepak Bhattasali. "It is an achievement unprecedented in human history."

But maintaining that momentum won't be easy. To do so, Deng's heirs must figure out a way to generate 15 million new jobs a year, even as they keep pulling apart old socialist structures. Their surest strategy is to shift people out of the countryside, where they are unproductive, and into cities like Shenzhen, where they can be efficiently employed. Currently two-thirds of China's population, or about 800 million people, live in rural areas. The government hopes to move 400 million into cities over the next 25 years. But accommodating an exodus of such staggering magnitude implies infrastructure on a gargantuan scale. Houses, roads, schools, railways, sewage systems, power plants—to build all those things, China will need far more resources than it has. Consider: China is already the world's largest steelmaker, producing 220 million tons of the stuff last year—more than the U.S. and Japan combined. But that still wasn't enough to keep pace with domestic demand, so China imported an additional 40 million tons. China is also the world's largest coal producer, but UBS Securities analyst Joe Zhang predicts it could become a net importer of coal within the next few years. Economist David Hale warns that China's ravenous appetite for commodities will force Beijing to strengthen ties to commodity exporters in the Middle East, Africa, and Latin America, and contemplate building a navy, potentially putting it on a geopolitical collision course with the U.S. "China's economic takeoff," Hale says, "has occurred so quickly that the U.S. and other countries have not yet fully come to terms with it. All are extremely sensitive to the risk of job losses from China's export growth, but they have not devised a strategy for coping with the larger consequences."

So far, China's rise has proved more of a boon than a bane to its industrial partners. Over the past eight years cheap imports from China have saved U.S. consumers more than $600 billion, according to Morgan Stanley, and significantly lowered parts costs for U.S. manufacturers. That, in turn, has helped Alan Greenspan keep interest rates lower for longer, making it easier for America's consumers to buy houses and for its companies to invest. And the charge that China is mercantilistic, focused solely on exports, just doesn't wash. Yes, China ran a $59 billion trade surplus with the U.S. last year. But for the most part, its U.S. trade gains came at the expense of rival exporters. Moreover, the U.S. is the only country with which China runs a significant trade surplus. For most other trading partners, China buys more than it sells.

SOME CHINA OBSERVERS—and sometimes the Chinese themselves—talk as if the country is hurtling down the development path blazed by Japan and South Korea, only at greater scale and speed. Many assume that it is only a matter of time before China has its own global brands, like Sony or Samsung. But such comparisons are misleading. China's development model differs from those of its neighbors, especially in China's treatment of outsiders. Planners in Tokyo and Seoul initially shunned foreign investment and sheltered domestic firms, and they continue to do so in many ways. China, by contrast, solicited foreign investment and trade from an early stage. It took in $54 billion in foreign direct investment last year—more than Japan has attracted since the end of the U.S. occupation. Foreign capital permeates vital sectors of China's economy in a way that would be unimaginable in Japan or South Korea. Its most successful carmaker, Shanghai Automotive Industry Corp., is supported by billions of dollars of investments from two foreign partners, Volkswagen and General Motors (see "Shanghai Auto Wants to Be the World's Next Great Car Company"). Indeed, nearly all of China's major auto firms have been married off to foreign partners. Foreign giants such as Nokia, Motorola, Philips, Intel, IBM, Hewlett-Packard, and Procter & Gamble have come to China in force. Until recently China insisted that foreign investors grant local partners a majority stake when investing in China. That's still true in automaking. But the dismal performance of early ventures has prompted Beijing to loosen restrictions. In a host of important sectors—from consumer electronics to retailing—wholly owned foreign-invested enterprises are becoming the norm.

Another difference is that, unlike Japan and South Korea, which ceded ownership largely to the private sector, China's communist planners have left major enterprises firmly in the hands of the state. China's dysfunctional financial system reinforces the dominance of state-owned dinosaurs. The state-controlled banks, which allocate nearly 90% of credit, dole out funds far more readily to state behemoths than to private ventures. State firms get preferential consideration for listing on China's stock markets, lucrative tax and regulatory breaks, and first dibs on government procurement contracts.

This is hardly the formula for spawning global giants, and it's telling that three of the four Chinese companies with the best shot at emerging as world-class companies (TCL, China's No. 1 TV manufacturer; Lenovo, the No. 1 PC maker; and Haier, the No. 1 white-goods maker) are so-called red-hat companies, where majority owners are public entities content to play a passive role. The fourth, Huawei, is employee-owned. Lenovo's chairman, Liu Chuanzhi, draws a contrast between his company, free to make its own business decisions but forced to listen to customers in order to grow, and Great Wall, a struggling state-owned rival launched with more money and assured contracts. "Eventually the government figured out that if it didn't let customers choose their own computers, Chinese industry couldn't develop. That helps us, because we can pick our own people and invest our money how we like." At most other large Chinese enterprises, however, even minor decisions about personnel, investment, and strategy require approval from government or party organizations. At a World Economic Forum meeting in Beijing in September, Li Rongrong, chairman of China's State-Owned Assets Supervision and Administration Commission, acknowledged that China's state-owned enterprises have "too many mothers-in-law" telling them what to do.

Yet for all the trappings of a centralized economy, China lacks a coherent central plan. The political system grants vast power to party satraps at the provincial and municipal level, who squeeze entrepreneurs and channel resources to value-destroying dinosaurs. The quirks of China's economy, George Gilboy argued recently in *Foreign Affairs*, have left it fundamentally dependent on foreign multinationals for continued growth. U.S. fears that China is emerging as a competitive threat, he wrote, "overlook important weaknesses in China's economic 'miracle' and the strategic benefits the United States is reaping from the particular way China has joined the global economy."

In many ways the incoherence of China's development model reflects Deng's personality. The disastrous results of Mao's agriculture policies left Deng suspicious of heavy-handed central planning. He saw foreign investment as a useful expedient for raising living standards, boosting China's global stature, and shoring up the party's legitimacy. But his agenda lacked a grand design. In a memorable phrase, he described the ad hoc approach as "crossing the river by feeling the stones."

Foreign firms, too, are feeling their way, with surprising surefootedness. A 2003 survey by the American Chamber of Commerce in China suggests that the Chinese operations of U.S. firms are thriving. Three-quarters of the group's 254 member companies said their operations were profitable, and nearly half said margins in China were higher than their worldwide average. Coke, which boasts more than 600 million Chinese consumers and a 55% share of China's soft drink market, says it has been profitable in China for the past nine years. Yum Brands, with 1,000 KFC stores and 120 Pizza Huts in China, dominates China's fast-food industry. Yum Brands CEO David Novak told FORTUNE earlier this year that KFC "makes almost as much money in China today as it makes in the U.S."

By contrast, Beijing's reluctance to relinquish control of large enterprises, while throwing open the economy to foreign competition, makes life tough for China's global contenders. Lenovo, formed in 1988, rose from an IBM distributor to China's top PC manufacturer in nine years. The firm still commands a 25% market share. But as executives set their sights on overseas markets, foreign rivals, including Dell, HP, and IBM, chipped away at their domestic lead. That forced Lenovo to postpone its global ambitions and focus on core markets at home. In an effort to undercut foreign competitors, the company recently unveiled a low-end model that sells for as little as $360. TCL grabbed headlines with a deal that gives it effective control over the TV-manufacturing arm of France's Thomson Electronics, making it the largest TV producer in the world. But while CEO Li Dongshen plots global strategy, he's battling to keep TCL profitable at home, where it is fending off competition at the high end from the likes of Sony, Sharp, and Philips, and at the low end from more than 20 Chinese rivals subsidized by cheap credit from provincial banks.

CHINA MARKED the centenary of Deng's birth in August with a cavalcade of films, TV documentaries, and ceremonies. Shenzhen rang in the anniversary with a concert of 200 grand pianos. But on the eve of an important September party conference, Deng emerged as a curious totem in a political tug-of-war between his two handpicked successors. Former President Jiang Zemin, who remains head of the party's powerful central military committee, has invoked Deng's memory to support his view that China's economy can grow much faster and that recent policy measures to rein in the economy are misguided. Hu Jintao, the current President, has expressed concern that rapid growth has resurrected a gaggle of old demons—inequality, inflation, crime, and greedy landlords—and instead lauded Deng for his willingness to surrender positions of formal authority. Their dispute about growth echoes disagreements that propelled Deng to Shenzhen a dozen years ago, but this time the stakes are different. Since Deng, China has felt its way too far into the river. The stones may be sharper and the current swifter, but there is no chance of turning back to shore.

From *Fortune*, October 2, 2004, pp. 86, 88, 90, 92, 94, 96, 98. Copyright © 2004 by Fortune Magazine. Reprinted by permission.

'LET US SPEAK!'

Social debate is opening up China ... but the Communist Party still dictates.

Chris Richards tracks the boundaries of the new political space.

You could not help but notice it. A huge red banner—always popular in Beijing—strung high over the entrance of Renmin University, welcoming NGOs to a meeting about the environment. Inside more than 200 people from 150 organizations from all over China (with a sprinkling of international representatives) talk for two days about strategies to raise public awareness of environmental issues and lobby government officials.

Within minutes of arriving, a group of activists are telling me about the Nujiang river: 'As big, as amazing as the Yangtze and the Mekong, but little known outside China.' And—at the time we are speaking—soon to be dammed. Wen Bo (who was the first Greenpeace worker in China and now represents Pacific Environment) suggests I write an article about it. 'The Government takes its international image very seriously,' he calls over his shoulder as he runs to answer his mobile phone. 'We need international coverage to bring pressure to bear.'

And so it happened. After that conversation in November last year, the NI was one of a number of international outlets that publicized the proposal to dam this recently listed World Heritage Site. By April this year, Premier Wen Jiabao had called a halt to the project for further assessment.

Here was a side of China not reported in the Western press. A forum that nurtures civil society and welcomes debate. A government sensitive enough to critics to reverse a major plan.

As the country hurtles towards a capitalist economy, the ability of the Chinese people to debate social and economic issues is beginning to blossom. Many will tell you that in the 55-year rule of the Chinese Communist Party (CCP) they have never felt so free to exchange views amongst themselves.

Go directly to jail!

But this new-found freedom to speak is fragile. China is, after all, still a one-party dictatorship. Any activists worth their salt can explain the clearly defined no-go zones. Challenging the supremacy of the CCP (the Chinese Communist Party) is off-limits. So are statements that undermine the unity of the Republic. Those who step outside these boundaries can expect a reaction that's swift and brutal.

The CCP has maintained stable government since 1949 by wiping out perceived opponents before their views gain any support amongst the people. Those who propose another Party or suggest alternatives to the authority of the CCP are destined for jail.

Since 1998, there have been at least 71 people detained for their use of the internet. Almost all have been found guilty of subversion and sentenced to between 2 and 12 years jail. Most of them can be linked to one of three categories: banned groups like the China Democracy Party (whose members are amongst those receiving the harshest sentences); criticism of a high-ranking Party official; or the 4 June 1989 Tian'anmen Square protest (when the State stepped in to shut down democracy demonstrators and in the process killed and injured hundreds).[1] Indeed, just the act of demonstrating in the Square now is enough to get you arrested.

Also heading for prison are those who speak about territorial independence from China in Tibet, Xinjiang and Inner Mongolia. Government reactions range from high-powered international diplomacy to direct internal repression. The blood that has flowed from the Tibetan independence movement is known worldwide (see page 19). The lesser-known struggle by the Uighur in the northwestern province of Xinjiang—just above Tibet on the map—has also left many dead. Of those Uighur whose fates are known (and many are not) 134 have been charged with separatist-related activities over the past five years. Twenty-nine received the death penalty while the rest are serving sentences from one year to life.[2]

In deciding what other groups constitute a threat, more unlikely contenders can be caught in the net. Falun Dafa (also known as Falun Gong) is a practice of purification through exercise and meditation. As a movement, it is about individual spirituality and health as opposed to social reform. Up until 25 April 1999—when more than 10,000 practitioners held a peaceful gathering in Beijing outside the Chinese leadership compound—Falun Gong was freely practised. Within two months of the gathering, the practice was declared illegal. The ability of the movement to mobilize such large numbers of people—rather than the beliefs of its members—is thought to be what the CCP has found so threatening.[3] The movement is now brutally suppressed (See article).

This subjectivity and uncertainty means there are no guarantees. While you may think that you're on the acceptable side of the public debate line, the Government may not end up agreeing with you. A 'freedom' like this—allowing views to be expressed that can be arbitrarily and instantaneously removed later—is no freedom at all. How could anyone argue otherwise?

Then I meet 'Dan'.

Redefining democracy

Dan says it's not as black and white as I paint it. He challenges me to step outside my usual frameworks to assess the rights that the Chinese can and do enjoy. Dan's an information technology executive who got a Masters degree in the US in the 1960s and later returned to work in China. He's not a Party man himself. But he puts a view that he says is controversial to Westerners like me; a view I hear time and time again. He says that—in judging whether the people have a voice in how they are governed—there is a form of democracy in China. No, it is not a representative democracy with directly elected politicians gathering in a Parliament or Congress. Nationally, he points out how difficult that would be. In a population of 1.3 billion, if you had, say, 1,000 elected representatives, each would need to represent the views of 1.3 million people. 'But amongst ourselves we do debate our views freely. We can have an impact at a local level and on government officials. We can be heard.' He then relates his recent meeting with CCP officials in which there was a healthy difference of opinion about policy issues, argued without adherence to a Party line.

> Let a hundred flowers bloom and a hundred schools of thought contend... Questions of right and wrong in the arts and sciences should be settled through free discussion*

Mao Zedong, February 1957

Other NGOs in Beijing also report how the bureaucracy and Party welcome a debate about issues and new ideas. And, as a window to Party policy, *China Daily*—the country's officially sanctioned English language newspaper—presently promotes discussion over a range (albeit limited) of social issues. Nowhere is this more prevalent than with the environment. For it is here that the Government needs all the help it can get. Of the world's 20 most polluted cities, 16 are in China. Western analysts now estimate that 300,000 people will die prematurely here from air pollution and that more than 20 million people will have to leave their homes because of lack of access to water or degraded land over the next 15 to 20 years.

'In fact, one of the nice things about having a one-party system is that you always have a range of different views in government so that you always have someone who is sympathetic to your views,' Greenpeace campaigner Sze Pang Cheung says. Sze thinks that political lobbying in China is easier than in the US where politicians have an eye on donations rather than issues. 'A lot of officials here are prepared to take our views very seriously. They give you an opportunity to be heard.' In addition to Greenpeace, international NGOs like WWF, Oxfam, ActionAid and Médecins Sans Frontières now openly work in China as a civil society starts to grow.

Civil society emerges

Indigenous NGOs have mushroomed: between 1965 and 1996 national social associations grew from 100 to 1,800 while local groups ballooned from 6,000 to 200,000.

The attraction of these organizations to the CCP is more about their potential to offer resources that can absorb the burden of a downsized government than it is about a desire to promote community participation or listen to the people. Since the 1989 protests in Tian'anmen Square, clamps on advocacy are tight. Organizations must register. To do this, they must have a sponsor: a government body or an organization authorized by government to oversee its day-to-day activities. The search for a sponsoring agency—called 'finding a mother-in-law'—is difficult, particularly for organizations that want a national profile and therefore must find a national sponsor. And even if a sponsoring organization can be found, security is tenuous: sponsors are authorized to unilaterally terminate the relationship if the sponsored group acts or speaks out of line.

For the CCP, it is a system that encourages social assistance to individuals while keeping down groups with 'undesirable' messages. The regulations can react to prevailing conditions. After a tightening of the system in 1998, by 2000 the number of social organizations plummeted to just under 137,000.[4]

Despite these constraints, the mere process of running such a huge number of civil society groups is starting to train people about a range of issues—the rights of women, people with disabilities, rural workers, the unemployed and children. Such skills and knowledge will increase the likelihood of their becoming effective advocates when the time is ripe.

> 'He who is not afraid of death by a thousand cuts dares to unhourse the emperor'—this is the indomitable spirit needed in our struggle to build socialism and communism

Mao Zedong, 1949

The communications revolution

Transnationals like to brag that China's integration into the global economy will help propel the Government to observe human rights. The argument is that free markets and free speech are travelling companions: if one develops, the other will naturally follow. This position has many flaws (see article). Nevertheless, there are a number of indirect consequences flowing from the opening-up of China's markets that should push China closer to a free speech climate.

First, the diaspora. According to official figures, more than 20 million Chinese went overseas last year. This record number included students, tourists, businesspeople and tens of thousands of workers. They are building highways and bridges in the United Arab Emirates, Jordan and Yemen, drilling for oil in Sudan and Venezuela, mining ore in Peru

---

* Many of Mao's thought and predictions about China—some written over 80 years ago—are surprisingly relevant today. a selection of them appears throughout this issue.

and Australia and picking fruit in England and Israel.[5] Those that return will have a different view of the world and changed expectations.

Second, as China embraces capitalism, an important reason for the people to accept a curtailment of freedom to speak is disappearing.

A postgraduate student at the Peking University who talked frankly and openly to me about a range of issues, nevertheless felt uncomfortable discussing human rights in public. She described it as the dominance of 'the Big I over the Little I'. She believed that the Government is justified in setting aside individual rights if it means that the collective good is promoted.

This feeling, still prominent, is nevertheless in retreat. As the market economy pushes the gap between rich and poor further and further apart, the belief that the CCP continues to champion the collective good is diminishing. Workers are no longer able to rely on an 'iron rice bowl'. Previously expected employment rights to job security, healthcare, housing, pay and pensions are receding as the number of people employed by state-owned enterprises falls. Rising in its stead is a new industrial workforce that gets rock-bottom pay, sweatshop conditions, and little (if any) education or pension rights (see article).

The Little I battles the Big I

The health system is already based on user fees. Recent research says that, as a consequence, as much as 30 per cent of poverty in China is directly attributable to medical bills.[6] And as the State provides less and less for the collective good, the justifications for sacrificing individual rights are also retreating. As a result, public resistance is becoming

more visible. Nicola Bullard (see article) points out that: 'In 2001 the Chinese Ministry of Social Security reported an average of 80 "daily incidents" but by December 2002 this had swelled to 700 per day.' And, over time, it looks likely there will be less community tolerance for harsh action being directed at those who publicly criticize the authority of the CCP. This will leave the CCP in a much more difficult climate in which to silence its critics.

The exponential growth in the market for Chinese people to communicate with each other must also help free up expression. Officially, the mainland has more than 300 million mobile phone subscribers. They sent a staggering 10 billion SMS (text) messages during this year's seven-day Spring festival, which is 7.7 messages for every one of China's citizens.[7]

It is a communications revolution that even the Chinese authorities will not be able to contain, giving new potential to individuals who are not yet organized into a group with common goals. He Xiaopei, a lesbian organizer in Beijing since 1994, describes the mobile phone as offering an immediately accessible, but largely invisible, way for tongzhi (homosexuals) to obtain information and support in a previously hostile environment. Volunteers who provided counselling and advice for a mobile hotline were confronted by personal problems that they hadn't expected, which provoked group discussions over a range of issues. Out of this: 'We have moved from being alone to helping others, from struggling for survival to seeking liberation, from rescuing ourselves to liberating others.'[8]

For women's groups, labour and human rights activists; for those who wish to start building a democracy

with Chinese characteristics in all areas where debate is not yet welcomed, the potential is enormous.

Then there is the power of the words themselves. Not just the words being heard through mobile phones. Also those in teahouses and kitchens, factories and bedrooms, spoken by farmers and workers. The thoughts previously hidden that are now being spoken. Having taken form, these words are slowly seeping out from the private into the public domain. A growing bulk of articulated opposition pressing against the boundaries that the State has erected.

Poetic justice waiting to be done.

Notes

1. Bobson Wong, 'The Tug-of-war for Control of China's Internet' in *China Rights Forum*, No 1, 2004.

2. The full list can be read in 'In Custody: Recent Arrests In Xinjiang' in *China Rights Forum*, No 1, 2004

3. Hu Ping 'The Falungong Phenomenon', *China Rights Forum*, No 4, 2003.

4. J Howell, 'Women's Organizations and Civil Society in China' in *International Feminist Journal of Politics*, 5:2, July 2003; S Liang 'Walking the tightrope: civil society organizations in China' in *China Rights Forum*, No 3, 2003.

5. *South China Post*, 24 June 2004

6. N Young, 'The physician will not heal himself', *China Development Brief*, July 2003.

7. South China Post, 23 June 2004; *China Daily*, 28 January 2004 and 5 February 2004.

8. He Xiaopei, 'Chinese Queer Women Organizing in the 1990s' in *Chinese Women Organizing: cadres, feminists, muslims, queers*, Berg, Oxford, 2001.

China's Voice Now Rises from the International Stage

Nicola Bullard

In scenes reminiscent of the Cultural Revolution—or a motivation session for sales reps—thousands of young Beijingers wearing matching sky blue tracksuits sit in row after orderly row repeating with frightening enthusiasm the words of their evangelical English teacher, a well-dressed 30-something Chinese with an American accent: 'I want to be somebody some day.' 'We will occupy the US, we will occupy Europe, we will occupy Japan. The three markets!!!' 'Make ... money ... internationally!'

For these privileged urbanites, the slogans of capitalism have long replaced the wisdom of Chairman Mao, and individual achievement—rather than collective endeavour—is the new path to reclaiming China's lost glory.

Though it is now an economic powerhouse, China is also a political powder keg. In the vast rural hinterland—home to 800 million—the peasants and workers, too, have discarded their blue Mao jackets in favour of tracksuits. But the rural classes who symbolized the Chinese Revolution are suffering (see A single spark). In the past five years, annual growth has exceeded nine per cent, but there are signs that the economy is 'slowing' and the Government has warned of a 'breakdown in public order' if growth falls below seven per cent. Modern-day China is teetering on the edge of a social cataclysm as the Communist Party attempts to balance its warm embrace of capitalism and private property rights with the cold facts of rural poverty, mass unemployment and growing discontent in its own backyard. The Chinese Government knows this and so does the rest of the world.

China's official aspirations are peace, prosperity, democracy and unity. It rejects 'one value' or being 'led by one country' and claims to base its foreign relations on mutual respect, sovereignty and co-operation between states: an emphasis clearly directed against the role assumed by the US as global leader.

Two-faced reflections

Even though these elements are clearly visible in China's international relations stance, the way that it is expressing itself both inside and outside the country is more schizophrenic. Partly, this is a realistic response to where the country finds itself. Within its borders, the developing and developed worlds live side by side. The US is pushing for China to have closer relations with the G8: the élite grouping of major industrial powers comprising the US, France, Canada, Japan, the UK, Germany, Russia and Italy. Yet China is also a leading member of the G20: the emerging coalition of developing countries within the World Trade Organization (WTO). It is an economic dynamo of capitalism—able to out-price and out-produce all competitors. Yet the Communist Party leadership explains that this free-for-all capitalism is a necessary step on the road to true socialism. Balancing these tensions—while advancing its own interests—are the hallmarks of China's engagement in international affairs.

In 1997, Jiang Zemin—who was then the President and chair of the Chinese Communist Party (CCP)—announced that China would take a more active role in the world by 'opening up in all directions...developing an open economy, enhancing our international competitiveness, optimizing our economic structure and improving the quality of our national economy'. This was the official signal that China was open for business.

It is now so integrated into the global economy that its economic health is a matter for global concern. With an economy the size of China's, national interests can easily become international concerns. China now accounts for 48 per cent of Japanese exports—spurring Japan's recovery after 10 years of recession. Any sign of a slow-down in China sends shivers through Tokyo's financial markets. In March this year, the People's Bank of China had a foreign exchange reserve of $439.8 billion: almost as much as the US foreign debt. In the past two decades, China has opened its doors to some $600 billion of foreign investment and almost 65 per cent of its exports are from 'foreign-invested enterprises'— Chinese subsidiaries of global transnational companies.

The transition to a market economy—which started in the late 1970s—culminated in 2001 when China joined the WTO after almost 15 years of negotiations. For the Chinese leadership, joining the WTO was a matter of pride: a sign that the country was ready to run with the 'big boys'. Many developing countries welcomed its arrival in the WTO, hoping that it would stand up for the interests of the global South. China has joined with India and Malaysia in opposing the kind of far-reaching investment agreement the developed nations want. What's more—in the WTO and elsewhere— China offers an important 'demonstration effect' for other developing countries because it vigorously defends its own interests. Last year, after the US accused China of dumping, it slapped duties on several Chinese products. China's diplomatic retaliation was immediate: it cancelled a high-level mission to the US to sign orders for agricultural products.

Together with India, Brazil and South Africa, China is a leading member of the G20—the group of Southern countries that built a common position on agriculture and stood up against the European Union (EU) and the US at last year's ministerial meeting of the WTO in Cancun. With China on board, the G20 represents 60 per cent of the world's farmers, giving it a powerful legitimacy in agricultural negotiations. China's economic weight and enormous internal market also ensures that the US and the EU take the G20 seriously. Most importantly, though, China's support for India in the G20 has been a critical factor in holding the group together, not because other developing countries disagree with India's positions, but because many were politically too weak to withstand the combined pressures of the US and EU. One Chinese analyst, Xu Weizhong, has predicted that 'the spirit of unity demonstrated by the developing countries during the WTO talks will impact on the entirety of North-South relations.'

Globalization's embrace

Most developing countries try to defend their agricultural sector. However, few have the political weight to succeed. Historically, India has carried the torch for agriculture in the South, with some support from Kenya and the Philippines. With China on their side—even if its specific interests may differ—developing countries now have a better chance of holding their ground. What's more, China has 70 staff in their Geneva mission working on trade issues alone and this provides a level of technical and negotiating expertise that even many Western countries would envy.

People of the world, unite and defeat the US aggressors and all their running dogs. People of the world, be courageous, dare to fight, defy difficulties and advance wave upon wave. Then the whole world will belong to the people.

Mao Zedong, November 1964

But China (like all nations) will stand up for the interests of the developing world only when these coincide with its own interests. In 2003, Thailand signed an agreement with China to eliminate trade barriers in fruit and vegetables. Six months into the agreement, Thai farmers were complaining that they couldn't compete with Chinese produce: Thai longans (fruit similar to lychees) cost 35 baht (85 cents) a kilo, while Chinese longans reach the Thai market at 20 baht (49 cents) a kilo. The story is the same for other fruits, onions and garlic. Thais complain that China is creating obstacles—such as delays at the border and food safety checks—which make it impossible for their products to enter the Chinese market. In this respect, China's approach to trade is no different from the US or the EU, both of which use WTO rules to maximize their own gains and back this with hefty economic weight.

Diplomacy rules

No-one is willing to put pressure on China to make trade or financial concessions that could jeopardize its continued growth. At the same time, the pressure on China to open up its trade even further will be a permanent feature of its WTO negotiations. For some Chinese critics, membership of the WTO is a disaster in the making. Predictions of a rural crisis and the decimation of Chinese industry (which have proved to be well-founded) have found their way into mainstream debates. Nevertheless the Government has continued its single-minded pursuit of international legitimacy.

Although China competes economically with many countries in East and Southeast Asia, it is showing leadership in trade, security and strategic affairs. Furthermore, Asian countries feel comfortable dealing with China, knowing that they are unlikely to be lectured about shortcomings in civil and political rights.

Unlike Japan, which even at the height of its economic power refused to assume political leadership, China has repeatedly responded to emerging situations in a way that guarantees it is now accorded the level of respect usually reserved for the US. For example, China has backed the 'ASEAN+3': an Asian regional trade agreement. This is a long-term strategy to build a regional trade block that could one day counter the Free Trade Area of the Americas (FTAA) and the EU.

China is also playing an important diplomatic role in the 'Six party talks' on South Korea and mediating between its traditional rival India and its traditional ally Pakistan. No doubt, China's present assessment is that regional stability—rather than ideological point scoring—serves its interests best.

Its 'engagement' in international relations now goes beyond the obvious spheres of the Security Council, the WTO and the UN, as it seeks to influence other international debates. This could be viewed as a strategy of China defining its own problems in the international arena rather than leaving the field open to its competitors and critics.

In April this year, China applied to join the Nuclear Suppliers Group and attended for the first time a meeting of the Missile Technology Control Regime: two groups it previously dismissed as US-dominated cartels controlling access to nuclear technology. By joining these small clubs, China is strengthening its relations with the US, Europe and Japan, and at the same time guaranteeing access to state of the art nuclear energy technology.

China also lobbied hard for a vice-presidency in the International Labour Organization—a move derided by labour rights activists who suspect its motive is to control the debate on international labour standards while at the same time deflecting criticism from its own labour conditions.

Officially, China denies any hegemonic or global leadership ambitions; yet economically and politically, it is positioning itself to be the rising power of the 21st century. So far, China's only voice on the international stage is the official voice of the Government. However, it is obvious that the impacts of rapid economic transformation are creating social divisions and dissent. Clearly, important things are happening in China that are not being reported either by the Communist Party or by the business press. While the 'conditions' may not be right for the growth of a democratic movement inside China, in the new era of the internationalization of the anti-war and anti-globalization movements, Chinese workers and peasants are the missing voices that still need to be heard.

Nicola Bullard works with Focus on the Global South, an activist research NGO based in Bangkok, Thailand.

LITTLE EMPERORS

China's only children—more than 100 million of them—make up the largest Me Generation ever. And their appetites are big.

BY CLAY CHANDLER

IT'S BACK-TO-SCHOOL DAY FOR 3-year-old Feng Qiyi, and things are off to a bumpy start. The bites inflicted by mosquitoes that feasted on him overnight have begun to itch. The smelly ointment smeared on him by his grandmother isn't helping. And now his grandmother and the family driver have deposited Qiyi at the Beijing Intelligence and Capability Kindergarten, waved goodbye, and left him to fend for himself. "Children," asks the teacher, "does everyone remember the baby-chick dance we learned before vacation?" Qiyi's classmates flap their arms and sing. The best performers are invited to the front of the class and rewarded with candy. Qiyi isn't picked. "I have candies at home!" he blurts. "Many, many candies!"

And so he does. As the only child in a well-to-do Beijing household that includes his father, his mother, and his mother's parents, Qiyi is used to getting plenty of candy, lavish praise from grownups, and pretty much anything else he wants. Indeed, nearly every aspect of Qiyi's short, comfortable existence has reinforced the notion that he is the center of the universe. That may not be the most rational view of the cosmos, but it is one shared by millions of other Chinese youngsters born since 1980, the year China's social planners issued a sweeping edict limiting each family to just one child. Beijing touts the one-child policy for its success in reducing poverty and raising living standards. Government demographers credit it with preventing nearly 300 million births over the past 25 years and lowering the average number of children per woman to two from more than six. But it is widely lamented that

the policy has had a nasty side effect: spawning a generation of selfish brats.

The Chinese have a special name for those tots: *xiao huangdi*, or "little emperors." They are regularly deplored in the state-run press. China's children are growing up "self-centered, narrow-minded, and incapable of accepting criticism," declared Yang Xiaosheng, editor of a prominent literary journal, in a recent interview in the *Beijing Star Daily*. Wang Ying, the director of Qiyi's kindergarten, concurs: "Kids these days are spoiled rotten. They have no social skills. They expect instant gratification. They're attended to hand and foot by adults so protective that if the child as much as stumbles, the whole family will curse the ground."

> **CHINA'S LITTLE EMPERORS** will emerge as a driving force of lifestyle and market trends—not only in Asia but in the U.S. and Europe as well.

The one-child policy has been loosely enforced in the countryside, where more than two-thirds of China's people live. In remote areas it's not uncommon to find farm families with as many as five or six children. But in cities one child per family remains the norm. Demographers estimate that of Chinese under age 25, at least 20%—about 100 million—have been raised in one-child households. That's only a sliver of China's 1.3 billion people. But for foreign companies hoping to capture the hearts and minds of Chinese consumers, little emperors are a crucial market vanguard. They're confident, cosmopolitan, and eager to try new things. And unlike their rural cousins,

they have the financial wherewithal to gratify their whims. An April survey by Hill & Knowlton and *Seventeen* magazine of 1,200 students at 64 universities in Beijing and Shanghai found that six in ten reported spending more than $60 a month on "unessential items"—a staggering sum given that monthly per capita income in those cities averages less than $250. Many analysts predict that as their purchasing power increases, China's little emperors will emerge as a driving force of lifestyle and market trends beyond China—not only in Asia but in the U.S. and Europe as well. Says Conrad Persons, a consumer-trends analyst at Ogilvy & Mather: "Get ready for the biggest Me Generation the world has ever seen."

The key to understanding this generation is to recognize that it is a breed apart. Everything is different for these kids; the sibling dearth is just the start. China's little emperors and empresses have come of age in an era of unprecedented prosperity. Their parents and grandparents endured years of famine under Mao's disastrous communal agriculture policies and the chaos of the Cultural Revolution. They remember the trauma of the crackdown in Tiananmen Square. But for the Chinese born since 1980, that's ancient history. For youngsters in Beijing and Shanghai—and even second-tier cities such as Dalian, Chengdu, or Kunming—each passing week brings a gleaming new shopping complex, restaurant, highway, or residential development.

And it's not just that they're better off. They're better informed. Although the state retains a firm grip on the media and has spent billions on technologies enabling police to snoop on Internet users, Chinese kids today know more about the world beyond the Middle Kingdom than any previous generation. They are avid techies, making ready use of mobile phones, the Internet, and electronic gizmos of every sort. They track the latest fashion fads in Tokyo, swoon over pop stars from Hong Kong and Taiwan, and watch Hollywood blockbusters on knockoff DVDs.

Consider Wang Qi, a 19-year-old hip-hop music producer scouting for a new pair of Air Force 1 sneakers at the Nike shop in Beijing's trendy Xidan district. Wang—who prefers to be addressed by his street name, "Jerzy King"—moved to Beijing three years ago from the coastal province of Shandong. A music school dropout who has never set foot outside China, he totes a mini-disc player loaded with Eminem, Puff Daddy, and Fabolous. On this particular day he's looking phat in a blue-and-white fleece jacket bearing the logo of the Toronto Maple Leafs. Where'd he get the getup? The U.S., of course. He spotted the jacket and matching pants on www.footlocker.com, paid for them using a Western Union credit card, and had them express-mailed to Beijing via the U.S. Postal Service. At last count Wang's wardrobe included more than 100 jackets and jerseys. Using an Intel-powered computer he assembled himself, he spends hours each day tracking the latest hip-hop styles over the Internet and does a brisk business importing and reselling American street wear (most of it originally manufactured in China) to Chinese friends.

China's little emperors are weaned on cheeseburgers from McDonald's, pizza from Pizza Hut, and fried chicken from KFC. Their enthusiasm for fast food is fattening their own bottoms as it fattens multinationals' bottom lines. In big cities one in five children under 18 suffers from obesity. The weight-loss business is booming. At the Aimin Fat Reduction Hospital in Tianjin, a former military institution that launched China's first weight clinic in 1992, doctors treat 200 patients, most of them under 25, with a daily regimen of acupuncture, exercise, and healthy food. Fifteen-year-old Liang Chen reports proudly that he has lost 33 pounds in less than a month at Aimin. But he can't stop reminiscing wistfully about his regular visits to KFC. (Indeed,

his favorite T-shirt is a souvenir from China's largest KFC store.) "I used to be able to eat an entire family-size bucket all by myself," he recalls. "Just one?" snorts his roommate, 14-year-old Li Xiang. "That's nothing. I used to be able to eat four buckets—sometimes five, if I didn't eat the corncobs and bread."

> **THEIR ENTHUSIASM for fast food is fattening their bottoms as it fattens multinationals' bottom lines. In big cities one in five children is obese.**

TV commercials starring Jay Chou, a pop heartthrob from Taiwan, have helped persuade millions of mainland teens to drink Pepsi. China's urban youngsters are easily dazzled by fast food, flashy clothes, and the glitter of foreign lifestyles, says Yi Wei, author of *Unbearable Happiness*, a book about Chinese youth. "This is a fragile generation," Yi argues. "They grow up sheltered, without any concept of sacrifice or self-control." Among parents, the nearly universal complaint is that young Chinese haven't learned to "eat bitterness," a common expression for enduring hardship.

That criticism isn't entirely fair. Little emperors bear the weight of an entire family's expectations on their tiny shoulders. In today's China, urban middle-class children come under enormous pressure to excel academically from as early as 5 or 6 years of age. Parents prod their offspring through a gantlet of costly lessons: piano, English conversation, martial arts, even golf.

Beijing insurance saleswoman Leng Yaqun has drawn up a daunting study schedule for her 13-year-old son, Bingyang, during his six-week summer vacation. It begins at 9:30 A.M. with an hour of homework assigned from school. Then comes an hour of extra math drills and an hour memorizing *The Analects of Confucius*. After lunch there's an hour for penmanship and an hour for reading (among the titles she has assigned: a Chinese translation of Tony Robbins's *Awaken the Giant Within*). The day wraps up with an hour for listening to recordings of classic texts in English, including *Romeo and Juliet*, Darwin's *The Origin of Species*, and the lectures of Herbert Marcuse.

"I tell him, 'You are so lucky,' " says Leng. "When I was his age, we had noth-

ing to study except the basic school texts. But Bingyang can buy all kinds of books. He can take extra courses. He can get a private tutor and learn about anything he wants from the Internet."

Bingyang does his best. But the Bard goes in one ear and out the other; he gave up on Robbins after the first few pages. "Too boring!" He'd just as soon use the computer to play videogames and the break from school to pursue his real passion: assembling model cars. Many parents confess that they are pushing their kids so hard out of a desire to compensate for the opportunities they never had. But it is also true that advancing through China's educational system requires hours of drilling. The nation's university system has places for only about half of those who apply. Chinese families spare no expense to help their children pull high test scores. Some rent air-conditioned hotel rooms so that their little Einsteins can study in comfort before exams. And as they venture into the workplace, new graduates are figuring out that their status as only children, combined with the inadequacy of China's public pension system, means that responsibility for caring for aging grandparents and parents will fall to them. China's little emperors are "born into a kind of fairyland," says Ellen Hou, a planner at the Shanghai office of global advertising giant TBWA. "But from the moment school starts, their lives become a struggle."

At Qiyi's kindergarten, one of Beijing's best, the struggle starts early. Founded in 1996 by a prominent Chinese educator, the school takes up to 400 pupils each year, from 18 months to 6 years. Tuition and other fees run $6,000, about double the income of the average Beijing household. In addition to instruction in math, science, art, Chinese language, and music, the school offers lessons in English, golf, and tennis. The school puts a premium on discipline, competition, and proper manners. To help young students grasp the virtues of self-denial, teachers offer them candy early in the morning with the promise that they'll receive a second piece if they can refrain from eating the first one before lunch. "By age 3, most students have learned to control their desires," says Wang. That is also the age by which they are expected to be able to recite pi out to 100 digits.

Some thrive in this sort of environment. The Intelligence and Capability Kindergarten's star pupil is 5-year-old Ying Rudi, the son of China's most popular television host and sitcom director and the grandson of a famous actor who was once the country's Minister of Culture. Rudi began piano lessons at age 3 and is now a celebrated prodigy. A poster in the school's front hall congratulates him for sweeping piano competitions this year in Beijing, Chicago, and Geneva.

AT 13, XU QIUSHI is a hardened junior achiever. She, too, is an award-winning pianist—and a formidable competitor in Korean kickboxing. On a recent evening she mesmerized guests gathered in the family living room with delicate renditions of Beethoven and Debussy, then—after a few mouthfuls of fried chicken and a quick costume change—launched into a boisterous demonstration of her head kicks. Xu hasn't the slightest doubt that she will be admitted to Beijing's elite Tsinghua University, and she is already contemplating postgraduate study in Paris. Her long-term goal is to become a diplomat—like her idol, piano-playing Bush security advisor Condoleezza Rice.

Twenty-one-year-old Li Cheng, by contrast, just wanted off the academic treadmill. Li hails from a long line of scholars. His mother and father are nuclear scientists, and his parents and grandparents, he says, "have more advanced degrees than you can count." Li has found his calling as a veejay with the Chinese subsidiary of Rupert Murdoch's Channel V. He says his family is appalled. But the pay is good, and there are other benefits. On a breezy evening in August, Li's career choice puts him before a crowd of 10,000 pumped-up fans at the Summer Shake, a raucous Channel V music festival on the outskirts of Shanghai. His appearance onstage, in a sleek tank top and

straw fedora, provokes a wave of high-pitched squeals. Offstage he is mobbed by teenage girls begging for his autograph.

Rebellion has also proved a shrewd career move for Chun Shu, a 21-year-old high school dropout who shot to fame two years ago with *Beijing Doll*, a sexually explicit novel recounting her search for love, truth, and the perfect punk band. *Beijing Doll* lambastes the education system for draining the passion from China's youth and pours scorn on Chinese who came of age during the Cultural Revolution for their mindless preoccupation with academic achievement. Before being banned, the book sold hundreds of thousands of copies and was embraced by disaffected students throughout China.

"Don't talk to me about how kids my age have more freedom," Chun rages, stubbing out a cigarette in a Beijing cafe. "The day China abolishes these stupid entrance exams, that's when you can talk to me about freedom." For all her invective against scholarship, Chun herself can be surprisingly cerebral. She has strong views about George Orwell, Henry Miller, and China's own Lu Xun. But she is apt to lurch from a discussion of great books into a rant about the relative merits of Converse vs. Nike, or Courtney vs. Avril. "Does anyone in America believe Avril is a true punk?" she wants to know. "The silky hair, the perfect skin, that little nose—real punks don't look like that."

But Chun also confesses doubts about her own punk credentials. "I used to think of myself as a lover of music, literature, and ideas, who cared nothing for fashion or style," she wrote in a recent posting on Sina.com, a popular Internet portal. But "rereading *Beijing Doll* today, I realize that, even more than most girls, I have been obsessed with material things. I could spend money like water just to buy a handbag with a fancy brand name. I am easily bewitched by television advertisements and can be as vain as a peacock. Am I less able to debate the ideas of Dostoevsky because I'm wearing beautiful panties?"

China's old guard doesn't begrudge Chun and other young rebels their choice of lingerie, as long as what they debate in

it doesn't verge into topics that are really taboo—like whether they ought to have the right to choose their political leaders. Authorities have taken solace in public opinion surveys suggesting that Chinese born in the 1980s, fed a steady diet of patriotic propaganda in public schools, profess more ardent devotion to their country than those born in the 1970s.

And yet the spread of consumer culture is quietly subversive. The fundamental idea of communism, after all, is the subordination of the individual to the collective; consumerism presupposes the reverse. Gilbert Lee of Research International, a Beijing marketing consultancy, advises foreign firms hoping to win over young consumers to play precisely to their yearning for self-expression, crafting messages that stress values of individuality, freedom, and physical attraction. Lee sees a stark divergence in the preferences of Chinese consumers born before 1980, who are likelier to seek out products that help them arrange their lives in a more secure and orderly way, and those born after the one-child policy, who are looking to project themselves, establish their uniqueness, and make a positive impression on others. In a society where children are indulged as infants and grow accustomed as adolescents to asserting their identity through spending decisions every day—what to wear, what to eat, what music to listen to, what to drive—how much longer before some also begin clamoring for a say in other things: property rights, taxes, the quality of public services?

Shen Jie, a sociologist at China's Academy of Social Sciences, says that all the hand-wringing about little emperors is overdone. They haven't gone soft, he argues, and aren't about to foment revolution. Like kids everywhere, he says, they're just trying to find their way: "If you judge Chinese kids today by the standards of yesterday, then sure, they come up short. They don't like to suffer. They aren't used to eating bitterness. But so what? Is that the main thing China needs right now—more people who are good at being miserable? These kids have other skills. They're creative and opinionated, and have the courage to do new things. Shouldn't that be grounds for hope?"

China Fears a Baby Bust

After 25 years of the one-child policy, the nation risks producing too few children. But many parents have decided one is enough.

By Don Lee

SHANGHAI—Zhang Xiaofeng, a 28-year-old who runs a steel business here, doesn't need anyone to tell him about the joys of fatherhood. He eagerly pulls out his wallet and displays pictures of his 2-year-old son, Chengqi, with his mother's big, round eyes.

Zhang often passes up nights out with his buddies so he can race home to play with Chengqi.

"I bathed him, fed him and changed his diapers. I did all those things," he says proudly.

But ask Zhang whether he and his wife want another child, and his jaw tightens. Raising another child would be tiring, time-consuming and expensive, Zhang protests.

He sums it up: "One is enough."

For the last quarter-century, China's one-child decree has been criticized by citizens and outsiders alike as draconian. But as the nation takes steps to ease its policy, with some cities encouraging certain families to have a second child, people like Zhang illustrate how difficult it will be for the government to root out ingrained attitudes.

Having only one child is now widely accepted, especially among urban residents. In Shanghai, China's largest city, a recent government survey of about 20,000 young people found that more than 80% preferred to have just one child. Another 5% said they wanted no children at all.

The findings worried officials all the more because this metropolis of 17 million was already grappling with plummeting births. Last year, about 57,000 babies were born in Shanghai, but there were nearly twice as many deaths. Such a large gap has profound implications for the future workforce and for an aging society. At the current rate, the city would face labor shortages, even with its sizable inflow of migrants.

Shanghai, with its affluence, fast-paced lifestyle and gleaming skyscrapers, isn't a typical Chinese city. But researchers believe that its demographic quandary typifies what other areas in China will confront in coming years: a society with too few children.

Shanghai Eases Policy

Keenly aware of that, Shanghai's Population and Family Planning Commission reformed parts of the one-child law last spring, making it easier for people such as remarried couples to have more children. Zhang and his wife can have a second child because they come from one-child households.

Shanghai officials added 11 exemptions to the one-child policy, including removing the waiting period for certain families. Also, in the fall the city scrapped a financial reward that had long been given to childless couples.

So far, the changes in Shanghai have spurred only about 100 more people a month to seek government permission for a second child. The commission's director said her office was prepared to handle 10 times that number.

"It's a big problem," said Zhang Henian, associate professor at the Shanghai Academy of Social Sciences, a government think tank. Rapid economic growth and the rise of urban society are major factors underlying Shanghai's low fertility rate, but Zhang said it's hard to reverse 25 years of heavy promotion of the idea that one child is best.

"In the past, the goodness of the one-child policy was overly stressed," he said.

Shanghai's prevailing attitude toward childbirth isn't representative of all of China. Couples living in some rural areas have long been allowed to have two or more children, and many continue to prefer larger families.

In the mountainous southern province of Yunnan, there were 17 births per 1,000 residents last year—compared with four for Shanghai, five for Beijing and 12 for the country as a whole. (The U.S. birthrate was about 14 per 1,000 residents.) To bring Yunnan's birthrate more in line with the nation's, the government is rewarding some families that stick to one child with a pension and cash for school tuition.

At the same time, other regions of China are experimenting with ways to encourage childbearing. Beijing's municipal government recently drafted new regulations that would increase time off and improve insurance policies for older women taking maternity leave. In east China's Zhejiang province, one city sharply lowered penalties for those who break the one-child rule, which are typically several times a family's annual income.

Such geographical disparities make it difficult for the central government to formulate a new national birth-control policy. At this point, Beijing hasn't spelled out what local jurisdictions can do, but it's understood they can't stray too far from the existing national policy.

Officials say the one-child law has reduced births by about 300 million and lifted living standards. China, with 1.3 billion people, remains the world's most populous country.

But the policy has been condemned for leading to female infanticide and forced sterilization, and it has produced a troubling gap in the number of boys and girls. Male heirs are considered desirable. Enforcement of the law has been uneven and often cruel. Last month, a couple in Jiangxi province complained that local officials destroyed their home after they were unable to pay a fine of 16,000 yuan (about $1,935) because their daughter broke the one-child policy.

Seeking a New Rule

Today, scholars say, there is agreement among academics and leaders of the Communist Party that the one-child rule is no longer good for China.

Two years ago, Beijing took an important symbolic step in softening the harsh language of the law, saying that those having unauthorized babies would no longer pay "fines" but a "social compensation fee." More recently, discussions of overhauling the family-planning policy have grown more intense, fueling speculation that the government will adopt something akin to a two-child law.

But no one knows when the one-child rule will be discarded. Most experts think it's several years away, and even then there would be no guarantee that a change would make a difference in places where the effects of the law are most problematic.

"Whether a new policy could be implemented is another issue," said Peng Xizhe, a Fudan University professor who is among about 300 scholars advising the Chinese government on population planning.

"A low family fertility rate is very difficult to raise," he said. "Even when you ask young people to have one child, they will refuse. That's a big change in social pattern."

Even many older Chinese who grew up with multiple siblings and had several children are convinced that one is enough.

Steven Liang's mother feels that way. That means when the 30-year-old engineer gets married, she won't be pressuring him and his wife to give her more than one grandchild.

"Because the one-child policy has been around for so long, we're all used to it," said the 63-year-old woman, who asked that her name not be used. "In my generation, two or three was a good number," she said. As she spoke in the lounge of a Shanghai theater, Liang sat beside her and nodded. "Nowadays, one child is good."

But for people like Jessica Zhang, even one is too many.

A 30-year-old editorial director of a Shanghai fashion magazine, Zhang said she and her husband had decided not to have a baby. Their reasons: They can remain the center of their home, focus on their careers and enjoy more free time. They don't have to deal with the rising cost of educating a child, and they can decorate their home as they wish.

"Of course I may feel lonely when I'm old and be envious of people with children," Zhang said. "But I will have earned much more happiness when I was young."

She also takes issue with those who believe that having kids will provide financial security. "It's a stupid idea that children will take care of you," she said.

But who will? That's a question many are asking nowadays in China. With improved healthcare and living standards, the elderly population has grown sharply in such places as Jiangsu and Zhejiang provinces and the city of Tianjin.

In Shanghai, 2.6 million seniors make up about 16% of the city's population, which far outstrips the worldwide average of 7%. Their swelling ranks are straining the city's pension and social service systems. At the end of last year, there were only about 450 senior nursing facilities in Shanghai, with enough beds for just 37,000 people. Although a higher birthrate won't solve this problem, more young people entering the labor force would generate taxes to help pay for health and social services.

Shanghai officials downplay the severity of the population imbalance, saying the city's troubles will be cushioned by its 3 million migrants. But surveys show that group, many of them young workers from hardscrabble rural areas, isn't more inclined to have children than Shanghai's registered residents.

In any case, it is unlikely that Shanghai will see many more couples producing more than one child.

That worries Zhang Qi, an assistant headmistress of a middle school in the city. She observes the students in her school—almost all from one-child families—and fears for the future.

"Every student thinks she's in the middle of the circle. They consider little of others," Zhang said. "I think it's a great harm to our nation."

Zhang, 43, has only one child, a daughter in her teens. She grew up with five brothers and sisters and yearns for another child. But she doesn't qualify for any of the exemptions to the one-child policy. She was willing to pay the social compensation fee but decided against it because she would feel immense pressure to give up her government job, as lawbreakers typically do.

"The condition in China seems to be improving to have another child, but I think it's impossible for me now," Zhang said, noting that violators of the one-child law are still scorned by co-workers.

On a recent Saturday afternoon, Guo Xiaoli sat beside her 7-year-old daughter, Jiayi, at a McDonald's restaurant in Shanghai's fashionable Huaihai Road shopping district.

Jiayi was among a handful of children at the two-story restaurant, which was jammed with customers.

Guo, 33, comes from the southern province of Guangdong, where people tend to have larger families. But her husband is a native of Shanghai, and he's dead set against that. Part of it is social pressure, she said.

"My husband's mother keeps saying, 'One is enough,' because everybody around us has just one."

As time has passed, Guo has come to agree with her husband. Child-rearing is costly, exhausting and frequently annoying, she said.

Sometimes, Jiayi pleads with her mom for a sibling to play with. For such occasions, Guo said she has a ready answer: "If you have a little sister, I will like her, not you."

Holding Up Half the Sky: Women in China

"In analyzing the impact of economic reforms on women's lives in China, the government would evidently prefer to discuss its achievements in promoting gender equality and not the chronic discrimination problems that remain. At the other end of the spectrum, Western feminist discourse on China often presents an alarming portrait, replete with stark statistics on the skewed male-female birth ratio, domestic violence, and increased trafficking in women." Exactly how far have women in China come? And how far will Beijing let them go?

SUSAN PERRY

First in an Occasional Series: Focus on Women

As the members of Gao Yinxian's family prepare to take a family photograph, thousands of years of Chinese culture are evident in the seating arrangements. Gao, whose husband has died, is the family's oldest member; she is seated in the middle. Her eldest son, who positions everyone for the photograph, is seated to her right, and her first grandson is to her left. Gao's daughter-in-law and granddaughters remain standing behind the seated members of the family: they have second-row status. They will not earn the right to be seated until their own sons assume the responsibilities of family leadership.

Although Gao's family lives in a remote corner of Hunan province, these gender dynamics are universal in China. Women have been relegated to second-row status throughout the country's history, with their social standing determined by their reproductive role and the clout wielded by their husbands and sons. Yet the last 50 years have seen the role of women in China change more than during the last 5,000 years. These changes result from a variety of factors, the most important of which have been the modernization of Chinese society and the 1949 Communist revolution. Women have moved out of the home and into agrarian and industrial production—and into the professional spheres traditionally reserved for men. In doing so, they have been both assisted by the government and hampered by recent economic reforms.

While women have benefited from an increasing array of new opportunities, their liberation remains incomplete. Since 1990 there has been a noticeable nationwide slowdown in the move toward gender equality. This stems in large part from the economic reform program that has transformed China from a poor, isolated nation into a burgeoning economic powerhouse. The reforms have widened the gap between the urban and rural populations in terms of access to education, health services, and technology. At the same time, booming private businesses and farming for profit have created a new class of urban and rural elites whose exploitative tendencies recall feudal China. The government now acknowledges that men and women have also been affected differently by the introduction of free market policies.

THE BEGINNING OF CHANGE

Most women rarely ventured beyond their own doorsteps in imperial China, their feet bound and their lives one of domestic drudgery. They had few possibilities of earning an income aside from household handicraft production and, in extreme cases, prostitution. Marriage was designed to control female labor; the young bride was immediately transferred to her husband's family home, where she assumed domestic and reproductive obligations that lasted until old age.

During the late nineteenth century, foreign incursions into China's coastal cities resulted in their partial industrialization and altered traditional marriage patterns. In the Canton delta, foreigners and Chinese alike owned textile factories employing thousands of young women. Their nimble hands and good eyesight were essential attributes for silk reeling, which earned them a cash income. Consequently, many of these workers practiced a form of marriage resistance or delayed marriage; they continued working for three years after their wedding, at which point they would move to their husband's village and begin childbearing.

The increasing modernization of urban China, symbolized by the 1919 May 4th Movement and its call for China's political, social, and cultural transformation, created new opportunities; schools and universities were founded to train female students as teachers, social workers, nurses, doctors, activists, and politicians. The nascent urban communist movement gained considerable female support because it called for a change in the traditional Confucian gender relations that relegated women to a subservient existence. Beginning in the late 1920s, the Chinese Communist Party promoted liberal marriage laws in its peasant base areas, which generated support for the movement among rural women. With the 1949 Communist revolution, both urban and rural women were expected to participate fully in the socioeconomic transformation of society. Because Marxism emphasized class as the agent for change, however, Chinese women were encouraged to enter the workforce to gain their liberation. They were mobilized to contribute to socialist construction, rather than fight for gender equality.

The upheavals of the Cultural Revolution between 1966 and 1976 made the majority of Chinese women eager to embrace Deng Xiaoping's brand of "market socialism" in 1978. Yet it was during the Cultural Revolution that many women came to the forefront of political decision making. During this period femininity was criticized as a "petty bourgeois characteristic." Female Communist Party members were expected to be militant and ambitious. Shrill but effective political propaganda emanating from Chairman Mao Zedong's wife, Jiang Qing, helped double the number of women in local, provincial, and national government. Later, many of these women admitted that they had felt compelled to serve the Chinese Communist Party and had entered politics with little enthusiasm.

For Chinese women this is a time of opportunity and exploitation

Today 14 percent of the Communist Party's membership is female, but only 7.5 percent of the Central Committee is made up of women. Female deputies constitute 21 percent of the National People's Congress, but only 9 percent of its powerful Standing Committee. Three female ministers have been appointed to the State Council, the country's main administrative body; they represent 7 percent of all ministers, commensurate with the world average.

The most influential female minister is undoubtedly Wu Yi, who heads China's Ministry of Foreign Trade and Economic Cooperation. She has reached the highest echelon of political power because of her competence and ability to weather political storms. While she might serve as an example to younger women working their way up through middle-level management and government positions in China, overall interest in politics appears to be waning. Many urban professional women appear unwilling to make the necessary sacrifices to climb the political ladder when their time could be better spent making money.

GENDER ECONOMICS

For Chinese women this is a time of opportunity and exploitation; those with a good education or strong motivation are most likely to succeed. Women managers have proved to be extremely effective at running large and small-scale enterprises in the new market economy. A recent study by American and Chinese scholars has shown that they score the same level of managerial motivation as male colleagues in organizational behavior tests, especially on the desire to exercise power and stand out from the group. Opportunities fostered by economic reform and the government's one-child policy, which effectively limits women's child-care responsibilities, have clearly had a positive impact on their managerial motivation. These women form an elite class, set apart from the masses by their university education and professional experience.

At the other extreme are women who work on the assembly line. China's export drive during the 1990s has been dominated by goods traditionally produced by female workers. In southern China women make up the bulk of the labor force in light manufacturing—especially in textiles, which is one of the country's major exports. Many of these workers are young peasants who have fled agrarian underemployment for cash jobs and a bit of excitement in the cities and towns. In the southern city of Shenzhen, for example, women make up one-half the migrant workforce.

Some of these women will return to the countryside with considerable savings. Nearly 50,000 women who have returned to their villages in Guangdong province are now the owners of individual or private businesses. According to one study published in China, "when [women] return to the village, not only has their appearance changed but they have also become stronger in character, are no longer afraid of going out in public, are more poised and independent. They have a relatively stronger sense of business and affairs."

Businesses created by the returning women workers form the backbone of the expanding private sector. Throughout China, women account for approximately one-third of the 14 million self-employed rural individuals engaged in commerce and service trades. In the villages, female employers favor female workers for services and light industrial production because, as one woman factory owner noted, they are "obedient, detail-oriented, and do not get into fights." They are also cheaper, earning less than 77 percent of the pay given to men doing the same work. Like their great-grandmothers in the textile mills at the turn of the century, these young peasant workers are a malleable workforce, with few dependents and a willingness to work long hours for low wages.

The new private sector frequently denies women health benefits and maternity leave. Working conditions are unsafe, since national and provincial safety standards are not enforced. In 1992, for example, dozens of women were killed when a fire broke out in their Guangdong toy factory; windows had been barred and doors locked to prevent them from leaving before their 12- to 15-hour day was over.

Women working in the shrinking state sector still receive health services and three months guaranteed maternity leave. Nonetheless, unequal pay and fewer chances for promotion are also chronic in this sector; women workers are perceived as less reliable than their male colleagues because of their household responsibilities. This perception is misguided, since urban Chinese men are far more likely to share household and child-care tasks with their wives than their American counterparts; according to one official study, men in the city of Shanghai perform up to 80 percent of household and child-care chores.

Unemployment is disproportionately high among women. National surveys indicate that the average laid-off worker in China is female (60 percent), works in an industrial enterprise (60 percent), is between the ages of 35 and 45 (67 percent) and has a junior high school level of education. Moreover, Prime Minister Zhu Rongji has targeted the largest female employer, the state-owned textile sector, for major structural reforms. In the next three years, new jobs must be found for 1.2 million female state textile workers.

The All-China Women's Federation, generally considered a mouthpiece for party policy, has been surprisingly sharp in its criticism of inadequate government programs for the retraining of unemployed women. The "Pioneers Project," run by the federation since 1995, offers professional retraining for nearly half a million women each year; 230,000 were placed in new jobs in 1997. Many of these women move into the service sector, seeking employment in fields as varied as hairdressing and computer programming. Economic security is considered so vital in today's rapidly changing social climate that several female college graduates interviewed for this article claimed that taxi driving and prostitution—the old stand-by in times of extreme duress—were the only jobs that would guarantee them a high cash income and no layoffs.

"FEMINIZED" AGRICULTURE

Nearly 80 percent of China's female population lives in the countryside, where gender equality has been slow to develop. The launching of the household responsibility system in the early 1980s as part of China's reform program has helped maintain rural gender inequality. The system disbanded agrarian collectives and allowed peasant families to grow crops both for themselves and for the state. In many respects, this policy has reinforced the traditional social structure, which also relies on female labor. Women continue the Confucian tradition of "marrying out": moving into their husband's household after marriage, where their status remains that of a secondary family member even after the birth of their first son.

With the integration of women into the agrarian workforce, all but elderly women and very young girls bear the burdens of farm work, child rearing, and domestic chores. Moreover, with the increase in migration to urban areas by men and young women, middle-aged married women have taken over the bulk of farm work in all but the poorest areas, resulting in a "feminization" of agriculture. The Chinese government estimates that women shoulder between 40 and 60 percent of the workload in the fields, while some Chinese scholars believe that women now perform between 70 and 80 percent of all rural work.

The feminization of agriculture may, in the long run, adversely affect traditional patriarchal values. Many women who become the heads of their household when their husbands depart as migrant laborers often retain their position once their husbands return.

WOMEN AND THE STATE

The government's one-child campaign is a long-standing example of effective state intervention in the lives of Chinese women. Introduced on a broad scale in 1980, this nationwide campaign has attempted to reduce China's galloping population growth by restricting couples to only one child. The government's draconian measures appear to have lowered fertility in urban areas. Women who become pregnant a second time must pay a substantial fine; risk losing their employment, housing, or health benefits; and may be required to undergo forced abortion or sterilization. Their husbands may also be penalized at work. These potential sticks, along with the carrots of pay bonuses and promotion eligibility that accrue to one-child families, have altered fertility behavior to the extent that the majority of city couples willingly have only one child.[1]

Although the campaign has altered fertility behavior, it has done little to educate urban women about their sexual identity. Despite the diversity of new material on sex-related issues and a great deal more discussion on the subject in the 1980s and 1990s, the dominant discourse on female sexuality still promotes women's subordination as a natural condition of their existence. This discourse is reinforced by the policies used in promoting the one-child campaign. Nonetheless, Chinese publishers have recently purchased translation rights to classic self-help books, such as *Our Bodies, Ourselves* (minus the American edition's sections on lesbianism and masturbation) or psychologist Françoise Dolto's texts on the family and child rearing, which indicates that they anticipate the growth of a new market.

Rural areas remain removed from elite urban discourse on female sexuality. The convergence of the household responsibility system, urban male migration, and the one-child campaign has put unprecedented pressures on peasant women. The household responsibility system, designed to encourage peasants to farm for their own profit, places a premium on the number of family members able to farm, while urban male migration and the one-child campaign reduce the number of laborers per household. Peasants obviously oppose both the coercive methods and the logic of the campaign. Couples continue to have more than one child in hopes of producing one or more sons, who are considered essential help with the heavy farm work and in providing a daughter-in-law to take care of parents in their old age. The exception to this trend occurs when peasant women seek employment in rural industries; since hiring priority is given to women with only one child, these positions have become a major family planning incentive.

Overall, peasant couples have proved adept at managing the economic penalties imposed on families with more than one child. Nonetheless, the psychological costs incurred by rural women are inordinately high. The All-China Women's Federation has publicly condemned family violence against women because of the birth of unwanted baby girls, the use of ultrasound to "select" a child's sex, and female infanticide. Peasants have also chosen more benign methods to circumvent the one-child policy, such as not registering the birth of a female. Without a birth certificate, however, the hundreds of thousands

AT THE GRASS ROOTS

THE MOST NOTABLE TREND in gender relations in China today is not the return of patriarchal traditions, but the sustained support women have begun to extend to one another across urban and rural landscapes. The United Nations Conference on Women, held in Beijing in 1995, was a watershed experience for the development of the women's movement in China. The push toward learning from and networking with women's movements worldwide was the single most important legacy of this extraordinary international gathering.

During the preparatory stages to the conference, slogans such as "connect the rails" with the international women's movement appeared. The government, however, attempted to rein in preparations once officials became aware of the possibility of human rights demonstrations at the accompanying International Forum for Non-Governmental Organizations. The forum was moved 30 miles outside Beijing, to Huairou, and the 47 panels to be presented by Chinese women were screened by the government through "rehearsals."

Nonetheless, many Chinese men and women observed for the first time how nongovernmental organizations function and the role these organizations can play in empowering women. More important, these men and women also managed to convince Chinese officials that the women's movement would not form an independent power base in China and hence was not a threat to the government.

Fresh from this experience, women have begun to explore how best to promote gender equality in China. Because the government has been unwilling to sanction the formation of independent women's associations, activists must find an officially approved umbrella group that is already registered, or strike out on their own to form an unofficial association.

Examples of innovative individual initiative abound. A banner embroidered by over a thousand rural women from the Shanxi area and displayed at the Huairou forum has been sold to benefit Shanxi rural development programs for women. A support group for female migrant workers in Beijing helps these women find jobs in understaffed city hospitals. A successful urban lawyer runs a legal consultation service out of her home to assist battered wives and their children. That so many women are willing to devote their time toward building an ad hoc grassroots women's movement indicates that sustainable gender equality will not be imposed from above, but demanded from below.

S. P.

---

of unregistered girls born annually will be unable to receive state health care, enroll in school, or vote as adults.

Despite a renewed drive to limit births in 1992, the Chinese government, because of widespread peasant discontent, has had to officially sanction in rural areas the birth of a second child in cases where the first child was female. The government also tolerates the small-scale, peaceful demonstrations that occur with increasing frequency in front of the gates to the State Family Planning Commission. In private, many educated Chinese women bluntly question how much state coercion is legitimate in trying to arrest population growth, and they criticize the government's heavy-handed methods.

Studies by Chinese scholars show that female education is the single most important predictor of fertility. China claims that 96.2 percent of its 7- to 11-year-old girls are registered for school. Yet, although education is theoretically available to all regardless of gender, female literacy rates continue to be lower than the national average. Of the 200 million illiterate Chinese, 70 percent are female. Often a woman burdened by farm chores will keep her daughter at home to assist with the multiple workloads. The children of urban migrants are also less likely to be schooled, since enrollment would call attention to their "illegal" residency.

In rural areas, the government has responded to this problem with the "Spring Bud" program, which is designed to put girls back in school by helping with their school fees, including lunch, textbooks, shoes, and eyeglasses. Many adult women have benefited from literacy programs sponsored by nongovernmental organizations or the All-China Women's Federation.

Little has been done to provide public schooling for the unregistered children of urban migrants.

At the other end of the spectrum, Chinese women who study through the master's and doctorate degree levels are outnumbered by their male counterparts—in some cases, by as much as ten to one. Professors complain that female graduate students are informed outright that prospective employers are interested only in hiring male graduates in their field. Still, female students are encouraged to pursue graduate work in the sciences, particularly in medicine. China has a higher number of female graduate degrees in engineering and other sciences than the United States.

A NEW IMAGE EMERGING?

Urban popular culture has focused new attention on women's issues. Once taboo subjects, including marital sex, divorce, and domestic violence, are now thoroughly explored in a growing body of women's radio talk shows and magazines, such as the favorite *Jiating* (Family) magazine. According to an official from the All-China Women's Federation, demand for weekly or monthly publications devoted to women's issues is growing. The federation and its branches currently edit 47 periodicals, up from just a handful during the Maoist years.

Television has also moved to accommodate the expanding female viewer market. The extremely popular show *Dajia tan* (Everybody's Talking) pits husbands and wives against one another in a mock trial setting in which they hurl accusations regarding one another's chauvinism, infidelities, and spending habits. The show's high ratings indicate that the topic offers more than entertainment value: Chinese viewers are exploring the boundaries of the "modern couple." Recently, a media partnership of Chinese businesswomen has proposed a television project titled *Women in the World,* a series of documentaries designed to introduce Chinese women to support systems as varied as model French day-care centers or Emily's List, the election fund for women running for political office in the United States.

MAKING THEIR VOICES HEARD

In analyzing the impact of economic reforms on women's lives in China, the government would evidently prefer to discuss its achievements in promoting gender equality and not the chronic discrimination problems that remain. At the other end of the spectrum, Western feminist discourse on China often presents an alarming portrait, replete with stark statistics on the skewed male-female birth ratio, domestic violence, and increased trafficking in women.

In determining how far China has come in terms of real gender equality, Chinese women themselves are likely to focus their attention on the development of pragmatic strategies for improving women's lives locally. Although outright criticism of central government policy is discouraged, women are making their voices heard by using those avenues for dialogue that the government has made available.

According to the State Council, the most pressing problems concerning Chinese women include their legal status, equal access to employment, involvement in politics, and the right to organize and participate in international women's activities.

Officially recognized nongovernmental organizations, along with the growing number of unregistered women's groups, have expanded the limits of debate, bringing in new issues, such as the feminization of poverty and reproductive rights. Their tactics, which encourage respect for national law, enable them to push for enforcement of existing legislation without appearing to threaten political orthodoxy.

Many Chinese activists believe that the seeds for real gender equality were sown at an NGO forum in Huairou during the United Nations 1995 Beijing Women's Conference. Since then a movement that encourages individual initiative has taken shape, with motivated urban and rural women working to assist those who have not benefited from China's economic expansion. Inspired by their own grass-roots experience and a growing interaction with women's associations worldwide, these informal women's networks signal the beginning of what may become a modern civil society in China.

NOTE

1. The one-child campaign has slowed China's population growth to a gross fertility rate of 2 births per female, while producing a statistical anomaly: China had 118 male births for every 100 female births in 1992. Since most nations produce 105 males for every 100 females born, this warped statistic indicates that 12 percent fewer girls are born in China annually than anywhere else.

SUSAN PERRY *is an assistant professor of international affairs at the American University of Paris. She co-organized a three-day conference on "Women, Culture, and Development Practices" at the French Senate with UNESCO and the OECD in November 1998.*

Home alone

Twenty-five years ago, faced with a dizzying population crisis, China banned its citizens from having more than one baby. The policy was a huge success—but what of the children who grew up without siblings? And what are the implications for society of millions of young people who never learned to share? Beginning the second day of our week-long special report, Catherine Bennett talks to Shanghai's 'little emperors'

Catherine Bennett

Soon after China implemented its one-child policy in 1979, reports reached the west of a new breed of plump, pampered creatures who had never learned to share. They were called Little Emperors, and nobody ever said "No" to them. It was as if our own country had decided to spawn millions of Prince Andrews. As these children have grown older, they have not, according to many bulletins, grown nicer. They are said to be in love with consumer durables and so obese, due to routine parental overfeeding, that they require regular sessions in fat farms. After the terracotta warriors, its army of spoilt tinies is now one of the most famous things about China. But like real emperors, these miniatures seem to avoid the vulgar gaze.

Because among the crowds loitering on the Bund, which once, according to Shanghai historians, swarmed with ragged, insistently begging juveniles, the number of promenading babies, each with a retinue of doting adults, can be counted on one hand. They are held up to be photographed against Pudong's brazen spikes and pinnacles. Elsewhere, in Shanghai's malls, parks and cafes, little children are rarer than British sparrows. In one big toy and baby shop there is not so much as a laden buggy, not even a glimpse of a pregnant stomach containing an embryonic emperor. The city looks as if the Pied Piper of Hamelin has just been through it. Or like a city with a very, very low birthrate. Professor Peng Zizhe, a demographer and director of the Institute of Population Research at Shanghai's Fudan University, thinks it may be as low as 0.7.

Although the contribution of a majority of larger, rural families keeps China's overall birthrate at an estimated 1.8, in urban Shanghai, the one-child policy is well on the way to becoming a half-a-child policy. The ruthless suppression of breeding may have succeeded almost too well. "Twenty or 30 years of propaganda and government implementation

of the policy has really changed people's minds about reproduction," explains Peng. "The problem for demographers and policymakers in Shanghai is not, 'Will these children have two or three kids?'—but whether they will have any kids."

Nowadays, he says, it is not uncommon for people in their late 20s still to be living like children with their parents: "They still get enough love from their mothers, so they don't need to create the solid marriage unit." The fact that any couple wanting two children must file an application to the people's government, supported by the relevant documentation, probably doesn't help create the additional citizens Shanghai needs to avert a future pensions and labour crisis. Citizens who pledge to reproduce just once are still rewarded with a Certificate of Honour for Single-Child Parents, and a lump sum at retirement.

So it is likely that most classes, in most city schools, will continue to be composed of only children: individuals who were once pathologised by the psychologist and birth-order obsessive Alfred Adler as typically selfish loners, prone to exaggerated feelings of superiority and liable to have trouble building close relationships. Outwardly, of course, you would never know. No one in the group of 14-year-olds I meet at No 2 Fudan Affiliated middle school is even fat. Do they live like little emperors? Maybe, given that many Chinese emperors succeeded the throne as children and suffered thereafter an oppressed, semi-adult existence, they do. "We don't sound like children, do we?" asks one boy, Zhang Zhe Yuan, who, like several of these children, has lived abroad. (Even those who haven't, speak to me in astonishingly fluent, expressive English.) In their uniform of blue and white tracksuits, with red scarves at the neck, they have a keen, active look, but they say they do nothing but study. "For us it's a very hard life now," says another child.

"The competition is intense." One girl, Xie Lu, lived in Leeds for a while. "It's so much pressure; the child has too much to live up to," she says.

Yes, the children say, their parents love them. "They put so much love on us that love becomes a reason to do everything," Xie Lu explains. In particular, love is the reason they must work hard at school. "One of the things that happens when you're an only child, the thing that happens in China," says, Zheng Xiu Yi, a boy who has lived in the States, "is that everything is focused on your grades, every aspect of expectation is focused on your grades. If you don't have good grades, you aren't a good child."

The good parent's job, accordingly, is to create a perfect studying environment. "Our parents sometimes do lots of things that we should be doing for ourselves because they want us to concentrate on our grades," says Zheng Xiu Yi. "For instance, my mom pushes my bike out of the door, presses the elevator button and waits for me to finish my breakfast and go out. It's just study, study, study, study and nothing else."

If they are, in every other respect, incompetent, it's not their fault. "The one-child policy is a good way to reduce the population," says Zheng, judiciously, "but everything has its good side and bad side, and what may become of this generation of people is that when we grow up, we may not know how to do laundry, or wash our socks, or tidy up our rooms."

They will, on the other hand, be academically able to a degree that is almost chilling when you imagine them competing against British teens, with their collections of indulgently marked GCSEs. The scores of Chinese students now studying in Britain present a misleading picture. Everyone in China knows that the cleverest, most ambitious students choose top universities at home first; the US—where there are scholarships—second; and English universities only third or fourth. Chinese students know our speciality is selling prestige for money.

At No 2 Fudan Affiliated, where entrance exams for high schools are coming up, the 14-year-olds say it is normal to work 14 hours a day. I must look disbelieving. "We do!" insists Zhang Zhe Yuan. "We wake up at 6.30 am, we don't have enough hours to eat, I skip my dinners for homework. We're not supposed to have this much pressure at school, because we're kids, we're children!" Another boy, taciturn until now, speaks up: "I think I'm just like a robot."

If they finish their school homework, their parents produce home-made extras. "They want more and more," says Zheng Xiu Yi. "You give 'em As, they want A-plus, you give them A-pluses, they want A-plus plus plus. Maybe I'll get 98%, and my dad goes, 'What went wrong?'" It is common, apparently, to score 100% in a school test, then be urged to try harder next time.

Sunny, a solemn girl who says her pocket money is stopped when she fails to excel, is one of thousands of Chinese children who find themselves the family raison d'être. "Chinese people always have new year dinner," she says. "When we eat dinner, my grandmother, grandfather, uncle,

aunt, parents and other people always ask me, 'How is your study?' It is the only topic. There is not any other topic. They like to have this topic for a long time, and I can't eat. When I get a bad mark in my examinations, they always say, 'If you don't get good marks, your future is dark'."

It is tempting to attribute some of this grumbling to general 14-year-old disgruntlement. Doesn't Britain also have its quota of pushy parents? But at another Shanghai middle school, Ying Chang Qi Weiqi, where large portraits of Mao, Lenin, Trotsky and Engels preside over the consuming struggle to come top, the fiftyish headmaster, Mr Jin Weiliang, says the parental pressure is "unprecedented in China".

These days, the headmaster says, the government's directive is to alleviate, rather than increase academic pressure on students. It is the parents, many of whom belong to the "lost generation", born into famine, their own prospects sacrificed to the cultural revolution or the economic upheavals which followed it, who focus relentlessly on the exam marks required for a successful life. And, in the absence of guaranteed state pensions, many of them are being realistic.

Once, they would probably have had six children, not just one. "If one child didn't do very well," says Jin Weiliang, "you could put your expectations on another child. Now, if you have only one child, it's succeed or fail. Parents value the success of their children more than their own success. To some extent, the success of a child's education decides whether the family is happy or not. So a very poor family, if their child is doing well, may be very happy—and even a rich family may be unhappy."

Ying, a 22-year-old student, must have made his parents happy by winning a place reading English literature at Fudan, one of the top three universities in China. He is already considering the best way to repay their investment. He remembers when he was 12 and won a free place at a good middle school, that his parents thanked him: "You save us money."

"It's a huge burden for us," he says, "to take care of our mother and father. At our age, you have to start thinking about that. When I get married, my wife and I will have to take care of four old people, so I am deliberately screening out certain jobs already. That's the kind of thing we have to think about, before we think about our own interests. There is a saying: if you are a good student, you earn money for your parents. This has become part of our consciousness."

How much this joyless, long-term planning is a response to China's exiguous welfare system, and how much it illustrates the only child's over-intense parental bond, it's too early to say. Longitudinal studies on some of China's only children are incomplete. Early findings are equivocal. What did we expect them all to be like—Churchill or Stalin? Most of the traits we associate with the only child are anecdotal, or were just made up by people like Adler or the early American psychologist, G Stanley Hall, who said: "Being an only child is a disease in itself."

For every person who believes that only children are loners, there's someone else to argue that they form stronger friendships to compensate for not having siblings. Susan Branje of Utrecht University, who has a paper in this month's Journal of Child Psychology and Psychiatry on "Perceived Support in Sibling Relationships", tells me that, "The differences are very small between children with and without siblings. If a child has a good relationship with parents, and companionship from friends, there don't have to be any negative effects from not having siblings."

Some have questioned whether the Chinese should even be concerned. "In countries all over the world you will find more and more single children," Professor Peng points out, rattling off a few low-fertility competitors—Japan, 1.2; Italy, 1.3; Spain, 1.3. "So sometimes the Chinese say, 'In other countries they do not worry about the mentality of the single child, why should the Chinese worry so much?'"

But the savage pruning of aunts, uncles and cousins from the family tree, creating families in which the interest and expectation of six adults all bear down on a single child, is utterly new. And it is in China, where the decline in fertility has been so ruthlessly enforced, that families once worshipped their ancestors. Back in the 70s, says Peng, while mathematicians extolled the one-child plan, some people did wonder what it might do to the Chinese tradition of the family. "But no one paid serious attention to that, because at that time China was in the aftermath of the cultural revolution, the Chinese had just experienced great destruction of Chinese traditional culture, so no one realised there would be further destruction of the traditional norms of Chinese society." He wonders about a lack of kinship and sense of belonging in these children with no relatives except parents and grandparents.

In the professor's experience, the students he teaches are less sociable than generations before them—but then, they grew up in changing times. "We really don't know whether this is caused by the single child [policy], or by the openness to the outside world." A Fudan student has a different explanation for his generation's reserve. "I think the cultural revolution largely destroyed the trust between people, that's why I personally don't feel so secure." Another 14-year-old says it's just that their parents see their friends as academic rivals: "Your parents think they are telling lies, and perhaps your friends are working hard, but they don't want you to know."

Whatever it is, exactly, that has made this generation so self-sufficient, the result is something which Mao would surely have recognised as a threat to the collective life of the party. "Liberalism," he wrote in his 1937 tract, Combat Liberalism, "stems from petit bourgeois selfishness, it places personal interests first and the interests of the revolution second … " And had his successors wanted to design a generation of people who behaved in this lamentable way, they could not have done better, surely, than enforce a policy which filled the cities and universities with striving only children. The population planners may even have created a generation which will one day render their own style of authoritarian birth control, enforced with fines, utterly unworkable. For rich people, those fines are irrelevant. Equally, it is hard to imagine some of the 14-year-olds I met in Shanghai meekly filling in an application form for a second child.

"I would say [this generation] are more innovative, more creative, they have a very strong self-identity," says the headmaster of Ying Chang Qi Weiqi. "They are more individual." And even with the compulsory Mao studies, which still go beyond school into university, it could be a challenge to keep them on the right path. "We have a lot of pressure and competition," says one of these 14-year-olds, pondering on the character of 100 million only children. "That makes us stronger."

~~Definitely~~ Probably One:

A Generation Comes of Age Under China's One-Child Policy

By Claudia Meulenberg

The Chinese population-control policy of one child per family is 25 years old this year. A generation has come of age under the plan, which is the official expression of the Chinese quest to achieve zero population growth. China's adoption of the one-child policy has avoided some 300 million births during its tenure; without it, the Chinese population would currently be roughly 1.6 billion—the number at which the country hopes to stabilize its population around 2050. Many experts agree that it is also the maximum number that China's resources and carrying capacity can support. Standing now at a pivotal anniversary of the strategy, China is asking itself, Where to from here?

China's struggle with population has long been linked to the politics of national survival. China scholar Thomas Scharping has written that contradictory threads of historical consciousness have struggled to mold Chinese attitudes towards population issues. China possesses a "deeply ingrained notion of dynastic cycles" that casts large populations as "a symbol of prosperity, power, and the ability to cope with outside threat." At the same time, though, "historical memory has also interpreted a large population as an omen of approaching crisis and downfall." It was not until economic and development issues re-emerged as priorities in post-Mao Zedong's China that the impetus toward the one-child policy began to build rapidly. During Mao's rule population control was often seen as inhibiting the potential of a large population, but in the years following his death it became apparent that China's population presented itself as more of a liability than an asset. Policymakers eager to reverse the country's backwardness saw population control as necessary to ensure improved economic performance. (In 1982, China's per-capita GDP stood at US$218, according to the World Bank. The U.S. per-capita GDP, by way of comparison, was about $14,000.)

> Had China not imposed its controversial but effective one-child policy a quarter-century ago, its population today would be larger than it presently is by 300 million—roughly the whole population of the United States today, or of the entire world around the time of Genghis Khan.

The campaign bore fruit when Mao's successor, Hua Guofeng, along with the State Council, including senior leaders such as Deng Xiaoping, decided on demographic targets that would curb the nation's high fertility rates. In 1979 the government announced that population growth must be lowered to a rate of natural increase of 0.5 percent per year by 1985. In fact, it took almost 20 years to reach a rate of I percent per year. (The overestimating was in part due to the lack of appropriate census data in 1979; it had been 15 years since the last population count and even then the numbers provided only a crude overview of the country's demography.) Nevertheless the Chinese government knew that promoting birth-planning policies was the only way to manifest their dedication and responsibility for future generations. In 1982 a new census was taken, allowing for more detailed planning. The government then affirmed the target of 1.2 billion Chinese for the year 2000. Demographers, however, were skeptical, predicting a resurgence in fertility levels at the turn of the century.

The promotion of such ambitious population plans went hand in hand with the need for modernization. Though vast and rich in resources, China's quantitative advantages shrink when viewed from the per-capita perspective, and the heavy burden placed on its resources by China's sheer numbers dictates that population planning remain high on the national agenda. The government has also stressed the correlation between population control and the improved health and education of its citizens, as well as the ability to feed and employ them. In September

2003, the Chinese magazine *Qiushi* noted that "since population has always been at the core of sustainable development, it is precisely the growth of population and its demands that have led to the depletion of resources and the degradation of the environment. The reduction in birth rate, the changes in the population age structure, especially the improvement in the quality of the population, can effectively control and relieve the pressure on our nation's environment and resources and strengthen our nation's capability to sustain development."

The Reach of the One-Child Policy

Despite the sense of urgency, the implementation of such a large-scale family planning program proved difficult to control, especially as directives and regulations were passed on to lower levels. In 1981, the State Council's Leading Group for Birth Planning was transformed into the State Population and Family Planning Commission. This allowed for the establishment of organizational arrangements to help turn the one-child campaign into a professional state family planning mechanism. Birth-planning bureaus were set up in all counties to manage the directives handed down from the central government.

Documentation on how the policy was implemented and received by the population varies from area to area. There are accounts of heavy sanctions for non-compliance, including the doubling of health insurance and long-term income deductions as well as forced abortions and sterilizations. Peasant families offered the most significant opposition; rural families with only one daughter often insisted that they be given the fight to have a second child, in hopes of producing a son. On the other hand, in some regions married couples submitted written commitments to the birth-planning bureaus stating they would respect the one-child policy. Despite this variation, it is commonly accepted that preferential treatment in public services (education, health, and housing) was normally given to one-child families. Parents abiding by the one-child policy often obtained monthly bonuses, usually paid until the child reached the age of 14.

Especially in urban areas it has become commonplace for couples to willingly limit themselves to one child. Cities like Shanghai have recently eased the restrictions so that divorcees who remarry may have a second child, but there, as well as in Beijing and elsewhere, a second child is considered a luxury for many middle-class couples. In addition to the cost of food and clothing, educational expenses weigh heavily: As in many other countries, parents' desire to boost their children's odds of entering the top universities dictates the best available education from the beginning—and that is not cheap. The end of free schooling in China—another recent landmark—may prove to be an even more effective tool for restricting population growth than any family planning policy. Interestingly, the *Frankfurter Allgemeine Zeitung* has re-ported that Chinese students who manage to obtain a university education abroad often marry foreigners and end up having more than one child; when they return to China with a foreign spouse and passport they are exempt from the one-child policy.

There are other exceptions as well—it is rumored that couples in which both members are only children will be permitted to have two children of their own, for instance—and it is clear that during the policy's existence it has not been applied even-handedly to all. Chinese national minorities have consistently been subject to less restrictive birth planning. There also appears to have been a greater concentration of family planning efforts in urban centers than in rural areas. By early 1980, policy demanded that 95 percent of urban women and 90 percent of rural women be allowed only one child. In the December 1982 revision of the Chinese constitution, the commitment to population control was strengthened by including birth planning among citizens' responsibilities as well as among the tasks of lower level civil administrators. It is a common belief among many Chinese scholars who support the one-child policy that if population is not effectively controlled the pressures it imposes on the environment will not be relieved even if the economy grows.

More Services, Fewer Sanctions

Over time, Chinese population policy appears to have evolved toward a more service-based approach consistent with the consensus developed at the 1994 International Conference on Population and Development in Cairo. According to Ru Xiao-mei of the State Population and Family Planning Commission, "We are no longer preaching population control. Instead, we are emphasizing quality of care and better meeting the needs of clients." Family planning clinics across the country are giving women and men wider access to contraceptive methods, including condoms and birth-control pills, thereby going beyond the more traditional use of intrauterine devices and/or sterilization after the birth of the first child. The Commission is also banking on the improved use of counseling to help keep fertility rates down.

Within China, one of the most prevalent criticisms of the one-child policy has been its implications for social security, particularly old-age support. One leading scholar envisions a scenario in which one grandchild must support two parents and four grandparents (the 4–2–1 constellation). This development is a grave concern for Chinese policymakers (as in other countries where aging populations stand to place a heavy burden on social security infrastructures as well as the generations now working to support them).

A related concern, especially in rural China where there is a lack of appropriate pension systems and among families whose only child is a daughter, is that it is sons who have traditionally supported parents in old age. The one-child policy and the preference for sons has also widened the ratio of males to females, raising alarms as the first children born into the one-child generation approach marriage age. The disparity is aggravated by modern ultrasound technology, which enables couples to abort female fetuses in hopes that the next pregnancy produces a son; although this practice is illegal, it remains in use. The 2000 census put the sex ratio at 117 boys to 100 girls, and according to *The Guardian* newspaper, China may have as many as 40 million single men by 2020. (There are several countries where the disparity is even greater. The UN Population Fund reports that countries such as Bahrain, Oman, Qatar, Saudi Arabia, and United Arab Emirates have male-to-female ratios ranging between 116:100 and 186:100.)

A Younger Generation: Adapting Tradition

However, the traditional Chinese preference for sons may be on the decline. Dr. Zhang Rong Zhou of the Shanghai Population Information Center has argued that the preference for boys is weakening among the younger generation, in Shanghai at least, in part because girls cost less and are easier to raise. The sex ratio in Shanghai accordingly stands at 105 boys to every 100 girls, which is the international average. Shanghai has distinguished itself over the past 25 years as one of the first urban centers to adopt the one-child policy, and it promises to be a pioneer in gradually relaxing the restrictions in the years to come. Shanghai was the first region in China to have negative fertility growth; 2000 census data indicated that the rate of natural increase was –0.9 per 1,000.

A major concern remains that as the birth rate drops a smaller pool of young workers will be left to support a large population of retirees. Shanghai's decision to allow divorced Chinese who remarry to have a second child is taking advantage of the central government's policy, which lets local governments decide how to apply the one-child rule. Although Shanghai has devoted much effort to implementing the one-child policy over the past 25 years, the city is now allowing qualifying couples to explore the luxury of having a second child. This is a response to rising incomes (GDP has grown about 7 percent per year over the past 20 years) and divorce rates. As noted above, however, many couples, although often better off then their parents, remain hesitant to have more than one child because of the expense.

The first generation of only children in China is approaching parenthood accustomed to a level of economic wealth and spending power—and thus often to lifestyles—that previous generations could not even have imagined. However, China also faces a rapidly aging population. In the larger scheme of things, this may be the true test of the government's ability to provide for its citizens. The fate of China's family planning strategy—in a context in which social security is no longer provided by family members alone but by a network of government and/or private services—may be decided by the tension between the cost of children and the cost of the elderly. There seems little doubt, however, that family planning will be a key element of Chinese policymaking for many years to come.

Claudia Meulenberg, *a former Worldwatch intern, received her master's degree from the George Washington University's Elliott School of International Affairs in May and now works at the Institute for International Mediation and Conflict Resolution at The Hague in her home country of the Netherlands.*

Rivers Run **Black**, and Chinese Die of Cancer

By JIM YARDLEY

HUANGMENGYING, China—Wang Lincheng began his accounting at the brick hut of a farmer. Dead of cancer, he said flatly, his dress shoes sinking in the mud. Dead of cancer, he repeated, glancing at another vacant house.

Mr. Wang, head of the Communist Party in this village, ignored a June rain and trudged past mud-brick houses, ticking off other deaths, other empty homes. He did not seem to notice a small cornfield where someone had dug a burial mound of fresh red dirt.

Finally, he stopped at the door of a sickened young mother. Her home was beside a stream turned greenish-black from dumping by nearby factories—polluted water that had contaminated drinking wells. Cancer had been rare when the stream was clear, but last year cancer accounted for 13 of the 17 deaths in the village.

"All the water we drink around here is polluted," Mr. Wang said. "You can taste it. It's acrid and bitter. Now the victims are starting to come out, people dying of cancer and tumors and unusual causes."

The stream in Huangmengying is one tiny canal in the Huai River basin, a vast system that has become a grossly polluted waste outlet for thousands of factories in central China. There are 150 million people in the Huai basin, many of them poor farmers now threatened by water too toxic to touch, much less drink.

Pollution is pervasive in China, as anyone who has visited the smog-choked cities can attest. On the World Bank's list of 20 cities with the worst air, 16 are Chinese. But leaders are now starting to clean up major cities, partly because urbanites with rising incomes are demanding better air and water. In Beijing and Shanghai, officials are forcing out the dirtiest polluters to prepare for the 2008 Olympics.

By contrast, the countryside, home to two-thirds of China's population, is increasingly becoming a dumping ground. Local officials, desperate to generate jobs and tax revenues, protect factories that have polluted for years. Refineries and smelters forced out of cities have moved to rural areas. So have some foreign companies, to escape regulation at home.

The losers are hundreds of millions of peasants already at the bottom of a society now sharply divided between rich and poor. They are farmers and fishermen who depend on land and water for their basic existence.

In July and August, officials measured an 82-mile band of polluted water moving through the Huai basin. China rates its waterways on a scale of 1 to 5, with 5 being too toxic even to touch. This water was rated 5. For fishermen, it may as well have been poison. "If I had wanted to, I could have gone on the river and filled a boat with dead fish," said Song Dexi, 64, a fisherman in Yumin. "It was smelly, like toilet water. All our fish and shrimp died. We don't have anything to live on now."

The Huai was supposed to be a Communist Party success story. Ten years ago, the central government vowed to clean up the basin after a pollution tide killed fish and sickened thousands of people. Three years ago, a top Chinese official called the cleanup a success. But the Huai is now a symbol of the failure of environmental regulation in China. The central government promotes big solutions but gives regulators little power to enforce them. Local officials have few incentives to crack down on polluters because their promotion system is based primarily on economic growth, not public health.

It is a game that leaves poorer, rural regions clinging to the worst polluters.

"No doubt there is an economic food chain, and the lower you are, the worse off your environmental problems are likely to be," said Elizabeth C. Economy, author of "The River Runs Black" (Cornell University Press, 2004), a study of China's environment. "One city after the next is offloading its polluting industries outside its city limits, and polluting industries themselves are seeking poorer areas."

China is facing an ecological and health crisis. Heavy air pollution contributes to respiratory illnesses that kill up to 300,000 people a year, many in cities but also in rural areas, the World Bank estimates. Liver and stomach cancer, linked in some studies to water pollution, are among the leading causes of death in the countryside.

"Over the past 20 years in China, there has been a single-minded focus on economic growth with the belief that economic growth can solve all problems," said Pan Yue, the outspoken deputy director of China's State Environmental Protection Administration. "But this has left environmental protection badly behind."

Too Poor to Flee, or to Get Well

Few places bear that out more than eastern Henan Province, which includes Huangmengying. The isolated region has tanneries, paper mills and other high-polluting industries dumping directly into the rivers.

One of the biggest polluters is the Lianhua Gourmet Powder Company, China's largest producer of monosodium glutamate, or MSG, the flavor enhancer. But the company's political influence is so vast that environmental regulators who have tried to challenge the company have done so in vain.

The Huai River basin has neither the history of the Yellow River nor the mystique of the Yangtze. Yet the Huai, with its spider's web of canals and broad tributaries, irrigates a huge swath of China's agricultural heartland.

Farmers once spent lifetimes tilling the same plot of corn or wheat. But in the past decade, millions of farmers, unable to earn a living from the land, have left Henan for migrant work in cities, leaving behind villages of old people and young mothers.

One of those mothers is Kong Heqin, 30, who was the last stop on Mr. Wang's cancer tour in June. She stumbled into her dirt courtyard, disheveled and groggy from an afternoon nap. Her face was bloated and her legs were swollen. She had already had three operations for cancer, and new tumors were growing in her large intestine.

Earlier in the year, doctors had prescribed chemotherapy. But treatments cost $500 a series, nearly a year's income. She had borrowed $250 to pay spring school fees for her two sons, and she worried that chemotherapy would drain the family's meager resources away from her children.

So she stopped chemotherapy.

"We've wasted so much money on medical treatment," she said. "I think the best thing would be to give up on it."

Her rising medical bills were one reason her husband left a few years ago for construction work in a northern metropolis, Tianjin. He returns twice a year to plant or harvest crops. On good months, he sends home $60, but Ms. Kong says months go by with nothing in the mail.

Her illness shapes family life. Her elderly mother tends her husband's fields because she is too weak. Her sons wash the clothes. She grows a ragged garden in her courtyard because the pesticides coating vegetables at local markets make her sick. The plate of boiled eggs on her dresser was a gift from sympathetic relatives.

Asked about pollution, she seemed confused, as if unaware of the concept. But she has noticed that her well water smells bad and has changed in taste. She knows that others are sick, too. "There's a family next door with a case of cancer," she said. "But they don't like to talk about it. People here are scared to talk about these things."

Epidemiological research for cancer in the Huai basin is scant. None has been done in Huangmengying. Nor does any scientific evidence prove that pollution is causing the rising cancer rate. What is clear is the wide range of pollutants, from fertilizer runoff to the dumping of factory wastes.

But Dr. Zhao Meiqin, chief of radiology at the county hospital, said cancer cases in the area rose sharply after heavy industry arrived in the 1980's and 90's. Before, the area had about 10 cases a year. "Now, in a year, there are hundreds of cases," she said, putting the number as high as 400, mostly stomach and intestinal tumors. "Originally, most of the

patients were in their 50's and 60's. But now it tends to strike earlier. I've even treated one patient who is only 7."

Dr. Zhao said most cancer patients came from villages close to the factories along the Shaying River, a major tributary in the Huai basin. Mr. Wang, the village party chief, also said the highest concentrations of cancer were found in the homes closest to the village stream, which draws its water directly from the Shaying.

Polluters Hiding in Plain Sight

Health problems began appearing slowly in the early 1990's. Mr. Wang said he learned that the water was severely polluted after an environmental official came on a personal visit. Farmers also began complaining that their fields were producing less grain because of polluted irrigation water.

Today, pollution corrodes daily life here. Farmers too poor to buy bottled water instead drink well water that curdles with scum when it is boiled.

Xiao Junhai is 57 but looks two decades older. In June, he shivered under a quilt in a dark room, summer flies flitting at his head, cancer knotting his stomach. He could not lift himself from his crude bed.

"I grew up drinking the water here, and I still drink it," he said. "I don't know what pollution is, but I do know it means the water is bad."

His daughter, Xiao Li, 24, anguished over the dilemma that her father's illness had thrust upon her. She says her father takes traditional Chinese remedies and eats rice porridge because the family cannot afford treatment. If she returned to her migrant job on the coast, in Hangzhou, she might earn enough money to pay for it. But no one else can care for him. So she has stayed.

"The water in the river used to be clean, but now it's black and changing colors all the time," she said. "The water is being destroyed."

The Lianhua Gourmet Powder Company is based in Xiangcheng, upstream from Huangmengying. It is the area's largest employer, with more than 8,000 workers, and the largest taxpayer in Xiangcheng.

For Henan Province, Lianhua Gourmet is a signature company, the biggest producer of MSG in China. An analysis by a Chinese credit rating service, Xinhua Far East, found that in 2001 the factory produced more than 133,000 tons of MSG and has plans to raise production to 200,000 tons.

Under any circumstances, the company's sheer size would translate into significant political clout. But Lianhua, basically, is the government. Lianhua is traded on the Shanghai Stock Exchange, but according to the credit analysis, its majority stockholder is a holding company owned by the Xiangcheng city government.

This type of government-controlled enterprise is not unusual in China, but the potential for a conflict of interest is glaring. The production of MSG leaves potentially harmful byproducts, including ammonia nitrate and other pollutants that are supposed to be treated to meet environmental standards.

A damning report last year by the State Environmental Protection Administration blamed local officials for lax enforcement. The report said Lianhua had dumped 124,000 tons of untreated water every day through secret channels connected to the Xiangcheng city sewage system. The water eventually flowed into the Shaying River, almost quadrupling pollution levels.

"This constitutes a grave threat to the lives and livelihoods of people downstream," the report stated.

Officials at Lianhua did not respond to repeated written and telephone requests for interviews. Neither did officials in Xiangcheng nor with Henan Province.

But one retired local Communist Party official said party cadres had always protected Lianhua. He said a son-in-law of a Lianhua chief executive once even headed the city's environmental protection bureau.

"There are a lot of officials who don't care about pollution," said the official, who asked not to be identified. "Some leaders are just interested in making money."

He said the company often broke promises about cleaning up. "What they said and what they did were different things," he said. "They even said they would stop production if they weren't able to meet pollution standards. But they never did that."

A Stream of Black Water

This June, a reporter saw a noxious liquid flowing from a waste outlet into a stream near a Lianhua factory on the outskirts of Xiangcheng. A sign above the outlet said, "Lianhua Company, No. 3 Waste Outlet." Another sign said the outlet was under the oversight of the city environmental bureau. The acrid smell was so strong that it was difficult to stand nearby.

Less than a mile downstream from the waste outlet, Wang Haiqing watched his seven goats chew on weeds. Mr. Wang lived on the other side of the stream, in Wangguo, and said several neighbors had contracted cancer or other intestinal ailments. He said his goats vomited if they drank from the blackened water.

To reach clean drinking water, he said villagers must dig wells 130 feet deep. Most cannot afford to do so.

"It's been so polluted by the MSG factory," said Mr. Wang, 60. "It tastes metallic even after you boil it and skim the stuff off it. But it's the only water we have to drink and to use for cooking."

The rains of June in Huangmengying had given way to boiling humidity by the middle of August. Mr. Wang, the village chief, wore shorts and sandals as he again walked beside the village stream. He said four more people had died since June, two of cancer.

But much had also changed in the two months.

The 10th anniversary of the government's promise to clean up the Huai had become a major embarrassment for the Communist Party. Roughly $8 billion had been spent to improve the basin, but the State Environmental Protection Administration concluded this year that some areas were more polluted than before.

China's press, often given freer rein on environmental issues, published critical articles over the summer. The newspaper operated by the State Environmental Protection Administration blamed local officials for allowing powerful companies, including Lianhua, to continue polluting. Even tiny Huangmengying got attention: a crew from state television visited in July. Officials, fearing a humiliating expose, hurriedly started digging a deeper well for the village.

But the gesture was dwarfed by what Henan officials did for Lianhua.

For more than a year, the company had been in financial trouble, suffering from bad investments and a slowdown in the MSG market. For months, banks pressured it for roughly $217 million in unpaid loans.

The Henan Province government stepped into the breach. The Henan governor, Li Chengyu, organized a meeting at Lianhua headquarters in July to devise a plan to save the company. The Henan government also gave the company more than $25 million.

"The government is confident and the business is confident that Lianhua Gourmet can be brought around," Mr. Li said, according to the Chinese financial press. "The banks should support Lianhua Gourmet."

The signal was clear. Henan's government would make certain Lianhua survived.

In Huangmengying, Mr. Wang again visited Ms. Kong, the young mother with cancer, who was also struggling to survive. Her resolve in June to forego chemotherapy had withered with her health by August. She was pale and coughing as she explained that she had again borrowed money for more treatment. She would leave in a few days.

But it meant that she could not pay her sons' school fees for the fall semester. Her husband could not find work and had no money to send. And the friends who had loaned her money said they could loan her no more. "I'm scared," she said.

Only an hour earlier, Mr. Wang had been walking to visit Ms. Kong when a woman rushed toward him and knelt in a formal kowtow, touching her lips against the dirt. Her husband had dropped dead. Doctors had examined the body and discovered a tumor. She needed Mr. Wang to help with funeral arrangements. He asked where she and her husband lived.

In a small brick hut, about 50 yards from the village stream, answered the woman, Liu Sumei.

Ms. Liu, 50, led Mr. Wang to a friend's home, where her husband's body lay in a coffin under a large poster of Mao Zedong.

Ms. Liu had not known her husband had cancer, only that he was in poor health. But in Huangmengying, she said, poor health is not unusual. "Every family has someone who is sick," she said. "All the neighbors."

Urumqi Dispatch

Purple Haze

BY JOSHUA KURLANTZICK

THE GRAND HOTEL offers some of the finest accommodations in Urumqi, the frontier capital of Xinjiang, the vast western province of China bordering Central Asia. A swanky first-floor bar swarms with Chinese businessmen dressed in expensive suits, sipping Johnnie Walker. A twentieth-floor fitness club caters to Chinese yuppies trying out gleaming new Nikes but never working hard enough to sweat out their hair gel. But there is one thing the Grand Hotel doesn't offer: a view. When I got to my eighteenth-floor suite, the bellboy showed me the room's amenities—satellite television, a plethora of little liquors—and proudly pulled back the drapes so I could get a good look at downtown Urumqi, a beautiful city. Unfortunately, I could see little through the gray air. Few buildings were visible, though I knew they were there, just outside my window. The bellboy smiled. "Nice view," he said.

Sad to say, he was right. Of the several weeks I have spent in Urumqi, that day was one of the clearest. Later that afternoon, some of the smog lifted, and I could see the stunning mountains surrounding the city, a rarity. Meanwhile, on the road outside Kashgar, a city southwest of Urumqi, mines and construction outfits belched smoke into the broad desert sky, making the air a thick particulate soup; when I ran a wet cloth over my face, it turned black, as if I'd been in a West Virginia mineshaft. My driver, and everyone else in a taxi with me, incessantly coughed and spit soot and phlegm on the car floor. And Xinjiang is hardly unique. For years, Western observers and some Chinese have worried about China's enormous problems: a sclerotic economy clogged by mountains of nonperforming loans; a rapacious gerontocracy allowing its people slightly more freedom while simultaneously cracking down on groups that organize against the state. But largely ignored has been perhaps China's biggest looming disaster: The Middle Kingdom is hurtling toward environmental catastrophe—and perhaps an ensuing political upheaval.

ALREADY, MOST CHINESE cities make Los Angeles look like a Swiss village. In Beijing last week, I choked on hot, dust-filled air—a normal occurrence in summer, when plague-like sandstorms from China's expanding deserts wash over the capital. In Guiyu, a city in southern China known for its electronics-recycling industry, the *Los Angeles Times* reports that peasants work atop mountain-sized piles of toxic refuse. Even in Shanghai, one of China's most progressive cities, I have passed water systems full of trash, oil, and feces, and many days the acrid, gray air has stung my eyes. The numbers are even more depressing. Sixteen of the world's 20 most polluted cities are in China. Two-thirds of China's cities don't meet World Health Organization air-quality standards. Reports released by Chinese health experts have suggested that just living in these places is like smoking two packs of cigarettes per day and that, in some cities, 80 percent of children may suffer from lead poisoning—80 percent! Beijing's air has more carbon monoxide than Los Angeles and Tokyo combined. By 2020, 550,000 Chinese will be dying prematurely of chronic bronchitis caused by airborne pollution.

Rural areas are in trouble, too. Between 1994 and 1999, China's Gobi Desert expanded by more than 20,000 square miles, moving within 150 miles of Beijing, reducing groundwater supplies—and causing brutal dust storms that often spread over much of Asia. "No country has ever faced a potential ecological catastrophe on the scale of the dust bowl now developing in China," Lester Brown, president of the Earth Policy Institute, a leading international environmental group, has said. According to *The River Runs Black* an outstanding new book by Elizabeth Economy, a China scholar at the Council on Foreign Relations, five of China's biggest rivers are "not suitable for human contact." China's wastewater pollution may increase as much as 290 percent by 2020.

In a country historically viewed in the West as attuned to the Earth—American acolytes of Buddhism often cite its respect for nature—how have things gone so wrong? Contrary to common belief, even traditional Confucian scholars saw nature as a force to be tamed, not preserved; and Mao famously pooh-poohed environmental problems, so Chinese leaders never built strong environmental institutions. And, over the past two decades, the country has gone through history's most rapid industrialization with virtually no controls; China's equivalent of the Environmental Protection Agency has a staff of only 300, one-fiftieth that of its American counterpart. Coal powers nearly 70 percent of Chinese industry, compared with less than 25 percent in the United States, and Chinese consumers are feverishly buying cars with lax emissions controls—the number of cars on the road is doubling every four years, creating unchecked congestion. Corrupt Communist Party officials profit from unregulated construction, China's tissue-thin legal system barely constrains developers, and the massive construction that results is destroying forests and contributing to desertification. In Kashgar, Chinese firms are madly tearing down old, eco-friendly mud-brick homes built around airy courtyards to build acres of flashy new glass and steel shopping centers. Kashgar contractors strew trash everywhere.

What's more, as unemployment has risen, Beijing has tried to keep the economy moving with a series of massive—and massively destructive—infrastructure projects, including the world's biggest dam and its longest bridge. (In 2002, state investment in the economy rose 25 percent even as foreign economists touted the liberalization of Chinese private business.) In Xinjiang, China has built an unnecessary railroad across the lightly populated province, as well as an economically unviable oil pipeline.

But the ecological destruction is spawning a grassroots environmental movement that might have a positive impact. Ecological destruction in Poland, the Soviet Union, and other countries in communist Eastern Europe helped spark political change. Economy notes how, in Krakow in the 1980s, Polish intellectuals, trade-union members, and environmentalists banded together to push the government to upgrade local factories that were heavy polluters—the same coalition of activists that later took on the state directly. In the USSR, meanwhile, huge protests developed in 1986 over a project to divert the Siberian River—a project that was causing massive pollution.

Similar things may be happening in China. Already tens of millions of farmers in the central provinces, sites of the worst desertification, stand to lose their farms. But, as the U.S. Embassy in Beijing notes in a report titled "The Grapes of Wrath in Inner Mongolia" (Inner Mongolia is a Chinese province), "China's 21st-century 'Okies' have no California to escape to." Without an unpopulated frontier for the migrants, they are coming to provincial capitals and to Beijing, where, like Poles and Russians in the '80s, they vent their anger at the government. China's Ministry of Public Security has admitted that the number of large protests in China almost tripled between 1993 and 1999 (the most recent statistics available), with many taking place in cities where workers fired from state industries are joining forces with migrating farmers who lost their fields. Similarly, recent battles between farmers and industry over water scarcity have led to large, often violent, antigovernment protests.

And, unlike religious groups like Falun Gong or small political organizations like the China Democracy Party that have challenged Beijing, the environmental movement is relatively well-organized. Though security forces harass anyone who might threaten the Communist Party, they have not cracked down as hard on the greens because, until recently, China's leadership did not consider environmental groups "political" organizations. "The environmental NGO definitely gets more space," Yi Wen, a program officer for the activist group Green River, told reporters last year. Taking advantage of this permissive attitude, more than 2,000 environmental NGOs have organized in China. These dynamic green groups, legal aid centers, and crusading media outlets have energized grassroots activism and suggested to ordinary Chinese that, ultimately, only democracy will bring down corrupt officials and solve environmental problems. Their message is getting through: According to Economy, Li Xiaoping, executive producer of "Focus," a Chinese investigative news program, says peasants now come to the "Focus" studios to beg them to investigate environmental problems caused by local officials. And one day, perhaps, a massive environmental meltdown will empower these green groups, as Chernobyl catalyzed Soviets. Maybe then I will be able to see out the window at the Grand.

Can China Keep the Lights On?

More cars on its roads, more air conditioners in its homes, more factories pumping out products. To keep its economy humming, China is scouring the globe for oil—and roiling the world's energy markets in the process.

By Clay Chandler

When Chinese troops opened fire on Soviet counterparts at a border checkpoint in 1969, the shots reverberated across the oilfields of Daqing. For much of the decade, a cadre of Chinese geologists and engineers, tutored by Soviet technicians, had struggled to coax crude oil from Daqing's frozen prairies. With the border skirmish, Sino-Soviet relations collapsed. Moscow ordered its experts home. Wang Qimin, then one of Daqing's youngest engineers and now its most senior, still bristles when he remembers the Russians' taunts as they departed. "They mocked us," he recalls. "They said, 'Your methods are too primitive. Trying to develop Chinese oilfields without Russian expertise is as futile as trying to learn how to make a shirt by studying a piece of cloth.'"

Wang and his compatriots proved the Russians wrong. In the years that followed they raised Daqing's capacity to more than a million barrels of oil a day, an astounding feat that enabled Mao Zedong to thumb his nose at the Soviets—and the rest of the world—for years. Today Daqing, China's largest oilfield, is a sprawling state-run colossus: 90,000 workers tending 50,000 wells linked by a maze of pipelines and storage tanks across an 800-square-mile expanse in the northeastern corner of the country. In the city of Daqing itself, hundreds of rusting pumps bob methodically—beside government office buildings, behind restaurants and karaoke bars, and in the midst of dingy housing blocks.

But China's moment of energy independence is over. After years of drilling, China's main oilfields are badly depleted. Costs per barrel have crept up steadily. At Daqing annual output has fallen 12%, to 50 million tons, since production peaked in 1997. As domestic reserves dwindle, China's economy, which grew at a sizzling rate of 9.1% last year, is shifting into overdrive. With the engines of consumption, business investment, and government spending all running full throttle, China, an oil exporter only a decade ago, has acquired a voracious appetite for foreign crude.

Since 1990 oil consumption has jumped by an average of 7% a year. Last year China guzzled 5.4 million barrels a day, eclipsing Japan as the world's second-largest oil consumer, though still far behind the U.S., at 20.2 million barrels a day. Imports topped two million barrels a day, a third higher than in 2002. The International Energy Agency (IEA) in Paris predicts that by 2030 imports will account for 85% of China's total oil consumption.

China's hunger for oil and other forms of energy poses far-reaching challenges for Beijing and the global economy. Already China competes with the U.S., Japan, and Europe for oil from the Middle East. Bear Stearns energy analyst Adam Clarke cites surging Chinese demand as the "primary factor" behind OPEC's success in preventing a collapse in oil prices in the wake of a U.S. victory in Iraq. Says Cambridge Energy Research Associates chairman Daniel Yergin: "Growth in Chinese demand is a fundamental trend driving world oil markets today. People used to talk about China's oil needs in passing. Now it's the No. 1 topic."

There's more at stake than just high oil prices. Security experts fear that China's determination to minimize dependence on oil from the Middle East, where the U.S. has strong ties to major producers, will tempt Beijing into oil-for-arms alliances with states the U.S. accuses of sponsoring terrorists. Environmentalists worry that to slow dependence on foreign oil, China will renege on commitments to curb reliance on coal. (China now counts on coal for 70% of its total energy needs and is the world's largest producer of carbon emissions.) Beijing might also embark on an expansion of its nuclear energy program—a troubling prospect given China's abysmal industrial safety record. Investors, meanwhile, fret that Chinese leaders will prod state-run energy firms to squander capital on unproductive overseas oil assets or overspend on grandiose pipeline projects to transfer energy from China's resource-rich western provinces to its booming eastern seaboard.

China's leaders know that in the long run, failure to come to grips with rising energy needs will jeopardize growth. For now, though, they're scrambling just to keep the lights on. Last summer an unexpected spike in electricity demand wiped out large swaths of China's power grid, plunging scores of cities into darkness and forcing temporary closure of thousands of factories. As many as two-thirds of China's provinces suffered severe power outages. In Shanghai last month shopping malls were ordered to turn down their thermostats. In Changsha, the capital of Hunan province, authorities have limited electricity use to three days a week.

The shortages highlight Beijing's reluctance to loosen its grip on the Chinese economy. Bureaucrats at the National Development and Reform Commission dictate who can build new power plants, how much they can charge for electricity, and how much they pay for the coal and diesel fuel needed to produce it. But planners have consistently misjudged demand, forecasting a 5% rise in electricity usage last year when actual growth jumped 15%. The NDRC has commissioned a flurry of new power-generation projects, among them the Three Gorges Dam. But it will be several years before those new facilities yield enough additional capacity to meet China's needs, and it's far from clear that new juice can be delivered flexibly and reliably after they come online.

The power shortages reflect the "complete incoherence" of China's approach to energy policy, says Philip Andrews-Speed, a former BP executive who heads an energy policy institute at Scotland's University of Dundee. China's leaders deem energy issues too important to leave to the market, he says, but they haven't figured out an alternative mechanism to make sure energy gets where it's needed, when it's needed. "There's no energy ministry, no center for decision-making, no one place responsible for pulling it all together," says Andrews-Speed. "Different strands of the energy question thread through a tangle of different offices. Companies and agencies fight it out at the highest levels." The result? "Systems break down, monopolists block distribution, power stations don't get built, and there are no price signals to guide investors on where to invest."

That's a fair assessment, agrees Chen Fengying, research director at the China Institute of Contemporary International Relations. "Each company and industry pushes its own priorities," she says. "No one authority is in charge."

Beijing's mismanagement of the nation's power grid hardly inspires confidence in official projections about China's oil needs. The government's Energy Research Institute sees oil consumption rising to seven million barrels a day by 2010. That squares with IEA estimates, but the agency's most recent forecast assumes China's GDP growth will slip to 4.8% in coming decades—a good two percentage points below the consensus of private economists. Analysts at Cambridge Energy put 2010 oil consumption at eight million barrels a day, while Bear Stearns's Clarke thinks the figure will be closer to 13 million. With per capita income rising above the $1,000 mark, he says, China has

reached a crucial "inflection point" and is poised for explosive growth in energy demand.

In China's largest cities, passenger car sales, which leaped 55% last year, have spawned near-permanent gridlock. Industry experts predict similar growth this year as car prices fall and Chinese banks and foreign carmakers expand financing programs for would-be drivers. Meanwhile consumers throng the aisles of stores selling appliances. And the spread of television, with its constant barrage of advertisements and foreign programs, has kindled a yearning for consumer goods in even the remotest rural villages.

China imports about 60% of its oil from the Middle East. That vexes China's leaders not only because the region is a bastion of U.S. influence but also because it's unstable. Worse, tankers carrying oil to China from the Middle East must pass through chokepoints that leave China vulnerable to blockade by the U.S. in the event of a showdown over Taiwan.

Beijing has worked frantically to secure alternative oil supplies— but with mixed results. Last May, Chinese President Hu Jintao traveled to Moscow to endorse a 25-year, $150 billion contract between state-owned China National Petroleum and Russia's Yukos Oil. The deal envisioned construction of 1,500-mile pipeline capable of moving as much as 600,000 barrels of oil a day from oilfields in eastern Siberia to Daqing. From there it could be pumped to the rest of China. But the plan foundered in October, when Yukos CEO Mikhail Khodorkovsky was hauled off to jail and charged with tax evasion. The fall of Khodorkovsky, who had clashed with Russian President Vladimir Putin, enabled Japanese officials to advance a counterproposal to build a longer, larger pipeline skirting China and connecting to the Pacific port of Nakhodka. Tokyo is offering $7.5 billion in financing for the project and has promised development loans for the Russian Far East. It's not feasible to build both pipelines, and Putin isn't likely to pick a winner until after Russian elections in March. But his aides have talked up the Japanese proposal in recent weeks, noting that it would give Russia access to markets in Japan, South Korea, and elsewhere in East Asia.

Hu also paid a visit to Kazakhstan to back another oil deal, this one calling for a pipeline to transport oil from the Caspian Sea to western China. But hopes for that project dimmed when a consortium of Western and Japanese oil companies blocked China's efforts to secure additional drilling rights in the region.

With the two pipelines looking increasingly like pipe dreams, China's oil companies are scouring the globe in search of oilfields they can buy. In recent years they have made investments in Sudan, Kazakhstan, Indonesia, and Australia. But so far those acquisitions have yielded only about 200,000 of barrels of oil a day in new capacity—a drop in the ocean of China's projected needs. They've earned China a reputation for overpaying for unproven energy assets and squandering capital that could have been spent buying oil on the spot market. Japan pursued a similar plant-the-flag approach to energy security after the 1979 Arab oil embargo. But the policy was a fiasco, leaving the government holding $20 billion in unrecoverable loans to companies that bought unproductive wells abroad. These days the Japanese, who count on imports for 98% of their oil needs, have come to live with the notion that they're better off buying oil than buying oil wells.

That's a proposition China's leaders aren't ready to swallow. Instead they have opted to inject competition in manageable doses. In the late 1990s, Beijing restructured the nation's oil production and refining institutions into two competing giants, each with drilling and refining assets and separate distribution networks. The two enterprises shed millions of unneeded workers, scrapped some of their least profitable operations, and hired Western banks to help dress up subsidiaries for listing on exchanges in Hong Kong and New York. A third company, China National Offshore Oil Corp., was given primary offshore drilling and development rights. Subsidiaries of the two onshore oil firms, PetroChina and Sinopec, fetched $7.5 billion in offerings to foreign investors.

One of those investors was Warren Buffett, who last year spent an estimated $500 million for a 13% stake in PetroChina. At PetroChina's Beijing headquarters, president Chen Geng says that although he has never met the Sage of Omaha, he welcomes his investment. Chen, who holds PetroChina stock options himself, proclaims Buffett's investment a "wise decision"—and with the stock price up 133% since his purchase, it's hard to disagree.

But the effort to instill private-sector sensibilities goes only so far. Between gulps from a jar of tea, Chen stresses that PetroChina's most important shareholder is the Chinese government, not Warren Buffett. Visitors to Daqing get a carefully orchestrated tour of "model pumping stations" and well-scrubbed research facilities. Even the tiniest attempt to depart from a preset itinerary meets resistance from official escorts.

Daqing's patron saint is Ironman Wang, an illiterate peasant revered for proclaiming that he would gladly sacrifice 20 years of his life to find more oil for China. In the Ironman Wang Museum, sepia photos show him leaping into a slurry pit to mix cement with his own body in an effort to plug a ruptured well.

Mao lionized Wang as a revolutionary hero, and in official folklore Daqing came to symbolize the idea that through hard work and a commitment to socialist principles, China's people could bend nature to their will—without help from outsiders. But those lessons are out of step with China's current predicament. To keep growing, China will need more than hard work and revolutionary zeal—it will need all the outside assistance it can get. "People got all worked up when we started importing oil in 1993," says Wang Qimin, the Daqing engineer hailed by many as Ironman's successor. "But I see it as a sign that we've matured. We're so big now, we grow so fast. As long as we make money from exports, why not buy oil from the Middle East?"

One place where China could benefit from outside help is in developing reserves of natural gas, which currently supplies only 3% of the country's energy needs. Foreign capital might also finance the extensive network of pipelines needed to transport that gas from western provinces to eastern factories.

Western oil giants have long angled for a role in China's domestic energy markets. But with the exception of WTO concessions granting some rights to operate gas stations in China, Beijing has held the Western majors at bay. Mark Qiu, a former investment banker who is now CNOOC's chief financial officer, argues that China "has no choice" but to reach out to Western multinationals. "They're the big producers," he says. "They have the discovery and production technologies. They're the ones who make markets. We have to create an incentive for their partnership—and that means letting them make a profit."

For now, though, the drivers of China's economy seem determined to stick to the self-service pump, even if they have to pay more for the privilege.

Feedback *cchandler@fortunemail.com*

China

Quest for Oil

The Middle Kingdom can't find enough oil to meet booming domestic demand—
and the world is paying at the pump

BY MATTHEW FORNEY

YANG HUA'S PASSPORT IS STAMPED WITH visas that would alarm immigration clerks around the world. He showed up in Indonesia two days after the Bali disco bombings in 2002. He's logged trips on a moment's notice to Iran, Yemen and Qatar, as well as to the U.S., Australia, Canada, England and Brazil. And Yang doesn't try to hide the substances contained in little glass vials that he brings home from his travels. They're lined up on the windowsill of his Beijing office, affixed with labels such as "SAUDI SWEET." Yang, it turns out, works for the China National Offshore Oil Corp. (CNOOC), and is responsible for the state-owned company's efforts to secure oil and gas supplies all over the globe. The samples of crude are souvenirs that testify how far he must roam in his search. "I'd like it if there was oil under Paris," he says, "but I spend my time in less comfortable places."

Yang's globe trotting reflects just how powerful China's thirst for fossil fuels has become. A booming but energy-inefficient economy and an emerging middle class in love with cars and other modern conveniences have caused energy demand in China to soar. The nation's oil imports have doubled over the past three years and surged nearly 40% in the first half of 2004 alone, pushing the country past Japan to become the world's second largest oil consumers, behind only the U.S.

Today numerous factors are driving up the price of crude, from chaos in Iraq to turmoil in Nigeria to hurricanes in the Gulf of Mexico. "It is neither fair nor accurate to blame China for most of the rise in oil prices," says Jeffrey Logan of the Paris-based International Energy Agency. But China's impact should not be ignored. Even if China's blazing GDP growth of 9.1% in the first three quarters of this year (compared with the same period the previous year) slows to

8% in 2005, as the Chinese Academy of Social Sciences predicts, the country is now a permanent player in the global competition for oil. "More than a billion Chinese are joining the oil market," says Bo Lin, an energy specialist at the Asian Development Bank. "How can prices go down?"

[Over the past three years, China's oil imports have doubled, surging nearly 40% in the first half of 2004 alone]

China has not always been so heavily dependent upon imported oil. The discovery in 1959 of the Daqing oil fields under the Manchurian grasslands meant the once largely agrarian country was for decades able to produce more crude than it required, a circumstance that the government celebrated as a political victory. (Study Daqing!" chanted legions of Red Guards during the 1966–76 Cultural Revolution.) Oil and gas discoveries in the South China Sea and Bohai Gulf, where drilling began in 1979, made China seem all the more invulnerable to oil shocks, and the country remained an oil exporter until 1993. Today, however, output from China's top four oil fields is in decline. By some estimates, the country's current proven reserves will be depleted in as few as 14 years. Meanwhile, largely untapped petroleum pools believed to lie beneath western China's desolate Tarim Basin, even with prices at $50 a barrel.

That doesn't mean China's robust economic engine will grind to a halt. The mainland meets more than two-thirds of its energy needs with coal and boasts the world's largest coal reserves. But to keep its economy racing ahead—and to ease some of the pollution that comes from burning coal—China's leaders are forced to seek ever-greater supplies of petroleum from overseas. More than half of China's oil imports currently come from the volatile Middle East, making oil security a growing concern in Beijing. China plans to build a strategic oil reserve, and the country has several pipelines planned that would theoretically receive supplies from fields in Russia, Central Asia and Burma. But China's state-controlled oil industry, consisting of three major companies (**China National Petroleum Corp.** [CNPC], **CNOOC** and **Sinopec**), and numerous overlapping bureaucracies has yet to develop a clear, comprehensive energy policy.

Leaders in Beijing want to avoid the fate of other oil-poor countries like South Korea, which buys all of its crude on the open market and is therefore exposed to sharp price rises. The way to do that is to invest in exploration and development in countries that have oil fields but lack the capital or technology to exploit them. When Chinese companies have a stake in oil coming out of the ground, even if it originates abroad, they'll have secured long-term supplies independent of the world's fickle prices. The process of overseas exploration began in 1997, when Premier Li Peng encouraged state-run oil concerns to look outside China's borders for investment opportunities, and in the past few years the search has ranged all over the world.

CNOOC, for example, signed a deal two years ago to extract a million bbl. of oil a day in Indonesia, and a year ago it signed a major contract to produce gas in Australia. In February, President Hu Jintao traveled to Gabon hoping to secure deals in Africa, and in June, he led a delegation from China's natural gas industry to Uzbekistan. Chinese oil executives have even begun courting Ecuador and Colombia. "Latin Americans feel frustrated that the U.S. has virtually ignored the region, so turning to China is prudent and will pay financial dividends down the line," says Cynthia Watson, a professor of strategy at the National War College in Washington.

China's stepped-up efforts have already created some friction. Earlier this year, **ONGC Videsh**, the overseas-investment arm of India's largest oil-and-gas producer, was on the verge of completing a deal that would have given it an 11% stake in a proven oil field in Sudan when China's CNPC swooped in with an offer that was reportedly 17% higher—snatched the deal away. "The Chinese are definitely very aggressive in the price they are willing to pay," says R.S. Butola, managing director of ONGC Videsh. Beijing has demonstrated that it is willing to face down international pressure to protect its energy interest. Last month the U.N. discussed imposing sanctions on Sudan as punishment for sponsoring genocide in Darfur. China—which has invested a reported $15 billion in Sudan oil projects and imports oil from there—threatened a veto. The Security Council passed a watered-down measure.

Despite that aggressiveness, China's overseas investments supplies just 5% of imports—the rest is purchased on the open market. Mainland oil companies have twice been foiled in their efforts to buy stakes in fields in Kazakhstan, and they haven't secured any significant drilling rights in Central Asia or the Middle East. The fields that Chinese companies have bought into are already mature, and many experts feel they've overpaid. "China has been singularly unsuccessful in its overseas ventures," says Jim Brock, a Beijing-based energy consultant. "They're trying to learn in a decade what it's taken big foreign companies a century to master."

The failure to secure a pipeline from Russia has especially frustrated China's leaders. Siberian oil is currently transported into China at great expense in trains and trucks. For years Beijing has lobbied its former communist brother for a pipeline to refineries in Daqing. During a visit to Moscow last month, Premier Wen Jiabao repeated China's entreaties but received no promises. Meanwhile, Japan has offered to pay for part of the multibillion-dollar pipeline—as long as it terminates in the Russian port of Nakhodka, near Japan. Moscow seems inclined to take Japan's offer, and "China feels betrayed," says Bernard Cole, an expert on China's oil needs at the National War College.

> By some estimates, the country's current proven oil reserves will be depleted in as little as 14 years

Even at home, Beijing has faced. In August, a consortium led by oil giant **Shell** pulled out of a just-finished gas pipeline—running 2,730 miles from the western deserts to Shanghai—after the firms decided their returns would be too small. A planned oil pipeline covering the same distance has also seen no takers. Then last month, Shell and **Unocal** backed out of a multibillion-dollar project to tap gas fields under the East China Sea.

So China's import requirements will continue to rise—putting upward pressure on world crude prices. If China's oil demand keeps growing at an average 7% a year, as it has since 1990, in less than 20 years will be consuming 21 million bbl. of oil a day, matching the current U.S. consumption. "The world has the oil," says Chen Huai of China's Development Research Center, a think tank in Beijing run by China's Cabinet, "and China has the money." The question is: How much is China—and the world—willing to pay for it?

—With reporting by Susan Jakes/Beijing

China's Africa Strategy.
Out of Beijing

BY STÉPHANIE GIRY

ALL SUMMER, THE U.N. Security Council debated whether to condemn the Sudanese government for supporting the murderous *Janjaweed* militias in Darfur. In July, it passed a weak resolution threatening sanctions against Khartoum. Then, in September, after the Bush administration labeled the massacres genocide, the United Nations passed another, similar, resolution threatening sanctions if the killing continued. Yet the United Nations did nothing more, even as the death toll rose in Darfur. On the campaign trail, John Kerry blasted the Bush administration for failing to push the United Nations to take a stronger stand. Editorials in nearly every major U.S. newspaper echoed Kerry's criticisms, denouncing the White House for watching Sudan burn.

Yet the Bush administration is getting more blame for Darfur than it deserves—and Beijing is not getting enough. Quietly but steadfastly, China's ambassador to the United Nations, Wang Guangya, has helped defang U.S.-sponsored drafts against Sudan, transforming language threatening to "take further action" against Khartoum into the more benign "consider taking additional measures." China then abstained from voting on even this weakened resolution, along with longtime ally Pakistan.

Beijing's goal? Probably to protect its investments in the Sudanese oil industry, including a 40 percent stake in a refinery pumping more than 300,000 barrels a day and a 1,500-kilometer pipeline from Sudan to the Red Sea. China's projects in Sudan are the pride of Beijing's new policy of prospecting for oil abroad—especially in Africa, where vast untapped fields could help fuel China, which recently became the world's second-largest oil consumer. In fact, by massively investing not only in African oil but also in African public works, telecoms, agriculture, and other sectors, China is trying to buy the hearts and minds of African leaders as part of a broader push to win allies in the developing world and boost its soft power abroad.

But China's efforts don't bode well for African democracy—or for Washington. As the diplomatic wrangling over Sudan shows, China's march into Africa will, at best, complicate African and U.S. efforts to bring good governance and human rights to the continent. At worst, it will hurt the fight against terrorism and weapons proliferation.

IN 1993, AFTER decades of self-sufficiency, Chinese domestic oil production could no longer satisfy demand, which had shot up because of the country's extraordinary economic growth. Since then, China has had to import more oil every year, from 6.4 percent of its total consumption in 1993, to 31 percent in 2002, to a projected 60 percent by 2020. As a result, Beijing has started to look for more oil abroad. In theory, China should be content to buy its oil from the world market. But, like the United States and other big consumers that don't trust the system to work perfectly, China has started courting producers directly. In 1996, Beijing set out to meet a third of its oil needs by sending state-controlled companies to prospect throughout the world.

So far, Beijing hasn't had much success in Asia. Deals for two pipelines, one from Russia and another from Kazakhstan, have stalled, leaving China with no direct conduit to any of its neighbors. Meanwhile, China's state-controlled oil companies haven't been able to displace well-established U.S. and European competitors in the Middle East. Despite fervent negotiations, the most Beijing has obtained from Saudi Arabia, the world's largest producer, is the right to prospect natural gas to sell to the domestic Saudi market.

Little wonder, then, that Chinese officials see Africa as El Dorado. A recent report by Washington's Center for Strategic and International Studies estimates that proven oil reserves in West Africa have doubled in the past decade to over 60 billion barrels. And, thanks to foreign investment, production in the region is about to take off. A decade ago, West African states (other than Nigeria) produced little. But Cambridge Energy Research Associates predicts that the region will turn out one in every five barrels of crude produced in the world between now and 2010, as well as vast quantities of natural gas. Except for Chad's inland fields, most of these resources lie offshore in the Gulf of Guinea, in an arc running from Nigeria to Angola. West African oil is of high quality—light, waxy, and low in sulfur—and is especially appealing to Chinese companies, which use refineries designed for domestic oil of a similar grade. Another bonus: Aside from Nigeria, West African producers do not belong to OPEC and are not subject to the cartel's production or export caps.

Chinese officials understand what is at stake. China already buys 25 percent of its foreign oil from Africa, and a recent report commissioned by Prime Minister Wen Jiabao urges Beijing to increase that share to 30 percent. (African oil accounts for 14 percent of total oil imports to the United States.) Beijing now hopes to get 50 million tons of oil from the Gulf of Guinea every year. In 2003, it bought ten million tons from Khartoum and more than $1 billion of petroleum products from Angola and South Africa each. Chinese oil firms CNPC and Sinopec recently struck deals with Algeria, Gabon, and Nigeria, and they are in talks with the governments of Niger, Chad, the Central African Republic (CAR), Congo, and Angola.

These projects are the handsome result of Beijing's aggressive new African diplomacy. In February, President Hu Jintao embarked on a highly publicized tour of Algeria, Egypt, and Gabon, with oil executives in tow. At roughly the same time, Sinopec inked its first agreement with Gabon, a nation also wooed by the Bush administration and U.S. oil majors. Gabonese Foreign Minister Jean Ping, whom Hu met earlier this year, is known in West Africa as a Sinophile. Ping could prove a good intermediary between Beijing and the tiny island-state of Sao Tome and Principe, an official ally of Taiwan that sits on major untapped oil reserves and was saved from an attempted coup last year by Ping's intervention. Sao Tome and Principe has also been aggressively courted by U.S. oil companies.

Beijing's oil-slicked diplomacy builds on goodwill generated by its past actions in Africa. In the 1960s and 1970s, China supported revolutionary movements in Africa, vowing to help protect it against U.S. and Soviet imperialism. Many of those former rebels have now come to power and are willing to repay favors to Beijing. What's more, unlike the United States or other Western nations, Beijing does business without setting conditions on human rights, transparency, or good governance. It requires only that trading partners not recognize Taiwan, and all but a handful of African states have obliged.

PREDICTABLY, THIS AMORALITY has been a boon to business. Beijing's state-controlled CNPC held onto its share in the Sudanese project, even after stockholders pressured the Canadian company Talisman to pull out because of Khartoum's human rights violations. (A 1997 sanctions package prohibits U.S. companies from investing in Sudan at all.) For African nations, doing business with U.S. companies "is complicated," says International Crisis Group special adviser John Prendergast, an Africa expert. "[The United States has] a lot of different compartments to [its] foreign policy. And many of them are really

irritating to countries like Nigeria and Angola." Authoritarian African leaders understand the difference. Unlike America's "tied aid," China's cooperation comes only "with mutual respect and regard for diversity," said Gabonese President Omar Bongo, who has been in power for three decades and is routinely condemned in State Department human rights reports.

Unencumbered by principles, Chinese companies are free to go where many Western firms cannot. Beijing moved closer to Nigeria in the mid-'90s, when, after the execution of writer-activist Ken Saro-Wiwa, the U.S. Congress considered blocking new investments by U.S. oil companies. Chinese companies positioned themselves in Libya well before the U.N. sanctions against Tripoli were lifted last year. Consider also Beijing's tactics in the CAR. When the European Union and international lenders refused last year to bail out the new authoritarian government until it restored constitutional order, Beijing stepped in, bankrolling the entire civil service. The move was savvy: Being in the CAR government's good graces won't hurt when, as energy experts predict, access to Chad's oil fields opens up on the CAR side of the Chad border.

Beijing's methods apply to more sectors than just energy. In 2000, the Chinese government launched the China-Africa Cooperation Forum, "a framework for collective dialogue" designed "to seek peace and development." Beijing could have said commerce, too. By the forum's second meeting, three years later, Beijing had signed 40 trade agreements and 34 investment treaties with African states and wiped clean a collective debt worth some $1.3 billion. Meanwhile, China has helped build roads in Equatorial Guinea, dams in Morocco, an airport and a nuclear reactor in Algeria, and government offices in Cite d'Ivoire, Djibouti, and Uganda. Beijing is even pushing the Chinese to travel to Africa and to buy African products at home.

In return, Beijing has won access to critical resources like natural gas, in addition to oil and vast new markets for its goods. Chinese trade with Africa has almost doubled since 2000; and, although it is still a fraction of China's worldwide exchanges, it's already about half of U.S.-Africa trade.

BEIJING'S TACTICS SHOULD worry Washington—not to mention average Africans. Increasing demand and shrinking domestic supplies are making the United States overly dependent on oil imports, and Washington's search for new suppliers in West Africa—which the Bush administration has called "the fastest-growing source of oil and gas for the American market"—seems to pit the United States against China. Yet, because Washington already has solid ties with the Gulf of Guinea's largest exporters and U.S. companies are better equipped than CNPC or Sinopec to perform the type of deep offshore drilling that will unlock the region's resources, China's hunt for African resources is not a direct threat to U.S. energy security.

It is, however, a threat to other U.S. interests on the continent. While Beijing courts African capitals for access to oil and gas, Washington is trying to convince African leaders to host U.S. military bases, battle terrorism, and emphasize human rights. After all, with growing concerns that unstable regions could become terrorist havens, Washington's commitment to democracy in Africa is becoming a security imperative. The Bush administration has committed $100 million to help eastern African states train their forces to better patrol borders, block terrorist financing, and reinforce aviation security. And, at an international counterterrorism conference in Algiers last month, U.S. ambassador for counterterrorism Cofer Black renewed a pitch for Africa's cooperation, arguing that "the struggle against terrorism is also in part the struggle for a better society." Meanwhile, through the Millennium Challenge Account, the Bush administration has offered aid to African nations that promote good government.

But China's march could scuttle Washington's efforts. In a searing 581-page report, Human Rights Watch recently argued that Chinese companies are complicit in Khartoum's efforts to displace populations in southern Sudan to clear the way for oil rigs. It also charges that China's oil purchases have enabled the Sudanese government to buy arms—sometimes from Beijing itself—fueling the violence in Darfur that Washington says it is now trying to stem. And who knows how much of the $1 billion in arms that Beijing sold to Ethiopia and Eritrea during their 1998–2000 war has migrated over the border into Sudan? Of course, it would be unfair to hold Beijing solely responsible for Sudan's abuses, or to absolve Washington simply for condemning the violence in Darfur. But, by bypassing sanctions, trading arms, and leveraging its influence with international organizations, Beijing has indeed helped prop up Sudan's oil industry and the government in Khartoum.

Likewise, Beijing could undermine U.S. efforts to stabilize President Olesegun Obasanjo's beleaguered and reform-minded government in Nigeria, a key ally in Washington's fight against terrorism. "It would really help the Nigerian [government] to have a business partner that didn't care about human rights," Prendergast says, citing Obasanjo's continued struggle with rebels who regularly disrupt oil production. "Sudan is providing a really good example of how China can help you," Prendergast says. "If I were Obasanjo, I'd be looking at that."

Some already are. Last month, the governor of Nigeria's Kaduna Province, the site of sectarian killings and an enforcer of sharia law, invited Chinese investors to set up businesses there. And the government of Angola, which has been pushed by the United States and the World Bank to improve transparency in the distribution of its oil money, has increasingly sought out Chinese investors. As a result, Beijing's unconditional investments across Africa could feed violence or prop up authoritarian leaders, much as they have in Burma, Laos, and other Asian nations where China has become the main external power. Just ask Darfurians where that can lead.

Stéphanie Giry is an associate editor at FOREIGN AFFAIRS.

Love your Liposuction.
Tout that Tummy Tuck.
Flaunt the Fake Nose.

If You've Bought It, Flaunt It
China has taken to plastic surgery with gusto.
Now the artificial beauties, many of whom seek a Westernized look, have their own pageant.

By Mark Magnier, LA Times Staff Writer

Organizers of a new beauty pageant here believe artificial breasts and surgically sculpted butts not only shouldn't be hidden away, they're something to brag about. Welcome to the brave new China, which is making history with what it claims is the world's first Miss Plastic Surgery contest.

"Naturals," with their God-given, pain-free looks, have no place here. This stage belongs to those who have suffered for their beauty and now live beyond the cutting edge. All nationalities are welcome, but contestants must show a doctor's certificate at the door.

At a news conference this month announcing the contest, to be held in early November, a host of beauty and cosmetic industry luminaries were trotted out, in a nation where plastic surgery is a runaway hit.

"To us doctors, altering beauty is a very natural thing," Zhao Xiaozhong, a medical professor and industry expert, told journalists. "When you do sports, you alter your muscles. We do the same thing through surgery."

Then came the moment everyone was waiting for: a peek at a genuine artificial beauty. "Down in front!" yelled one cameraman as Lu Xiaoyu, 23, took the stage to the oohs and ahhs, applause and neck-craning of several dozen reporters.

"I hope this contest helps people learn about plastic surgery," the former farm girl from Hebei province said. "I hope to see a day when it's so commonly done we'll no longer use the term 'artificial beauty.'"

"Do you have scars, and will you show them at the pageant?" one inquiring mind wanted to know.

"I'd be willing to," Lu responded as several cameras flashed.

Lu, who has large eyes and curly brown hair—thanks to dye and a permanent—said she had wanted plastic surgery since childhood. Everyone else in her family was good-looking, and she felt like the ugly duckling.

A few years ago, she left her hometown—with its back-breaking farm work and peasants—for Tianjin, a city just outside Beijing. There she landed a job in a beauty salon, where she decided to revamp her nose, add dimples and creased eyelids, and otherwise reshape her face, all with the help of her employer.

"I feel like I'm living in a dream," she said. "Now my parents say, 'We finally look like we're from the same family.'"

Since the contest was announced, Lu has been joined by more than 30 Miss Plastic Surgery hopefuls from as far away as New York, Malaysia, South Korea and the vast reaches of China, all keen to nab the title and the $1,200 prize money.

Although that's hardly enough for a tummy tuck, the real lure is publicity—the pageant will be televised and the winner is promised a role in a planned Chinese TV drama in which every actor or actress boasts man-made charms.

"We'll draw a line, though," said Zhao Chaofeng, a planner with Beijing Culture & Media, a pageant and beauty products company that is sponsoring the contest. "The director won't need cosmetic surgery. "

A lot of the credit for the world's first plastic surgery showdown goes to Yang Yuan, a leggy 18-year-old with hennaed hair. She created a stir last year when she underwent 11 surgeries, at a cost of $13,000. Her goal was to enter the Miss Intercontinental beauty contest, also sponsored by Beijing Culture, but she was disqualified when someone noticed her picture in a before-and-after surgery ad.

A tuck here and there is all right, Beijing Culture argued, but Yang had changed her entire face. Yang wept, cried foul, then called her lawyer and filed a $6,000 lawsuit charging psychological damage. The court showed little sympathy.

But faster than you can say "Botox," the pageant company created this contest, built around the sculpted-body arts. "We want to give them a forum of their own," organizer Zhao said.

Having knocked down the door for artificial beauties, Yang ultimately decided not to cross the threshold.

"I'm not entering this contest," she said as she sat in the clinic where she had her face redone, dressed in denim hot pants and a sleeveless brown shirt, her long hands fingering two late-model cellphones. "The committee is treating me like a ball they can kick around."

Despite the high-profile no-show, interest in the pageant has far exceeded expectations, befitting a country that has embraced plastic surgery with gusto. The industry has surged from almost nothing a few decades ago to a $2.4-billion business that is growing by an estimated 20% a year.

The most eager to be altered reportedly are women in their 20s—hoping to supercharge their careers—and 40s—hoping to remain young—although more and more men are taking the plunge. And though a surgical overhaul is costly, modest procedures can cost a couple of hundred dollars or less.

Driving the growth are higher living standards and a more global outlook, beauty experts say, as well as pent-up demand stemming from the communist history that condemned individual beauty as a bourgeois affectation.

Shi Sanba, 54, president of Beijing's MengNiHuan surgical clinic, recalls rubbing red paper on her lips during the Cultural Revolution to simulate lipstick, risking a self-criticism session.

"In those days, there was no such thing as beauty," she said. "Having breasts was shameful, so we made little tight bras to keep them hidden. Everything was about revolution."

Today, many young women who choose to have cosmetic surgery want a more classically Western look. Creased eyelids, thinner noses and larger breasts are among the biggest sellers.

This has prompted traditionalists and women's rights groups to fret that China is losing its soul in the headlong embrace of all things foreign.

"We're losing diversity in the rush for a global beauty standard," Renmin University sociologist Li Lulu said. "If everyone starts looking the same, it will be a pretty dull world."

In China's highly competitive society, cosmetic makeovers are often seen as a way to earn more money, get a better job, even find a wealthy husband.

"I'm not looking for a sugar daddy, but I hear good looks may boost your salary by 30%," Yang said.

The quest is not without peril. In a case reported this spring in the newspaper Heilongjiang, businessman Jian Feng married a woman from Qingdao without realizing she'd had a surgical overhaul. When an "amazingly ugly" child arrived in 2003, Jian accused his wife of infidelity, then divorced her, obtaining a $120,000 settlement for misrepresentation and "lost opportunities"—namely, that he could have married someone else.

Exploding demand for surgery has created a thriving business in gray-market procedures, including many performed by fake doctors in the back rooms of beauty salons. Approximately 200,000 botched surgeries were reported during the 1990s.

"Driven by profit, many nonmedical groups are stealing into this industry, ruining our reputation," said Wang Jigeng, head of the plastic surgery division of the Chinese Medical Assn.

At the MengNiHuan facility in Beijing, which says it uses only qualified surgeons, several young women entering the storefront clinic were ushered into a consultation area furnished with small wicker tables, fake-grass carpet and a mock palm tree.

As the customers fidgeted nervously, young assistants ran them through picture books, touching their eyes, noses, temples as they explained what needed to be cut, moved or filled in. On a shelf, a model skull, its brain cavity empty, sat beside bottles of herbs and a half-full pack of cigarettes; nearby, a staff member wielded a fly swatter with aplomb as her cellphone rang with a jazzed-up version of the "Nutcracker" Suite.

Owner Shi, dressed in Hawaiian-print pants and a red scarf that complemented the tropical decor, believes in leading by example. She's had 20 to 30 operations and foresees a day when getting plastic surgery will be as routine as having a tooth filled.

She declared that all the work she had done makes her look 20 years younger, whipping off her scarf to reveal a long hairline scar from a face lift. Aside from a slightly rippled area above her mouth, there was little evidence of what had been done.

"I really bring joy to people's life," she said. "Designing people's faces is a bit like playing God, but if someone's born ugly, why shouldn't they be beautiful?"

A few feet away, a nervous-looking Amie Xu, a 20-year-old college communications major, sat with her mother. She had just returned from a year abroad and wanted her jaw broken and rebuilt to narrow her face before next semester. She was also considering a few changes to her nose, chin and eyes.

"They say it will be about $6,000," she said as her thin-jawed mother looked on approvingly.

Several miles away, at the Evercare Medical Institution, belly dancer An Li was conferring with chief surgeon Zhou Gang about a nose and jawline correction, an eyebrow transplant, breast augmentation and liposculpture.

She's betting the $12,000 in work will further her career.

"I hope to dance in Las Vegas someday," the petite 25-year-old said, demonstrating a few gyrations in the consultation room. "In my line of work, beauty is only skin-deep, and I hope to work at this job until I'm much older—maybe 33."

The dancer is too late to enter this year's Miss Plastic Surgery pageant—contest rules require that all work have been completed by May to prevent rapid-fire surgeries, which can be dangerous—but An says she'll consider entering next year.

Next in line for a consultation was Sonny Lu, a 24-year-old male translator interested in a sex change. Liu Tong, Evercare's medical director, reviewed Lu's face, then showed him a photo album labeled "Deep Emotion," filled with before-and-after shots of male breast jobs.

"Your nose is a little long, the cheeks a little high, and the chin would need to be bent out," Liu said. "But look at these little delicate arms. They're wonderful! I've done more than 10 breast operations, but broad shoulders are a problem."

"Have you considered modeling?" a nurse chimed in.

But even as contest organizers create a pageant for artificial beauties, they are drawing the line at transsexuals, fearing this would be too much for the Communist Party.

"Then again, maybe transgender people will get their own contest someday," said Sonny, who plans to have his work done in installments.

The pageant will judge contestants in several categories, including swimsuit and evening gown, talent and personality. Entrants must show a certificate from a recognized plastic surgery hospital, but promoters say no job is too small as long as you're even a little bit artificial.

In a country where just about anything can be pirated, the organizers say they'll do their best to ensure that the certificates are real. That includes working with the plastic surgery branch of the medical association, which happens to be a sponsor.

Organizers also plan a seminar on the problem of unlicensed, unauthorized and fake plastic surgeries, as well as a debate between natural and artificial beauties on which is better.

It's all great publicity for the surgeons, who appear to have their sights on a wider audience at a time when China seems to be taking over every other global industry.

"We're already seeing Koreans and Japanese come over for operations," and Europeans and Americans could follow, clinic owner Shi said, running down her price list, which ranges from $1,200 for a nose job to $3,500 for extensive breast work. "We're cheaper and faster, so there's a real global market here, provided we boost our reputation.

"We'd need to become more adept at Western faces, though," she added after a moment's thought. "We won't get anywhere trying to make foreigners look Chinese."

Lijin Yin in The Times' Beijing Bureau contributed to this report.

The Latin Americanization of China?

Land reforms aimed at raising rural incomes and promoting urbanization could accelerate the crisis already building in China's cities. If urban legal and social reforms fail to keep pace, China could face intensifying conflict between a burgeoning class of have-nots and an entitled minority, a consolidated alliance between political leaders and business and social elites, and a host of other social and political ills familiar to many Latin American states.

GEORGE J. GILBOY AND ERIC HEGINBOTHAM

Wide-ranging liberal market reforms have produced rapid gains in China's overall economic growth over the past two decades. Yet rural policy since 1978 has been rent by opposing influences: the state recognizes the growing plight of farmers facing market reforms, but it refuses to accept rural migrants as full members of urban communities. Today, however, China's leaders are deepening land reform programs in the countryside. Reformers hope this will spur consolidation of land into larger, more efficient agricultural holdings while encouraging inefficient farmers to divest their land, leave the countryside, and help fuel healthy industrial growth by selling their labor in China's burgeoning cities.

As Karl Polanyi, the author of the 1944 study, *The Great Transformation*, could have predicted, this process is not going smoothly. Although Polanyi was describing the enclosure movement and subsequent social, economic, and political crises in eighteenth-century England, many common themes are now being played out in China's own great transformation, including worsening inequality, rising expectations, and increasing conflict and violence in the countryside.

Yet the current crisis in the countryside is only a precursor to the deeper and more fraught crisis that is growing in China's cities. China's economic reforms have created what Sun Liping of Tsinghua University calls a "cloven society." The new rich and powerful now live in walled, guarded villas and modern apartment complexes, enjoying vast differences in wealth, power, and rights from the swelling ranks of the rural poor and urban dispossessed. The latter are composed of millions of migrant workers living in shantytowns, alongside the growing numbers of urban unemployed and low-income residents who are being forcibly removed from the city center to make way for new real estate development. This second, developing crisis is not only a crisis of infrastructure and incomes—the hardware of urban life. As millions of peasants seek a permanent home in China's cities, it is also a battle for identity and entitlements—the critical software that makes urban society workable. These "urban rights" include

legal status and accompanying access to jobs, education, health services, insurance, and social welfare benefits.

The outcome of this second crisis, though it will certainly involve increasing scope and intensity of conflict and confrontation, need not be endless discord or regime collapse. China's tumultuous reform process could see the creation of new, more liberal legal and social institutions. Transforming migrants into urban citizens with equal rights and allowing social groups to organize and articulate their own interests would both improve the ability of the government to govern effectively and minimize long-term threats to stability and economic development.

But other outcomes are also possible. The state could refuse to allow liberal institutional innovation and slip into a modern form of authoritarian corporatism in which political leaders might seek to channel social energies toward nationalist ends—the "revolution from above" about which Barrington Moore warned. Or alternatively, China could catch the Latin American disease, characterized by a polarized urban society, intensifying urban conflict, and failed economic promise. Indeed, despite aggressive efforts to make the state more responsive and adaptive, the speed with which social cleavages and conflicts are growing today arguably makes this last outcome easier to imagine than the others.

THE SUFFERING COUNTRYSIDE

China's rural areas are now deep in crisis, with sluggish income growth, peasants burdened by excessive taxes and fees, and local governments overstaffed, in debt, and unable to provide adequate services for peasant families. Rampant corruption among local officials has combined with these factors to incite increasing levels of peasant organization, protest, and violence. This crisis is not new, but it is reaching a new scale and intensity. In a 2004 Chinese Academy of Social Sciences (CASS) survey of 109 of China's top sociologists, economists, managers, and legal experts, 73 percent of the respondents identified the "three rural problems" (*san nong wenti*) of agriculture, peasants, and rural areas as China's most urgent challenge. Combined with other issues such as corruption, the intensity of the rural turmoil led more than half of the respondents to see a systemic crisis as "possible" or "very possible" within the next 5 to 10 years.

Small-scale inefficient agriculture and the relative decline of township and village enterprises are contributing to a widening rural-urban income gap. Average annual rural income stands at just $317 today, and the gap between urban and rural income has grown from 1.8:1 in the mid-1980s to 3:1 in 2003. Between 2000 and 2002, incomes fell in 42 percent of rural households in absolute terms. And according to a July 2004 government report, the number of farmers living under the official poverty line of about $75 per year increased by 800,000 in 2003, the first net annual increase in absolute rural poverty since economic reforms began in 1978. At the same time, farmers suffer from a disproportionate tax burden while receiving fewer services; according to the State Council's Development Research Center, the urban-rural income disparity soars to between 5:1 and 6:1 when entitlements, services, and taxes are included in the calculation.

Unsurprisingly, organized rural protest is on the rise. Actions range from tax evasion and blocking roads and railways to the assault or kidnapping of officials and even to riots that have involved hundreds or thousands of people. Even so, the nature of rural protest and of the state's response to it limits the possibility that rural conflict alone could threaten regime stability. As Yu Jinrong of CASS notes, when rural residents do engage in collective action and protest, they often seek alliances with central government officials against local officials, rather than seeking broad-based systemic change. Yu argues that today's peasants are not the revolutionary "peasants of Mao." They are seeking legitimate political organization to defend legitimate economic interests, and, he warns, suppressing their aspirations and organization carries significant political risks.

Beijing has been highly attuned to rural problems for the past several years, and has taken steps to address them. In particular, the central government has had some short-term success in reducing the peasant tax burden by cracking down on illegal local fees and converting fees to more transparent taxes. It also has moved to share a larger amount of central revenue with local governments. The central government has created more safety valves for expressions of rural discontent, clamped down on abuses by local officials, explained policies to peasants, paid out monies to mollify protesters, and allowed village elections (although it has also simultaneously removed considerable tax and fiscal power to the higher township level, not subject to elections).

These measures are, however, also creating a strong sense among Chinese citizens that they have "legal rights." Rural residents increasingly refer to these "rights" in their protests—a potentially significant development for the future of Chinese politics. And, despite the government's success in localizing, suppressing, or conciliating potential rural threats, the leadership does not believe that such measures represent a real long-term solution to the *san nong wenti*.

THE GREAT ENCLOSURE

Many key Chinese policymakers and social scientists believe the solution to the rural crisis lies in a more radical approach: a combination of land reform, industrialization, and urbanization. Wang Mengkui, the director general of the State Council Development Research Center, argues, "Too many people and too little land makes large-scale production difficult and is therefore the greatest problem for farmers to increase their incomes." The consolidation of larger farms and the movement of farmers to the cities will go far toward solving the rural problem, he asserts, and as an additional benefit of urbanization, "large numbers of migrant workers [will] supply cheap labor, thus helping to enhance the international competitiveness of Chinese industries." Pan Wei, an influential government adviser and Beijing University professor, also argues that Beijing should encourage a rapid acceleration of peasant migration to urban centers, proposing that China should develop an additional 100 cities of 5 million people or more over the next 30 years, either by building new cities or expanding existing ones.

The migration from country to city, already massive, is accelerating. In part, this is being driven by illegal land seizures and the conversion of farmland to industrial and recreational use. In November 2003 the Ministry of Land and Resources reported more than 168,000 cases of illegal land seizure, twice as many as in the entire previous year. According to the State Statistics Bureau, China lost 6.7 million hectares of farmland between 1996 and 2003—three and a half times as much as the 1.9 million lost between 1986 and 1995. The trend continues to accelerate, with some 2.53 million hectares, or 2 percent of total farmland, lost in 2003 alone. According to the 2004 *Green Book of China's Rural Economy*, for every *mu* of land (approximately 0.07 hectares) that is transferred to nonagricultural use, about 1 to 1.5 farmers lose their land. According to official statistics, some 34 million farmers have either lost their land entirely since 1987 or own less than 0.3 *mu*, and the new surge in land transfers almost certainly indicates acceleration of that process.

The government has met with some success in curbing the transfer of farmland for nonagricultural purposes during 2004, but a more sustained, legal, and probably larger-scale shift in rural land tenure patterns is in the offing, in this case driven by the central government's efforts to rationalize agriculture and raise rural incomes. The landmark Rural Land Contracting Law (RLCL), which took effect in March 2003, is the latest means toward that end. Under the post-1978 household responsibility system, land remains owned by the village, with use rights allotted by village leaders to individual households. The lack of secure land tenure periods and the frequent use of "readjustments" by village leaders (that is, reapportioning land between households) inhibited improvements to the land, transfer of land-use rights between farmers, and the emergence of commercial-scale agriculture. The RLCL mandates written contracts between farmers and villages, and sets the period of land tenure at 30 years. It includes clear provisions for the farmer's right to transfer land rights to others. And, to give potential buyers confidence that their land-use rights will be respected, it prohibits "readjustments" except in extreme cases (for example, natural disasters). No doubt, enforcement of the RLCL will be inconsistent. But the central government appears committed to the task and will almost certainly continue to sharpen land-use legislation. Indeed, the agenda may be expanded to ease rules on mortgages and to push the household-based tenure system toward an individual-based system, two measures that would substantially speed the transfer of land-use rights.

Among well-heeled urban young people, the phrase "You're so farmer!" (Ni zhen nongmin!) has gained currency as a playful expression of disgust.

If successful, land reform will accelerate China's internal mass migration. But the impact of both illegal seizures and land reform will not be limited to an increased rate of migration. The composition of the "floating population" also will be affected. Many of those who previously crowded onto trains for the cities went in search of higher incomes and were, in fact, adding one income to the family effort since their wives, husbands, or parents continued to work the farm in their absence. Today, an increasing number of people are moving with families in tow, no land or homes behind them, and no guarantees ahead.

China's best-known business and economics magazine, *Caijing*, has called the recent spate of rural and urban land seizures by alliances of local officials and real estate developers a new "enclosure" (*quandi*) movement, consciously echoing the process that sped urbanization and was so disruptive and violent in eighteenth-century England. But for many peasant families, legal transfers under the RLCL will have a similarly dislocating effect. Rural reform is incomplete without also guaranteeing the assimilation of China's migrants as full, productive members of urban society.

A Bigger Crisis Tomorrow

Speeding China's urbanization trades one social and political problem for another that is potentially more severe. The problem of poor farmers working small plots becomes that of poor migrants working dangerous jobs with few rights and virtually no social security safety net. The scale of China's urban landscape is already daunting: 166 cities of more than 1 million people (the United States has 9) and 500 million official (that is, without counting migrants) urban residents. Urban population growth is already at 2.5 percent per year (versus 0.8 percent for India), and the government expects 300 million people to move to China's cities and towns between 2004 and 2020. Because most of China's migrant workers retain their *shenfen*, or personal status, as farmers in their home locality, they are cut off from access to urban services, social security, and effective legal protection. This problem could worsen unless the next

generation of migrants who have lost their land—either through illegal seizures or through the legal operation of a land-use rights trading system—are granted rights and benefits that will allow them to fully join urban society. The current plight of China's migrant workers offers a glimpse of the obstacles that must be overcome.

Migrant workers without municipal *hukou* (registration) cannot participate in regular job markets. When they do find work, their rights under Chinese labor law are frequently violated. Their wages are withheld for months or years. The government estimates that China's 100 million migrant workers are owed $12 billion in back pay. Mandatory safety conditions often go unmet. According to *The China Youth Daily*, in one urban area alone—Shenzhen and the surrounding Pearl River Delta region—industrial accidents claim more than 30,000 fingers from workers each year. The standard payout for such injuries is $60 per finger, but many employers refuse to pay any compensation. According to government officials, nearly 70 percent of migrant workers have no form of insurance. And most live in shantytowns outside the cities, where whole neighborhoods are subject to clearance and destruction on little notice and with little or no compensation.

The impact of this ambiguous, floating status falls disproportionately on children and other weak dependents who travel with workers. Currently, the floating population includes an estimated 3 million children aged 14 and under. According to a 2004 government report, pregnant migrant women and their children suffer mortality rates between 1.4 and 3.6 times the national average. Of migrant children between the ages of 8 and 14, some 15 percent do not attend school. Most of those who do attend pay high fees (often $100 or more) to enroll in improvised, substandard private schools. Pressures associated with payment—and shouldering an entire family's hopes for the future—have prompted a rash of student suicides and even murders. Although problems associated with migrant workers have been apparent for some time, the rapidly accelerating trend toward landlessness and the consequent growth in whole families on the move make specific problems associated with dependents new in magnitude if not in nature. And, while services in the countryside were also poor, China's underclass in the cities will perceive injustice more keenly as they see the benefits that the new rich enjoy every day.

In theory, even urbanization advocates understand that, as Wang Mengkui, the State Council official, put it, "urbanization requires institutional innovation." To date, efforts have been limited to protecting migrants against some of the worst abuses. In a major government work report in March 2004, Prime Minister Wen Jiabao declared that the government would "basically solve the problem of default on construction costs and wage arrears for migrant rural workers in the construction industry within three years." The Ministry of Labor and Social Security said this year that it will oblige construction and manufacturing firms to provide health and life insurance to millions of migrant workers. And the central government has encouraged municipalities to give migrants greater and cheaper access to public schooling, though, as with most measures, no central funds are earmarked. In the first sign of state and party support for a broader defense of rights for the urban poor, reformers in the

National People's Congress are now drafting a law that amounts to a "Bill of Welfare Rights" for China's internal migrants.

OBSTACLES TO REFORM

Despite the rhetoric and regulations, real progress has been limited, and the gap between rising consciousness of rights and the ability to act on and realize these rights is growing. The most obvious problem is money, or, as a group of Chinese scholars noted in a major new book, *China's Urban Development Report*, the question of "who will pay the bill for China's urbanization." The scholars' answer is simple: urban industry. But the construction industry, which is most relevant to the migrant economy, has resisted paying its existing obligations on time, much less shouldering additional costs. With local governments profiting from the construction industry and officials making their reputations based on building and development—not to mention widespread bribery and corruption—the incentives to overlook violations remain powerful.

Nanjing University's Pan Zequan, writing in *Strategy and Management*, argues that pervasive discrimination against migrants is not simply an inherited evil now being attacked and reversed; rather, it is built on consciously erected systems and policies and is regularly "produced" and "reproduced." Although progress has been made in some areas, Pan's contention that a dynamic struggle is under way rings true. Certainly, discrimination against migrants works to the advantage of—and is convenient for—those who already hold entitlements in the cities. Material interests are reinforced by strong local identities and prejudice against rural "outsiders" (*waidiren*)—a phrase invariably used in reference to migrants. Eastern urbanites frequently explain to Western visitors that *waidiren* are of "low quality" (*suzhi di*) and say they feel less in common with domestic migrants than they do with foreigners. Among well-heeled urban young people, the phrase "You're so farmer!" (*Ni zhen nongmin!*) has gained currency as a playful expression of disgust.

Given hostile interests and culture, it is not surprising that measures to lessen the hardships of migrants often meet with obstruction. Despite Beijing city officials' recent order to public schools to admit the children of migrant workers and to cut discriminatory tuition fees imposed on them, many schools continue to exclude migrants by claiming to be filled to capacity when, in fact, a survey by the Beijing Education Department showed 35,000 vacancies. Members of the floating population face discrimination even in death. In a recent incident in Luzhou, an explosion in a city gas pipeline killed several people. The families of city residents were compensated with $17,000—those of migrant workers were given $5,000. Although they lived and worked and died in the city, the migrants were still classified as peasants. An official justified the difference with the claim that "the cost of living in the countryside is lower."

Although China's leaders continue to view the rural problem as the nation's greatest threat, a 2004 survey by the Chinese Academy of Social Sciences singles out migrants as the great economic loser of the post-1978 era. When asked which of eight groups had benefited most from China's economic reforms, a distinguished panel of experts was unanimous on only one item: "migrant workers" were worse off than any other group. The least fortunate of them join what a leading government researcher, Zhang Xiaoshan, has described as a new class, the "three havenots": people with no land, no jobs, and no access to national income insurance.

THE FUTURE: LIBERAL, FASCIST, OR DICKENSIAN?

Broadly speaking, three possible 15-year outcomes to the dual rural-urban crises are possible: liberal, authoritarian corporatist (or fascist), and botched. None of these outcomes is preordained. We would argue, nevertheless, that China is now groping its way at least tentatively toward the first, though the pace of social change and the difficulty of overcoming entrenched interests may ultimately make the third most likely.

Progress toward a liberal outcome would see the village election system strengthened and expanded at least to the township level. Land reform would proceed but with land reforms matched by commensurate and simultaneous urban reforms that protect new arrivals in cities and second-generation migrants, and that permit employment beyond construction and road sweeping. In urban areas, the *hukou* system (already being revised) would be eliminated and public services made equally available to all people living and working in a given region. An awareness of legal rights would develop, along with the means to actualize them. Ultimately, individuals, regardless of their status, would be allowed to organize in groups free from direct state control to defend their interests.

Movement can be seen in most of these areas. Village-level democracy—imperfect as it is—is bringing greater accountability to the countryside. Nationally, progress in building legal institutions and, especially, fostering a "rights and accountability" culture has been made on a broader front. Although the road ahead is still much longer than that already traveled, the state seems prepared to countenance a judicial system that will be used to mediate interests as part of the local political process, not simply to administer justice. Legal awareness has been aided by the central government's emphasis on "rule of law" in its own battles to control provincial and municipal governments. And peasants are responding. The State Council Development Research Center reports that an increasing proportion of official petition and protest letters cite legal rights and protections as the basis for the complaints.

In the cities, the state has tolerated, if not encouraged, the rise of a few new independent social organizations. Writing in *The China Journal* in January 2003, Benjamin Read analyzed the development of urban housing associations focused on gaining control of management and improving service quality in upscale real estate developments. These groups capitalize on the government's recent promotion of notions of certain "rights" to property and consumer protection. They have fended off attempts at government co-optation and are promoting a sense of common identity among their members in addition to pursuing claims against negligent or corrupt real estate developers. While Read cautions that it remains to be seen whether these groups can sustain their current autonomy, they offer tantalizing evidence of the kind of ad hoc, innovative interest-intermediation groups that could become the basis for more permanent social and political institutions. Yet, by their very nature, these new associations highlight the disparities in income and social rights between China's haves and have-nots—such open organization and representation are not tolerated in migrant shantytowns. Nor have they successfully emerged among poor city residents who are forced to move from older downtown buildings demolished to make way for new developments. Despite the caveats, however, all this adds up to substantial progress toward a more liberal future.

Unfortunately, other social and political possibilities are also readily apparent. Observers such as Michael Leeden and Jasper Becker have argued that China, far from becoming more liberal, has moved in the opposite direction, toward fascism. Benchmarks for movement in this direction would include the consolidation of society into state-dominated and controlled hierarchical organizations; administrative, rather than judicial, mechanisms for social conflict resolution; the strategic use of anticapitalist and anti-foreign rhetoric; and the heavy involvement of the military in propaganda and social work.

In fact, this largely describes elements of China today. Yet all of these features are becoming less true of the Chinese state, rather than more. Private industry is growing relatively faster than state industry. New self-organized groups are cropping up faster than the state can effectively co-opt or suppress them. The media are more robust, independent, and commercial, with ever-shrinking restrictions on what can be reported. The legal system is growing stronger. And the military is distancing itself from its socioeconomic functions as it has been reduced in size and professionalized. In most key dimensions, China is currently headed away from authoritarian corporatism, not toward it.

There is, however, a third possible trajectory: a "Latin Americanization" of China in which the state could fail to develop institutions capable of adequately addressing China's new social crisis. The speed of social change and the explosive growth of social conflict may outstrip the state's ability to respond. Political leaders could settle into a collusive relationship with business and social elites. A semipermanent have-not class might engage in a constant and economically costly low-level war with the entitled minority. For many Chinese scholars and government officials, Latin American-style social and political problems are now an explicit frame of reference for what China might face if it fails to reverse social trends in the near future.

Despite movement toward a more adaptive, liberal future, the downward spiral toward failure may in fact be just as likely in the mid-term. Some indicators already point toward this outcome. The 2004 report of the Politics and Law Commission of the Communist Party found that the number of incidents of "social unrest" or "mass incidents" rose 14.4 percent in 2003, to 58,000 nationwide. The number of people involved rose 6.6 percent, to 3 million. In the cities, the "floating population" accounted for "up to 80 percent of all crime." Evidence from numerous urban areas suggests that avoiding the Latin Americanization of Chinese society and a descent into low-level class warfare will require more than partial measures designed to mitigate the worst suffering of migrants—it will require making them full citizens.

THE NEED FOR SPEED

China's leaders are intently focused on the nation's rural crisis and the growing gap between urban and rural quality of life. Their proposed solution to these problems—land reforms aimed at promoting mass migration and rapid urbanization—is likely to speed the arrival of a second crisis, pitting migrant families against entrenched urban interests in a struggle for rights and entitlements. Those urban interests are themselves powerful forces, including alliances of municipal officials, real estate developers, and construction industries, alongside a new wealthy urban class and existing ranks of urban poor and unemployed. Yet many migrant farmers, some accustomed to voting for their village leaders, and now promised new protections by the government, bring a new "rights consciousness" with them when they move to the cities. Their expectations of fair treatment and access to benefits such as insurance and health care can only be ignored at the government's peril.

While it is struggling with difficult but familiar rural conflict, China's leadership is less well endowed to deal with the coming urban social challenge. The Chinese system does have remarkable strengths, not least the practice of conducting pragmatic economic and political experiments in individual locations and then embracing successful methods nationwide. It is entirely possible that liberalizing interim solutions could become more permanent institutions, as they did in the England that Polanyi described. But with government plans calling for the market-based "enclosure" of China's rural areas, and several hundred million migrants likely to move to the cities over the next two decades, Beijing is in a race against time.

GEORGE J. GILBOY *is a research affiliate at the Massachusetts Institute of Technology's Center for International Studies.*
ERIC HEGINBOTHAM *is a senior fellow at the Council on Foreign Relations.*

Where the Broom Does Not Reach, the Dust Will Not Vanish

To whose voices is the Chinese Communist Part listening—the capitalists' or the workers'?

Chris Richards

SHE is only 36 years old. Yet when you look at her, you know that every part of her is wearing out. Her whole body sags.

Sometimes she works all day and through the night into the early hours of the morning. After all, there are many costs that come with living in the city. And she has many people in her family to support.

Li Siuling is a migrant worker: just one of 130 million who have left their rural communities to find work in the prosperous eastern cities. The money that she and others like her send back now makes up more than half of the income of the peasants and rural workers in the Chinese hinterland.

Suiling was only 15 when she left Anhui province, not far from Shanghai. It was soon after her father had died. On offer in the city were the menial, dangerous or difficult jobs that are always poorly paid. At first she worked in a restaurant, making 100 yuan ($12.10) a month while the city citizens working with her earned five times that amount. Then she sold clothes, earning 350 yuan ($42.35) while her urban colleagues took home 900. It was just one of the prejudices that she has had to face.

There are four others. After 20 years—like most Chinese citizens who move away from the household where they are officially registered—she still pays an annual fee for a temporary residence permit. Effectively, it registers her as a second-class citizen within her own country. It deprives her of a vote in local elections. It deprives her children of the right to free education. It exposes her to harassment from a police force strapped for cash. Indeed, the police came to her house around midnight once, demanding to see her permit. When she produced it, they tore it up in front of her. Then they fined her for not being able to produce her documents, and jailed her when she wouldn't pay.

The fifth prejudice cursing migrant workers has largely escaped her. Many who have lived in the city for as long as she has done are often unmarried and socially dislocated. But she has a husband—a migrant worker like her, from her province.

Li Siuling lives in Beijing. In Guangzhou, the capital of the southeastern province of Guangdong, I sit at a table with seven representatives of the All China Federation of Trade Unions (ACFTU) to find out whether her story is a common one here, too. Yes, they say, except that after seven years of working in Guangzhou, migrant workers can apply for permanent residence and the social entitlements that go with it.

They tell me that in March last year, Sun Zhigang—a young man from a northern province who worked in a Guangzhou garment company—was taken to a police station because he didn't have a temporary residence permit. He later died from the beating he received there. The public outcry that followed has led to reform of, but not an end to, the system of citizen registration.

The new face of labour

Migrant workers now form the country's main industrial workforce. There are an estimated 20 million of them presently in Guangdong alone. And they are in high demand. China is the factory of the world's manufacturers, and this province is one of its main production lines.

The transnational brands pouring from Guangdong's factories read like a corporate who's who. Economic growth here is staggering. It took Britain most of the 19th century to increase its per capita income by two and a half times, and Japan from 1950 to 1975 to improve its by six. Guangdong's per capita income since 1978 has increased 60-fold. A promotional book about the province says that by 2002, the amount of the province's import and export transactions with countries around the world totalled as much as that of Russia and twice that of India.[1]

The road from the coastal city of Shenzhen (another economic power-zone within the province) to Guangzhou is just over 200 kilometres: a virtually uninterrupted line of factories. A grey haze folds over the horizon. At the halfway mark, the factories seem to take up all the space that the eye can see—blocks after blocks after yet more blocks of grey cement and grey corrugated iron line up into the distance. Inside the larger factories 20,000, 30,000, even 70,000 people work. To an entrepreneur, this must be nirvana. To me it looks like hell.

Capitalism speaks

Taking this journey, the conclusion that corporations have freedom to speak and be heard in this country is inescapable. Local authorities here have been given special powers to govern their own economies. They have long ago learnt the language that attracts investment: special economic zones, tax havens and neglected labour regulations.

So too the Chinese Communist Party, whose government has nurtured this new culture. In November 2002, Jiang Zemin—the country's President at the time—announced that entrepreneurs would henceforth have both membership and a voice in making decisions within the Party. This only formalized what had already been happening. In the *Forbes* magazine list of China's 100 richest in 2002, a quarter said they were CCP members,[2] and nine were delegates to China's law-making body, the National People's Congress.[3]

> **The most outstanding thing about China's 600 million people is that they are 'poor and bland'.... On a blank sheet of paper free from any mark, the freshest and most beautiful charactes can be written, the freshest and most beautiful pictures can be painted.**

Just as big business is now being officially heard in Government, workers' representative committees and trade unions have a legal right to be heard in management. The catch is that there's only one recognized workers' voice: the ACFTU. Other labour unions must be affiliated to it in order to be recognized.

The ACFTU and the CCP are closely intertwined. At the union table in Guangzhou, I'm sitting opposite Chen Weiguang, who chairs the Guangzhou Trade Union Council and is also the convenor of this region's CCP (ACFTU branch). He smiles at any suggestion of conflict, and says the two should work as one. As for the CCP's ties with capital and management: 'On the surface there is conflict, but these must be resolved.'

Inside the larger factories 20,000, 30,000, even 70,000 people work

Ma Xiao Xin, a manager with Panasonic, is also sitting at the union's table. He outlines his company's 'particular attention to labour relations'. Whenever there are issues affecting workers, the company approaches the union. He describes a proactive union response—for instance, negotiating a drop in the temperature in a too-hot factory by 4 degrees. So who is the union representative in the factory he manages? He is. When I express surprise, he explains: 'When we wear the union hat, we have to fight for workers' rights.' International investors would be hard pressed to come up with a more attractive way of doing business with unions.

What should one make of such systems? China's paramount leader during the late 1970s and 1980s, Deng Xiaoping, dubbed the country's new capital-driven economy as 'socialism with Chinese characteristics'. He explained: 'It does not matter whether the cat is white or black; if it catches mice it is a good cat.' In other words, the primary concern should be the betterment of the people. And if the system that delivers the goal is driven by a free market economy, then that's considered appropriate in a socialist government agenda.

According to the World Bank, China's economic transformation has helped the Communist Party Government to deliver people from starvation. The Bank estimates that over the past 20 years between 300 and 400 million people have been lifted out of poverty.[4] In Guangdong province alone the GDP per head in 1978 was less than a dollar a week and well under the international poverty line. By 2000, its 87 million people enjoyed an average annual GDP income of 12,973 yuan (a little more than $30 a week): three times that amount for those who were living in the provincial capital, Guangzhou.

This city—with its row upon row of high-rise buildings reaching to the clouds—looks like any modern capital in the rich world. Its streets throng with an emerging middle class. But like any city, Guangzhou's buildings and streets hide the underclass that capitalism creates to survive: the groups of migrants who will do the menial and dirty work shunned by the locals. They sweat it out in nearby factories and dormitories.

On the factory floor

Overwhelmingly, rural migrants work the factory production lines. In some, 80 per cent are migrants. In others, 90 per cent. Most are young women under 25 years of age: as workers they are cheap and compliant. They come here willingly, most knowing that they will work 10 or 11 hours a day, six days a week. In busy times, some will regularly work all night. Tired and bleary, they are accidents waiting to happen.

I visit a chemical factory, passing by a sign that warns workers that if they use the lift 'inappropriately' they will be fined (a common way for employers to regulate their employees in these provincial factories). The sign is on the way to the factory floor where male workers with scant protection in stifling heat pack chemicals into bags. I see a toy factory where rows of migrant workers sit at tables, applying glue and paint with fine brushes to tiny resin soldiers. Again, the heat is stifling: the smell from the solvents inescapable. The workers have no protection. When I talk to one of the employers, he explains that there is ventilation 'and they have a break every 5 hours'. I wonder what the continual whiffs of glue must do to their brains.

These are the better places to work.

On average these workers earn less than 600 yuan ($73) a week; up to 800 ($97) with overtime.[5] The system maximizes employer control. During busy times temporary employees—who can be sacked when production falls—make up a large proportion of the workforce (40 per cent is not uncommon). Accommodation is tied to employment, giving employers extended control over the lives of workers outside their factories and minimizing the opportunity for employees to find out about other—more appealing—workplaces.[6] The workers live together in dormitories—some attached to the factory, some in nearby towns. In the less restricted ones, there are only six people to a room. All are sex-segregated.

Even if the workers were given the choice (and few are), the workers could not afford to live in a house with a partner as a family. Exhausted after long workdays, many have never been to nearby towns. The factories and the dormitories define their lives.

Industrial action appears a remote prospect. Less than 30 per cent of foreign factories in Guangdong have an ACFTU representative. Indeed, it was not until last year that membership of the ACFTU was even open to migrant workers.

In any event, a priority for both the CCP and the ACFTU is courting capital. Their policies for the modernization of China depend on it. To attack the chronic underemployment in the countryside (where 30 million still have inadequate food and clothes) surplus rural labour needs to be sucked into the cities. But to support the rural influx, jobs must be attracted—and lots of them. China's labour laws must be the trade-off. Both the Chinese Government and the ACFTU know that if factory wages and standards are raised, capital investment will go elsewhere—to other parts of China, or worse still, to other countries.

Left to their own devices, with no ability to form their own associations, there'll be no real change for migrant workers—in the short term, at least. But as the Chinese saying goes: Water supports a boat; it can also capsize it. China's rural workforce was 488 million in 2003, with over 300 million of them underemployed. As a consequence, the number of rural workers leaving the countryside to seek work in the cities is increasing exponentially: estimated to grow to 200 million in the next decade. Just as the peasants and farmers provided the vital support needed for the Communists to

take power, the migrant workers they have become can take it away.

Resentment and frustration are spilling over. There have been sporadic strikes and walkouts. Reports of vandalism, sabotage and violence by workers are seeping past the factory doors and on to the desks of NGOs. If the legitimate grievances of migrant workers remain unresolvable, they are likely to pose a long-term problem for the stability of the Government.

Occasionally there is an encouraging sign from the CCP. Xiong Deming, a woman in a remote village in Chongqing province, complained to Premier Wen Jiabao last October that her husband, who was working on a nearby construction site, had not been paid for several months. This is a common tale: in the construction industry alone, payments in arrears owing to workers have accumulated to 30 billion yuan ($3.6 billion). After Premier Wen's direct intervention, the money owed to Xiong's family was paid that night. It sparked a Government-led campaign to help migrant workers retrieve the arrears owed to them.[7]

The ACFTU too is working to improve living standards: running model dormitories and setting up health schemes for the workers.

But these responses dance around the central issue. To pull a few up to riches, the money that is being made through the sweat of the migrant workers pushes many down to appalling pay and conditions.

To an entrepreneur, this must be nirvana. It looks like hell

It is still not too late. The CCP's public embrace of the market economy does not mean that it needs to

trash socialist principles completely. It could resist the present economic orthodoxy overseen by the World Trade Organization, the World Bank and the International Monetary Fund that the state should refrain from any intervention in the market. It could then reach into the factories and dormitories with laws and inspectors to guarantee better, safer working conditions and a more equitable share of income. However, if it accepts the prevailing capitalist view and takes a 'hands-off' approach, the voices of the migrant workers are likely to remain unheard in the crowded factories and dormitories where they are presently trapped.

Notes

1. Jin Huikang, *Aspects of Guangdong*, Cartographic Publishing House of Guangdong Province, Guangdong, 2002.
2. J Chan 'Chinese Communist Party to declare itself open to the capitalist élite.' *World Socialist Web Site*, wsws.org, November 2002.
3. China's 100 Richest 2002, *Forbes*, forbes.com, 24 October 2002. This was a fall from 2000, when 12 of the 50 richest were delegates to the Congress—see R Hoogewert 'China's 50 Richest Entrepreneurs' *Forbes*, forbes.com, 27 November 2000.
4. World Bank President, James Wolfensohn, speech to a Poverty Reduction Conference in Shanghai on 26 May 2004.
5. Research conducted by Kai Ming Liu, Executive Director of the Institute of Contemporary Observation, Shenzhen.
6. Pun Ngai 'The Precarious Employment and Hidden Costs of Women Workers in Shenzhen'. The Chinese Working Women Network, September 2003.
7. *China Daily*, 19 January 2004.

DISCONTENT IN CHINA BOILS INTO PUBLIC PROTEST

By David J. Lynch

WANLI, China—When the local government announced it was going to confiscate their homes and businesses to make room for a new development, residents of this village in southeastern China fought back.

They complained to five levels of local government. They sued officials in charge of the relocation effort. They journeyed to Beijing to petition central authorities. And when all that failed, they staged a sit-in on the grounds of a local factory that they had built with their pooled savings.

All but one of the demonstrators were women, many middle-aged or older. But that didn't stop more than 100 police officers, backed by thugs led by a notorious local criminal, from busting through a locked gate and beating them into submission.

"They bent my arm and threw me on the ground. I started to cry when I fell on the ground," says 54-year-old Huang Yaying, who suffered a broken arm.

After a half-hour melee in which 15 people were injured, authorities seized control of the four-story liquor factory. And in the weeks since the Aug. 1 incident, relocation crews have returned again and again to knock down nearby homes and businesses. Today, blue metal construction fences line Wanli's narrow streets, shielding piles of cement and brick rubble.

The continuing siege in this village of 600 people illustrates that beneath a veneer of authoritarian rule, economic strength and Olympic success, popular discontent is bubbling in China. Last year, there were 58,000 "mass incidents" across China, according to government statistics cited in the Chinese magazine Outlook . That's more than six times the number of protests and demonstrations the authorities admitted to in 1993.

The increasing willingness of individuals to confront state authority is powerful evidence of the emergence of what political scientists have labeled "rightful resistance." As China's economy grows freer and more tumultuous, these government-tolerated grievances are starting to hint at what a more open political environment might entail. They illustrate both the partial loosening of restraints on popular action and the very real limits that remain.

The disturbances—by farmers upset over high taxes, laid-off workers demanding overdue pensions and residents outraged over improper land grabs—have been carefully limited to economic issues. Even the most ill-educated peasant understands that the Communist government would swiftly crush any organized demand for political liberalization.

That doesn't mean China's rulers regard the complaints with indifference. Last month, Zhou Yongkang, minister of public security, listed such protests as one of several factors potentially threatening the country's stability.

Once a rarity, examples of public disobedience are now plentiful. Earlier this year, 2,000 workers and retirees at a textile plant in Suizhou in central Hubei province took to the streets to demand unpaid benefits. In November 2003, thousands of people in Zoucheng, Shandong province, stormed a government building, smashing windows and office equipment, after a sidewalk vendor was accidentally run over when city officials tried to enforce a new policy against such sales. And in March of 2002, tens of thousands of workers in Liaoyang massed against corruption and unpaid wages in the longest-lasting protests in China since the 1989 "Democracy Wall" campaign.

"They are very concerned about it," says Murray Scot Tanner, a China specialist at the Rand Corp, a think tank. "Unrest has gone up every year by at least 9% or 10% since they started keeping these numbers in 1993."

The central government in recent years, however, has relaxed its view of people who balk at official actions. Beijing established "letters and visits" offices to receive citizen complaints and seems to understand that the dislocation caused by China's unprecedented capitalist evolution, not foreign agents, explains much of the discontent.

At the same time, Chinese protesters also have become adept at choosing their targets. To increase their chances of success, they usually argue that unjust local officials are failing to implement the central government's policies. "When you protest and wave the central government's policies as your shield, it provides some sort of political protection," says Minxin Pei of the Carnegie Endowment for International Peace.

Since taking office two years ago, Communist Party General Secretary Hu Jintao and Premier Wen Jiabao often have spoken publicly of the need to address legitimate citizen complaints. Ensuring that the ruling party is more responsive to those being left behind in China's pell-mell capitalist surge is likely to be emphasized at a high-level party meeting in Beijing later this week.

Still, local officials don't always handle citizen disputes with the sophistication Beijing would prefer. And, as Wanli demonstrates, they don't hesitate to employ brute force to get their way.

'Take him out and beat him'.

Wanli once was a farming village outside of Fuzhou, the bustling capital of Fujian province. In the 1980s, the Cangshan district government confiscated almost 500 acres of village farmland to use for a new university, hospital and other institutions. As their vegetable farms kept shrinking, villagers decided in 1985 to pool their savings and invest in a series of shops, factories and six-story apartment buildings. Four years later, the enterprise, similar to village-level commercial ventures sprouting throughout China, was formalized by the local government as the Wanli Group.

By the mid-1990s, rent from these properties was the villagers' main source of income. By then, continued development in the area had turned the former farming village into an urban neighborhood on the outskirts of Fuzhou.

In 2000, villagers began hearing that the local government planned to evict them to make way for unspecified redevelopment. It's a familiar story in contemporary China. Roughly 20 million of China's 900 million farmers already have lost their land to commercial projects, according to the state-run Xinhua news agency. Often these disputes pit developers and their government allies against some of the least powerful people in China.

Almost everyone in Wanli was opposed to the project, fearing it would eliminate the businesses that are their only source of income and leave them without enough money to buy new homes elsewhere. With their farms gobbled up by China's insatiable urban development, many residents survive on the 330 yuan ($39.85) they receive each month as their share of income from the village enterprise.

Villagers say they receive 100 yuan ($12) per month in compensation for some of the businesses that already have been torn down. And they have been offered 1,300 yuan per square meter ($14.59 per square foot) for their homes. That's less than half the amount they say they need to buy comparable new housing.

But as is routine amid China's construction binge, local officials rejected the residents' complaints about improper treatment and insisted they evacuate. Villagers, displaying a surprisingly strong faith in the central government, blame corrupt local officials for their predicament. Earlier this year, the villagers sued the Cangshan district government office that is directing the relocation. But the case has languished. "We think it's useless. The government and the court, they're the same thing," Li Hua says.

On March 18, the first wrecking crews arrived in the neighborhood and began tearing down several buildings. In their wake, one despondent resident, Jiang Biquan, 45, died after drinking a bottle of agricultural chemicals. His suicide did nothing to slow the redevelopment. Several weeks later, about 300 police officers and men in civilian clothes surrounded the village and beat residents trying to halt additional demolition efforts.

On July 17, a group of villagers flew to Beijing and spent 11 days fruitlessly making the rounds of relevant ministries. One official at the Ministry of Construction even telephoned the Fujian Construction Bureau to inquire about the situation. But demolition work continued. "The Cangshan district government doesn't care what the law says. Their attitude is, 'We want this land and you have to give it to us,'" villager Li Wu says.

On Aug. 1, the dispute finally boiled over. Villagers had learned that the relocation office staff was preparing to seize the liquor factory to serve as its headquarters during the final stages of clearing Wanli. Relocation officials told villagers their existing office suffered from poor feng shui, meaning the building's design lacked harmony.

Local residents, who had invested as much as 10,000 yuan each in the village enterprise, feared the loss of their financial stake. So a few minutes past 8 a.m., about 20 people assembled inside the compound, which is ringed by a cement wall.

About 8:30 a.m., Shen Li, an official with the Cangshan district government, arrived and demanded to be let in. The villagers refused to unlock the gate, so Shen clambered over the wall and ordered an underling to break the lock with a hammer. "We tried to stop them from breaking open the lock. Some of us surrounded the lock," says a woman who gave only her surname, Jing, saying she feared retaliation from officials.

At that point, officials began roughly pulling the women away. As the skirmish intensified, says Jiang Bibo, the only male villager inside the compound, "You can't beat people like that!"

Villagers say Shen responded by ordering another man to punish Jiang, who was sitting nearby, saying, "Take him out and beat him."

Jiang, 50, was punched in the face and hurled to the ground. As he lay there, wrapping his arms protectively around his head, several men kicked and struck him repeatedly, witnesses say. Jiang was later hospitalized with a concussion and internal bleeding. "If it wasn't for another woman who went over and covered Jiang Bibo with her body, somebody could have died that day," Jing says.

Operating alongside the police that day were 14 recently released prisoners headed by Zhao Zhenguang, a local mobster. With ambulances parked nearby, the authorities seemed prepared for violence, witnesses say. One of the vehicles was needed to ferry Pan Lanfang, 56, to the hospital. The local resident had gone to the factory to look for a friend, not to protest. She ended up badly beaten.

Efforts to reach Shen at his office and on his cellphone for his account of the August 1 events were unsuccessful. Authorities in Beijing say they were unable to provide any immediate comment on an individual relocation dispute.

Retaliation by officials alleged.

Despite central government directives to take a more subtle approach to quelling protest, such violent repression remains all too common, analysts say. "The use of hoodlums, really violent criminals, against peaceful civilians is quite prevalent," says Pei, the Carnegie analyst.

Chinese authorities have ordered local police to defuse protests without violence, according to Tanner. But the message doesn't appear to be getting through. "In the last couple of months, I've heard several cases of very deliberate use of harsh forces by the police and very undisciplined use of force," he says. "That's absolutely the sort of thing that, officially, they've been discouraging."

Following the attack, 60 villagers went to the Fujian province's Public Security Bureau to file a complaint. They also penned an open letter to authorities in Beijing seeking an investigation. Human Rights in China, a New York-based group, released an account of the dispute on Aug. 25. There has been no formal response to the villagers' letter.

But on Aug. 2, the day after the assault, relocation office officials staged a banquet at the factory compound for the police and other government officials who evicted the demonstrators. And in the intervening weeks, several villagers say, provincial officials have threatened them with prosecution for talking about the case with foreign reporters or other outsiders. "Everybody here is in great danger now. They could just put us in jail for seven or eight years. You never know," Zheng Rong says.

Already, one villager, whose home was demolished earlier this month, has been detained along with his wife and sister. Officials reportedly are pressing him to admit he helped organize recent protests, which would leave him open to prosecution. As part of their strategy to quell protest, Chinese officials customarily punish organizers more harshly than mere participants.

Officials are keeping the pressure on in other ways. In early September, as the temperature hovered around 97 degrees, they interrupted the public water supply for several days. Villagers also say their sleep has been interrupted by people knocking on their doors before dawn.

Amid continued clearance work, desperation is growing. On Sept. 8, Jiang Zongzhong, a retiree in his 70s, had had enough. As workers moved to demolish his home, the elderly man tried to kill himself by igniting a propane tank used for cooking fuel. Witnesses say a fireman knocked him to the ground, slapped him in the face and removed him from the home. The structure and the possessions of a lifetime were flattened.

Still, the villagers continue to profess faith in the central government in Beijing.

"We hope the real Communist Party can send us a Bao Qingtian," Li says, referring to a 12th-century Chinese imperial official renowned for his honesty and integrity. "We hope they send somebody like that to give us justice."

BEIJING'S AMBIVALENT REFORMERS

"The party has implemented various modest reforms in recent years. Some are designed to allow the party to implement its policy agenda more efficiently. Others aim to make it more responsive to a changing society, or at least to appear so. All are designed to perpetuate the Communist Party's rule, not necessarily to make China more democratic."

BRUCE J. DICKSON

China's leaders have been exceedingly cautious about embarking on extensive political reforms, and not without good reason. There is no guarantee that reform efforts will succeed, or that China will be better or more easily governed as a consequence of reform. There is certainly no guarantee that the Chinese Communist Party will survive as the ruling party if it initiates fundamental reform of the political system.

The leadership is acutely aware that even good intentions can have disastrous consequences: when Soviet Communist Party leader Mikhail Gorbachev launched his reforms in the Soviet Union, he did not envision the collapse and dissolution of his country, and yet that was the result. Even though the immediate causes of the Soviet collapse are not as salient in China (economic stagnation, separatism, populist leaders, Gorbachev himself), the country's leaders are concerned that political reform could lead to the same fate. With few examples of authoritarian parties sponsoring democratization and surviving as the ruling parties of their countries, the Chinese Communist Party is still searching for a suitable role model to emulate.

What kinds of reforms are necessary to keep the party in power, and what reforms would jeopardize its tenure? These are questions that bedevil the current "fourth generation" of leaders just as they did their predecessors. Both Deng Xiaoping and Jiang Zemin, leaders of the second and third generations, respectively (Chairman Mao, of course, was the first generation leader) believed that economic modernization had to precede political change, and took the Soviet collapse as a cautionary tale. The current leadership, symbolized by party General Secretary Hu Jintao and Prime Minister Wen Jiabao, has not made its full intentions clear, but it has not yet shown any inclination to experiment with bold political reforms.

Even though China has not experienced the kinds of democratization that most observers have in mind when they look for signs of political reform, the party has implemented various modest reforms in recent years. Some are designed to allow the party to implement its policy agenda more efficiently. Others aim to make it more responsive to a changing society, or at least appear so. All are designed to perpetuate the Communist Party's rule, not necessarily to make China more democratic.

ADAPTING THE PARTY TO THE NEW AGENDA

At the beginning of the post-Mao period, Deng Xiaoping and other reformers recognized that their goal of modernizing China's economy with "reform and opening" (*gaige kaifang*) policies could be undermined both by remnant Maoists, who did not support their policies, and by veteran cadres, who were not qualified to carry out reform even if they supported it. For their economic reforms to be successful, therefore, they changed the party's policies regarding recruiting of new members and appointments of officials. After removing the Maoists from their posts and easing the veteran cadres into retirement, they transformed the composition of the cadre corps and rank-and-file membership.

At all levels, party members and officials have become on average younger and better educated. To make the Communist Party younger, leaders assured that roughly two-thirds of new recruits each year were no older than 35. To prevent local officials from remaining in office indefinitely, the party instituted a two-term limit on all posts and required officials to retire once they reached a certain age and were not promoted (65 for provincial-level officials, younger for lower levels of the bureaucracy).

The emphasis on education is even more apparent. Whereas less than 13 percent of party members had a high school or better education in the late 1970s, by the time of the sixteenth party congress in November 2002, the figure had risen to 53 percent. Within the central committee—which comprises the party's top 150 to 200 leaders—the proportion of those with a college degree rose from 55 percent in 1982 to 99 percent in 2002. Improvements in the education qualifications of local officials from the provincial to the county level were even more dramatic: in 1981, only 16 percent had a college education; 20 years later in 2001, 88 percent did.

The current generation of leaders is often referred to as "technocrats," meaning they hold bureaucratic posts and have technical backgrounds in the sciences and engineering. As Li Cheng has noted in the *China Leadership Monitor*, all nine members of the Politburo's standing committee, the very top elite of the party, are technocrats, as are eight of the ten members of the State Council, China's cabinet. Below this top level, however, the growing dominance of technocrats has abated. According to Li, the proportion of technocrats on the central committee dropped from 52 percent at the fifteenth party congress in 1997 to 46 percent at the sixteenth party congress in 2002. Similarly, among provincial party secretaries and governors, the proportion of technocrats has declined from about 75 percent in 1997 to only 42 percent in 2003. Moreover, *none* of the provincial leaders appointed after March 2003 (the likely candidates for the fifth generation of leaders) are engineers. For this younger generation of leaders, educational backgrounds in economics, the social sciences, humanities, and the law are increasingly common.

Many Chinese now believe that economic success is based on personal connections with party and government officials, not individual initiative or quality work.

While it is dangerous to infer political preferences from academic backgrounds, this change in the composition of local leaders is a trend worth following. Many observers have predicted that the technocratic background of fourth generation leaders makes them more disposed to practical problem solving than bold experimentation. But the next generation of leaders comes from very different formative experiences. They have different educational backgrounds, have had greater exposure to international influences, and have enjoyed deeper experience in local administration. They may be more inclined to not just make the political system work more efficiently but to change it to make it more responsive to societal demands.

For these leaders, the key political event was not the Cultural Revolution (they are too young to have had their careers affected by those tumultuous years between 1966 and 1976 as previous generations of leaders did) but the popular demonstrations of 1989. The challenge facing the party is not to undo the mistakes of the Maoist period or to achieve rapid economic growth, but to prepare the political system for the consequences of modernization. How the party addresses this challenge will largely determine whether China will become more democratic, and more important, how systemic change may come about.

The best example of the party's change in recruitment strategy is the proportion of worker and farmers in the party. In 1194, they comprise 63 percent of all party members; by the end of 2003 their proportion had dropped to 44 percent in a party that had grown to over 68 million members. In less than 10 years, these representatives of the proletarian vanguard had become a minority in the Chinese Communist Party. The party now focuses on educational credentials and professional accomplishments in its recruitment strategy. Increasingly, that has meant turning to the urban entrepreneurial and technological elites.

These changes in the composition of the rank-and-file party members, local officials, and top leaders were designed specifically to promote economic reform. In this regard, the party reform was certainly successful. Over the past 25 years, China's economy has grown by 8 percent annually, lifting per capita income from $190 in 1978 to $960 (according to the World Bank, using current dollars) and shifting the bulk of economic activity from agriculture to industry, commerce, and services, and from the state sector to the nonpublic sector. Rapid economic change has also led to the emergence of new social groups, and the party has switched from excluding them from the political arena to actively incorporating them. It has co-opted these new elites by recruiting them into the party, appointing them to official posts, and creating corporatist-style organizations to integrate state and society.

THE RISE OF THE RED CAPITALISTS

In August 1989, soon after the end of popular demonstrations in Tiananmen Square and elsewhere, the Communist Party imposed a ban on recruiting private entrepreneurs into the organization. Party leaders were concerned that the economic interests of businessmen conflicted with the political interests of the party. This was not just Marxist paranoia: several prominent businessmen publicly supported the Tiananmen demonstrators and later fled the country to avoid arrest. Although the ban remained in place for more than 10 years, it was not very successful in keeping entrepreneurs out of the party. As I noted in my recent book, *Red Capitalists in China: The Party, Private Entrepreneurs, and Prospects for Political Change*, local officials had an incentive to reach out to entrepreneurs even if central leaders disapproved. Creating economic growth is now a key criterion for career advancement, and throughout the 1990s most economic growth and job creation came from the private sector.

Because of this, local officials in some communities—but by no means all—began to recruit successful entrepreneurs into the party despite the formal ban. The percentage of private entrepreneurs who belonged to the party—a group known as "red capitalists"—grew from around 13 percent in 1993 to more than 20 percent in 2000. Not all were new party members, however; most red capitalists had been members before going into business, but about one-third were co-opted after becoming successful businessmen. Orthodox Marxists harshly criticized the emergence of red capitalists and warned that it would spell the demise of the Chinese Communist Party.

As the economic clout of the private sector has grown, the political roles filled by private entrepreneurs have also increased. In addition to being members of the party, many entrepreneurs also belong to local party committees, the major decision-making bodies in China, further integrating them into the political system. At the sixteenth party congress in November 2002, entrepreneurs were among the delegates for the first time, although none were named to the central committee. When China's legislature, the National People's Congress, met in spring 2003, 55 entrepreneurs were selected as deputies. Entrepreneurs have been asked to serve in local legislatures in even larger numbers: over 17 percent of entrepreneurs belong to local people's congresses, and 35 percent belong to local people's political consultative conferences, a body designed to allow discussion between the party and other local elites. Many have been candidates in village elections, and most of the successful candidates are also party members, showing the party's desire to keep all political participation under its control.

In addition to bringing new social strata into the political arena, the party has also developed institutional ties with a variety of new social organizations. These groups are designed to be the party's bridge to society, allowing it to monitor what is occurring without directly controlling all aspects of daily life. China has tens of thousands of civic and professional organizations and hundreds of thousands of nonprofit organizations, such as private schools, medical clinics, job-training centers, and community groups, that provide a variety of social welfare services.

This vast number of organizations may create the foundations of a fully developed civil society, but at present they do not enjoy the kind of autonomy normally expected to be found in the groups that compose a civil society. All organizations must be formally registered and approved, and have a sponsoring governmental organization. They are also not supposed to compete with each other for members or for governmental approval. Where more than one similar group exists in a community, they may be pressured to merge or disband. These restrictions on social organizations suggest a corporatist logic to state-society relations, with controls over which organizations can exist and what kinds of activities they can engage in.

At the same time, many of these new groups are unlike the Communist Party's traditional "mass organizations," such as the All-China Federation of Trade Unions, which are seen by their nominal members as tools of the state rather than representing members' interests. Many professional organizations are not simply transmission belts for the party line, but instead are able to provide tangible benefits for their members. As Scott Kennedy shows in his forthcoming *The Business of Lobbying in China*, a variety of business associations have sprung up, often industry-specific, organized from the bottom up, and active at lobbying the state and in some cases unilaterally setting industry standards and regulations. These associations are more autonomous, more assertive, and less interested in simply representing the

state's interests. This may complicate the party's strategy of creating new institutional links to monitor and control the private sector, but at present these business associations limit their activities to issues within their sphere and are not involved in larger public policy issues.

That behavior remains the key to success for both individuals and organizations: do not stray into political matters, and do not challenge the Communist Party's monopoly on political power. While much has changed in China in recent years, this basic political rule has not. Yet most of these new organizations are more inclined to succeed within the existing boundaries than try to change them. This also is a function of civil society: not just to challenge the state, but to find ways of working with the state to pursue common interests. Most writing on contemporary China focuses largely on the conflictual nature of civil society, but the potential for cooperation is just as important and certainly more prevalent today.

COURTING NEW ELITES

As economic development created a more complex society, with new social strata that did not fit neatly into old class categories, third generation leader Jiang Zemin and his colleagues recognized that they relied on these new elites to maintain rapid economic growth and could not continue excluding them from the party. Beginning in early 2000 and culminating in his speech on the eightieth anniversary of the Chinese Communist Party's founding on July 1, 2001, Jiang laid out a new definition of the party's relationship with society, which became known as "the important thinking of the 'Three Represents.'" According to this formulation, the party no longer represented only farmers and workers, its traditional base of support, but now also incorporated, first, the advanced productive forces (referring to entrepreneurs, professionals, high-tech specialists, and other urban elites); second, the most advanced modern culture; and third, the interests of the vast majority of the Chinese people.

This was a very inclusive definition of the party's role, and while often ridiculed as an empty slogan, it indicates a serious effort to update the party's relationship with a changing society. It acknowledged that what brought the party to power in 1949 was substantially different from what the party faces in the twenty-first century. If the party's guiding ideology no longer fit China's economic and social conditions, then the ideology needed to be updated—but not abandoned altogether. The party still goes to great lengths to show how its ideology remains consistent with its Marxist origins, even if China's few remaining ideologues believe the party has already abandoned its traditions and betrayed its revolutionary goals.

After Jiang's Party Day speech in 2001, in which he recommended lifting the ban on recruiting entrepreneurs and other new social strata into the party, and after the sixteenth party congress in 2002, when the "Three Represents" was added to the party constitution, large numbers of these "advanced productive forces" were expected to join the party. That did not happen, but it is not clear why. It may have been that local officials were not enthusiastic about this new policy and resisted implementing it. While some local leaders had ignored the ban on recruiting private entrepreneurs into the party, other leaders adamantly believed that capitalists did not belong in the Communist Party.

After the "Three Represents" became official party doctrine, a small number of cities was chosen to experiment with recruiting members from among these new urban elites. The public media did not report on the results of these experiments, indicating little progress was made. The party's organization department issued new directives on recruiting private entrepreneurs in 2003, but the message was ambiguous. Local party committees were advised not to be so eager to recruit new members that they lowered the standards for party membership, nor so strict that they did not let in any. Without clearer guidelines, the adoption of the "Three Represents" slogan was not fully integrated into the party's recruitment strategy.

However, the lack of progress may have been due to declining interest among private entrepreneurs themselves in joining the party. The number of red capitalists has continued to grow in 2003—up to 30 percent of entrepreneurs were party members—but most of the growth has come not from new recruitment, but from the privatization of state-owned enterprises. As these enterprises were converted into private firms, their former managers, almost all of whom were party members, became owners of private firms, automatically becoming red capitalists. Other entrepreneurs, however, seem to have lost interest in joining the party. Some claimed that they did not want to belong to a party that seems increasingly corrupt. Others did not want to be subject to party scrutiny of their business practices.

In a more general sense, party membership has become less valuable for many entrepreneurs. When the party was more ambivalent about the private sector, membership was useful for promoting business interests, such as securing loans, finding new investors, limiting outside competition, and above all protecting them from predatory actions of local officials. Reports of the confiscation of private property and financial assets remain common, showing that many local officials are more concerned with profiting from the private sector than promoting it. As the party's commitment to the private sector grew, and the interests of businessmen became better protected in party policy as well as laws and regulations, party membership became a less valuable commodity for private entrepreneurs. Still, the slowdown in co-opting entrepreneurs into the party—which seemed to be the main motivation behind the "Three Represents" in the first place—remains something of a mystery.

While many no longer believe Marxism remains a relevant doctrine in contemporary China, there is no doubt that Leninism remains the guiding influence in the political system.

Even so, in the years after 1978, the party has steadily become younger, better educated, more professionally experienced, and more diverse as the farmers and workers, the traditional mainstays of the party, have been replaced by entrepreneurs, high-tech specialists, managers, and other new social strata. These changes have reinforced the commitment of party members and officials to the "reform and opening" policies. As a result, party adaptation has been generally successful by one measure: the changes have allowed the party to pursue its new goals more efficiently. However, a more challenging test of the party's adaptability is whether it is responsive to the changing wants and needs of society, and here the results have been more ambiguous.

RESTORING BALANCE

Under the leadership of Jiang Zemin, the Communist Party had a distinctly elitist orientation, emphasizing the first of the "Three Represents": the advanced productive forces, which are primarily the urban entrepreneurial and technological elites. In recent years, private entrepreneurs in particular have become more assertive in seeking political and legal protection of their economic interests, and the party has been very responsive to their interests. To further symbolize the party's commitment to the private economy, in November 2003 it decided to revise the state constitution to protect private property and to promote the interests of the private sector.

This increasingly close relationship between the party and the private sector has created the widespread perception that the benefits of economic growth are being monopolized by a small segment of the population while the rest of the Chinese people are being left behind. Many Chinese now believe that economic success is based on personal connections with party and government officials, not individual initiative or quality work. As people come to believe that the benefits of the economic reform policies are unfairly distributed, the legitimacy of the party's policy of letting some get rich first is jeopardized.

In response to this perception, the new leadership of General Secretary Hu and Prime Minister Wen has shifted the focus away from the elitist orientation of the Jiang era to the third of the "Three Represents": the interests of the vast majority of the Chinese people. Hu and Wen, along with many others, concluded that the pendulum had swung too far in recent years, favoring the elites over the general population. They now want to create a new image for themselves and the party. This can be seen in Hu's speech on Party Day in 2003. Like Jiang just two years earlier, Hu concentrated exclusively on the "Three Represents." But whereas Jiang had emphasized the advanced productive forces, Hu mentioned the new social strata only once in passing. Instead, he focused on the "fundamental interests of the vast majority of the people," a phrase he repeated 13 times. In doing so, he was not rejecting an important symbol of the Jiang era, but he was reinterpreting it to signal a shift in priorities.

Hu and Wen have done more than simply speak on behalf of the majority. They have also shown their support—or at least their sympathy—for the disadvantaged in their public appearances and activities. During the 2003 Chinese New Year, Wen visited and shared a meal with miners. During the 2003 SARS crisis, Hu and Wen visited SARS patients in hospitals. They fired the minister of public health and the mayor of Beijing for covering up the extent of the epidemic. On World AIDS Day in December 2003, Wen visited and shook hands with HIV/AIDS patients, the first top leader to recognize China's AIDS crisis. In January 2004, the Communist Party issued a new policy directive on improving rural conditions that included policies aimed at alleviating income inequality. The Hu-Wen team has also tried to alleviate regional inequalities by promoting development in the northeast rustbelt and the less developed western provinces. This effort was begun under Jiang but expanded under Hu and Wen. Experiments with local elections, also started under Jiang, have continued with the fourth generation. In recent years, there have been elections for party secretaries, township leaders, urban neighborhood committees, and other positions.

MIXED SIGNALS

Along with hints of change came signs of the enduring features of the political system. The doctor who exposed the SARS cover-up and became a national hero, Jiang Yanyong, was taken into custody by military officials in June 2004 and held for six weeks for advocating a reassessment of the official verdict on the 1989 Tiananmen demonstrations. Although the extent of the AIDS crisis has been gradually but not yet fully acknowledged, HIV/AIDS victims still rarely get the treatment they need and official culpability in the spread of the virus has yet to be admitted, much less punished. AIDS activists, most notably Wan Yanhai, have been harassed and imprisoned, and reporters who have tried to expose the policies of local governments that allowed the virus to spread have been fired and their stories suppressed. Residents of "AIDS villages" in rural Henan, where the AIDS virus has spread widely through blood donations that use unsanitary practices, have been beaten, arrested, and had their homes destroyed for seeking medicine and financial assistance from higher levels of government, for meeting with journalists to publicize their plight, or for attempting to gain the attention of investigating groups visiting China from the World Health Organization.

Efforts by top leaders to compensate the disadvantaged continue to be hampered by the failure of local leaders to act on new initiatives. For example, Wen may order local leaders to pay IOU's and unpaid wages to specific individuals in specific cases when they come to his attention, but similar cases that do not get singled out are rarely addressed. Local governments are themselves often starved of cash and cannot be as generous and proactive in identifying and addressing the many injustices that exist in their jurisdictions. And candidates in local elections are still either Communist Party members or independents; no new polit-

ical parties have been allowed to form, and there has not even been official discussion of such a possibility. Efforts to create the China Democracy Party went for naught, as petitions to register the party were denied and the activists who were behind the effort were arrested and sentenced to jail terms of more than 10 years.

At the same time, Hu and Wen seem determined to shift away from the elitist orientation within the party. There is now frequent media coverage of Politburo meetings. Hu reported on the work of the Politburo to the most recent central committee meeting in fall 2003, and lower-level party committees are also expected to give regular reports to the bodies that formally elected them. Hu also canceled the annual meetings in the resort city of Beidaihe, which have traditionally been held each August to decide major policy and personnel issues. Because they are more informal than Politburo meetings and central committee plenums, they have been frequently used by senior party leaders to influence decision making, even after these officials have formally retired from office. The decision to cancel the meetings gives greater emphasis to the formal meetings in Beijing, and may curtail the informal influence of retired elders.

These changes are designed to promote the transparency and accountability of top-level decision making and to give greater weight to formal processes over informal politics. While the party has described these changes as improving inner party democracy, a dubious claim to be sure, they should at least be recognized as creating greater institutionalization in the Chinese system, which by itself would be a generally positive trend.

But these changes occur within clear limits. Reports on Politburo meetings reveal little beyond the topic under discussion and the theme of Hu's remarks to the group. Work reports by themselves do not provide for much accountability, and in any event the central committee only "elected" the Politburo after top leaders agreed among themselves who would belong to it. And media coverage of the November 2003 central committee plenum highlighted again the party's secretive nature. Although the media reported that the central committee had approved major constitutional revisions, they did not report on the content of those revisions. Speeches were given by top leaders, but the texts of the reports—including Hu's report on the work of the Politburo—were not published.

These mixed signals are the result of several factors. First is the leadership's ambivalence about pursuing any one course exclusively, with the danger that concessions to some individuals or groups may be used as a precedent for others to make claims against the state, or might raise expectations that more expansive political reforms are being considered. Second is the fragmented nature of political authority in China. Not all actions are the result of coherent decisions by unified leaders; they are also the result of different parts of the state taking actions that other parts of the state, and other leaders, may be unaware of or even oppose. Third is the consequence of political decentralization. Policies announced in Beijing are not immediately or even inevitably implemented by local governments. Finally, the transition from the third to the fourth generation of leaders is still incomplete. Jiang retains his post as chairman of the Central Military Commission, the Communist Party's top body for military matters, and continues to intervene in domestic and foreign policies—he was reportedly behind the detention of Dr. Jiang (no relation). Hu and Wen, perhaps recognizing that time is on their side, have not directly challenged Jiang's interventions even when they run contrary to the new leaders' preferred direction. Which of these causes is behind each zig and zag is often difficult to determine by outside observers and even by the victims and beneficiaries of these steps and missteps within China.

BENEVOLENT LENINISTS?

While many no longer believe Marxism remains a relevant doctrine in contemporary China, there is no doubt that Leninism remains the guiding influence in the political system. There is still no organized opposition of any kind, and no public lobbying for policy change is visible outside the economic realm. But for those who do not choose to challenge the Communist state—and this involves the vast majority—the party is increasingly less pervasive and less intrusive. This is not to sug-

gest that the party is seen as legitimate, much less popular. But it points to a fact that is often overlooked in most criticisms of China: although freedoms of all kinds are sharply delimited, and not well protected by law, it is nevertheless true that the degree of mobility, expression of ideas, and access to information is increasing, not contracting. When compared to the freedoms enjoyed, even taken for granted, by citizens of democratic countries, this progress seems halting and minuscule. But when compared against China's own past, the changes are dramatic.

Whether they will be sufficient to forestall popular demands for more significant change, and to prolong the Communist Party's tenure as China's ruling party, remains a key question in Chinese politics. In short, it is still not clear if we are seeing a more benevolent form of authoritarianism or signs of more significant political reform yet to come.

BRUCE J. DICKSON *is an associate professor of political science and international affairs at George Washington University. His most recent book is* Red Capitalists in China: The Party, Private Entrepreneurs, and Prospects for Political Change (*Cambridge University Press, 2003*).

Nationalist Fervour Runs Amok

By Geoffrey York

Beijing—By taxi and bicycle, the young men arrived near the Japanese embassy. They carried loudspeakers and sirens and giant red Chinese flags. They wore shirts with anti-Japanese slogans.

It could mean only one thing. The Chinese patriots were on the prowl again.

"Flag-holder, come to the front," shouted an organizer as the patriots began to march. "Hold it higher, so that everyone can see!"

They waved their flags and unfurled two red banners denouncing Japan, and marched down the street to the embassy. They chanted slogans, listened to an emotional speech, sang the national anthem and delivered a petition to the embassy mailbox. "Japan, apologize for your crimes," they shouted.

Among the marchers was one of their chief organizers, an earnest 29-year-old computer programmer named Lu Yunfei. His boyish face, with his conservative haircut and wire-rim glasses, is the new face of China's resurgent nationalism: a well-organized movement that exploits Internet technology to launch petitions and verbal attacks against Japan and the United States.

As communism slides into irrelevance, the new nationalists are emerging as a powerful force in China, with ominous implications for its neighbours.

After a soccer match between China and Japan in Beijing this summer, hundreds of angry Chinese men chanted "Kill the Japanese" as they pelted Japan's team bus with plastic bottles and forced Japanese fans to hide behind a police barricade for hours.

A few years ago, Mr. Lu spent his weekends strumming his guitar and singing karaoke at Beijing nightclubs. Today, he spends all of his spare time—up to 50 hours a week—on his work with the Patriots Alliance, a network of nationalist activists with close to 100 volunteer workers and 79,000 registered supporters on its website.

He has postponed his wedding to his fiancée three times in the past year because he is so busy with the alliance. He has helped organize more than 10 public protests in the past two years—a stunning number in a country where such gatherings are normally illegal.

The nationalist mood seems to be gaining strength every year here. The schools are filled with "patriotic education" classes. Young people are organizing boycotts of Japanese products. Web petitions against the Japanese government are attracting millions of supporters. The Japanese are routinely denounced as "devils" and "little Japs" in chat rooms on the Chinese Internet, and one bar in southern China went so far as to post a "Japanese not welcome" sign.

A few years ago, optimists had hoped democracy would be nurtured by China's growing personal freedoms and its new Internet culture. But in reality, it is the national-ists, not the democrats, who have scored the biggest victories from the relaxed atmosphere.

"All of our events are very exciting," Mr. Lu boasts. "Everything we're doing is unprecedented. Nobody has done it before. People nowadays have more freedom to express their feelings and put them into practice."

While most of Mr. Lu's activities are aimed against Japan, he is also quick to vent his hostility against the United States. He fully expects a war between China and the United States, and he vows to be the first volunteer in the battle against the Americans if there is a war over Taiwan, which Beijing regards as a renegade province. "I love my country deeply," he says. "I have a stable job and a good income, but if a war happens, I'll go to the front lines without hesitation."

In some countries, a small protest by 20 people at an embassy might be a routine event. But in China, where everything is carefully regulated, the protest at the Japanese embassy last month was highly significant. The protest was held quite openly, with full police knowledge, on the third day of the annual conference of the Communist Party's central committee, at a time when police were strictly banning all protests by anyone else.

Thousands of petitioners and protesters in Beijing had been rounded up by police during the Communist Party meeting to avoid any embarrassment to the political elite. Yet even as arrests continued, the Chinese patriots were allowed to carry out their demonstration freely, under the noses of police officers who carefully supervised the event and even escorted one of the organizers inside the embassy's fence to deliver his petition.

It was further evidence of Beijing's semi-official approval of the new nationalists. The Communist leaders are seeking to harness Chinese nationalism as a unifying force, a sentiment that can be tapped by authorities to build loyalty, to quell opposition, and to fire the passions of young people who might otherwise drift into dissent.

"With the decline of Communist ideology as a source of legitimacy, the Communist Party depends even more on nationalism to legitimize its rule," wrote American scholar Peter Gries, author of the book China's New Nationalism, in a forthcoming issue of the China Quarterly.

The nationalism has become so virulent that it even questions China's economic reforms. Several recent Chinese books have denounced globalization as an evil U.S. plot and an attempt to imprison China. Chinese trade negotiators were called "traitors" for leading China into the World Trade Organization. One analyst said the nationalists were reviving the spirit of the Boxers—the xenophobic rebels who slaughtered hundreds of Christians and others with alleged foreign connections in a famous revolt in 1900.

Mr. Gries argues that the Maoist "victor narrative," which glorified Communist victories over Japanese and Western imperialism, has been replaced since the mid-1990s by a new "victim narrative" that emphasizes China's suffering at the hands of Japanese and Western occupiers during a "Century of Humiliation."

While the anti-American feelings of many Chinese first exploded into the world spotlight in the violent protests against the U.S. bombing of the Chinese embassy in Belgrade in 1999, most of the nationalist hostility today is aimed at Japan.

Even after decades of normalized relations and billions of dollars in Japanese economic aid to China, there is a mounting fury against Japan that goes beyond the historical memories of the Nanjing massacre and the Japanese invasion of the 1930s. In one recent survey, 53 per cent of Chinese respondents admitted to hating the Japanese.

"After a quarter-century of unprecedented economic growth, most Chinese no longer fear Japan, and a long-suppressed anger at Japan has resurfaced," Mr. Gries wrote. "Chinese animosity towards Japan is unquestionably out of control.... The Japan-bashers are ascendant. A winner-takes-all, show-no-mercy style reminiscent of the Cultural Revolution is prevalent."

The mood was symbolized by a 2001 incident in which Chinese actress and model Zhao Wei was photographed in a short dress with a large imperial Japanese flag imprinted upon it. The dress was created by an American designer and the photo was taken during a fashion shoot in New York, but none of that mattered to the enraged Chinese patriots, who used bricks and bottles to assault Ms. Zhao's house. One man assaulted the actress on stage during a New Year's Eve show, pushing her over and smearing her with excrement. "I don't think I did anything inappropriate," the man said later.

When liberal journalist Ma Licheng wrote an article in an academic journal in 2002 that criticized the anti-Japanese mood and called for rapprochement between China and Japan, he was immediately condemned as a "traitor." He suffered death threats, and his home address and telephone number were posted on the Internet, along with a call to burn down his house. He was obliged to quit his job and move to Hong Kong.

Other incidents swiftly followed. In 2003, the Patriots Alliance organized the first-ever attempt to send Chinese activists to land on the Diaoyu islands, a cluster of small volcanic islands controlled by Japan but claimed by China. (They are called the Senkaku islands in Japan.) The alliance also organized a Web-based petition against a Japanese bid to provide high-speed railway technology to China.

There was another outpouring of anti-Japanese rhetoric on the Internet in 2003 after a group of Japanese businessmen hired hundreds of Chinese prostitutes during a visit to southern China. And a few weeks later, when three Japanese students performed a risqué drag show during a university party in the city of Xian, there was an angry demonstration by 7,000 Chinese protesters who demanded an apology for the perceived insult to their national dignity.

Then came the violence of the China-Japan soccer match this summer, where fans burned Japanese flags, smashed the car window of a Japanese diplomat, waved theatrical swords, and sang martial songs with lyrics such as, "A big knife chops off the heads of the Japanese devils." China was forced to mobilize 12,000 soldiers and police officers to prevent rioting.

All of this is adding fuel to the campaigns of nationalists such as the Patriots Alliance. In an interview in a Beijing coffee shop, Mr. Lu wore a black T-shirt with a nationalist rallying cry in small white characters. "Protect the Diaoyu islands," the slogan proclaimed.

Despite a strict ban on political activism in China, the patriots have been free to display their slogans on their chests. "Nobody tells us not to wear such T-shirts," Mr. Lu says.

He sees the rise of China as a rightful return to its place as the dominant force in the world.

"The United States is afraid of our long history," he says. "China has had over 5,000 years of history, and our backward period was only the past 200 years. During the rest of our history, we were the world's leader."

TIANANMEN: VICTORY FOR CAPITALISM

Richard Spencer

'Tell me,' a Chinese friend asked me the other day. 'Why are you so interested in poor people?' Urban sophisticate that she is, she couldn't quite figure out the point of the story I was writing, about a poverty-stricken village in the middle of nowhere. I wasn't entirely sure myself, actually, but rather than go into the difference between Telegraph and Guardian readers I decided to try a bit of home-spun egalitarianism. 'Well, put it this way. You have a smart flat, a smart car, you go to smart restaurants, just like folk like you in Britain. But then there are your fellow Chinese who have nothing, not even running water. And you're a communist country too, right? That's interesting, isn't it?' The look on her face was halfway between bafflement and scorn. 'My goodness,' she said politely, and left it at that.

China is, as most people are aware by now, a funny sort of communist country. Journalists are often invited to press conferences with top leaders like Wen Jiabao, the friendly Prime Minister who visited Britain last month, and lectured about the virtues of stability and modest living. Chinese newspapers are less reserved, blasting away at American hegemonism, and extolling the virtues of Mao Tse-tung Thought, Deng Xiaoping Theory, and Marxism with Chinese characteristics.

Capitalism it's got big, but the one thing you rarely see discussed is just how right-wing China is. Not just right-wing for a communist country, with an injection of American investment here and there, but full of gut-instinct right-wingery; right-wingery of the sort that regards China's newly enormous disparity of wealth as perhaps natural, at worst a necessary evil; right-wingery that regards authoritarian, paternalistic control, particularly of the poor, as a duty of government; right-wingery of the sort that elsewhere has student radicals marching in the street crying fascist.

China had its own student radicals crying fascist once, of course. That was 15 years ago to the day as I write this, but when the tanks rolled into Tiananmen Square we all supposed the nation was being saved for socialism, and when the tanks didn't roll a few months later in Eastern Europe the contrast appeared to prove the point. Looking back on that time, however, it's pretty fair to say that we all got things wrong. Most people thought the Tiananmen massacre just delayed the inevitable, that shortly the regime would fall, democracy would arrive, and everyone would agree what a big mistake sending in the troops had been. Peasants would discover the voting booth, as they did in South Africa; apparatchiks would mingle with oil barons, as in Russia; it would all be very exciting and end-of-historyish. But in fact the opposite turned out to be true: not only is the regime still there, but it sees the Tiananmen Square crackdown as a monumental success story for China. It's just that it wasn't making the country safe for socialism, it was making it safe for capitalism.

When I say success, that might seem callous. But look at it from the point of view of Deng Xiaoping, the paramount leader. He thought China was about to fall into a second Cultural Revolution, only worse. He painted a picture of civil war, of refugees flooding China's prosperous Asian neighbours. But then he added, to those old revolutionary friends of his who held the country in their hands, 'The momentum of reform cannot be stopped. We must insist on this point at all times.' Opening and reform, the buzz words of the Deng era, were the key to it all. And who can deny they have happened? In the years since, the proportion of the economy owned by the state or collectives has plunged and the economy has grown by 9 per cent a year. Deng met his target of quadrupling GDP in two decades. Underdeveloped countries which did throw off the shackles of dictatorship, like the Philippines and Indonesia, now look on with awe. Hong Kong is full of educated, free-thinking middle-class people from the Philippines; they work as maids, because there are no jobs at home.

Some will rightly point out that if China is capitalist it's not capitalist in the sense that Wall Street bankers understand. Half the economy—over half, some say, depending on how you define and count—is still in the hands of the government and its state-owned enterprises.

When you see all these big Chinese firms listing on the Hong Kong and New York stock exchanges, they are usually pretty peculiar entities—artificially created subsidiaries of big state businesses which still have the controlling share. The listings are capital-raising exercises, rather than privatisation as we know it. Some go so far as to laud the 'Beijing consensus' as a Third Way alternative to Washington/IMF extremism.

Nevertheless, privatisation is happening, and even these listings couldn't happen if many state firms were not doing what privatised firms have to do to survive—merging, modernising, laying off workers.

Over the last half decade, anywhere between 20 and 40 million workers have been made redundant by the old state sector. You can see now more clearly where Tiananmen Square comes in. In Britain, 30,000 miners caused headaches. In Russia, the country China itself holds up for horrified comparison, all those new poor did outrageous things like, when they got the chance, voting for a return to the old communist ways. China couldn't have that.

And those numbers are small compared with the estimated 300 million rural peasants who, to make China's economy and agriculture efficient, will one day soon have to leave the land and come and find work in the cities. China is the World Trade Organisation now, too, and the thought of 300 million Chinese peasants discovering that their rice is more expensive than American rice and becoming anti-globalisation protesters is an alarming one. They might come to the city to riot, not work. Luckily, China still has a hukou, or residency permit system, akin to the old apartheid pass system, enabling unemployed peasants to be turfed back to where they came from before they cause trouble.

Probably the most visible success story in China in the last decade has been the cash flowing in from the multinationals and Taiwanese and Hong Kong businessmen. Foreign direct investment in China is now the highest in the world: a vote of confidence in its economic ordering if ever there was one. Opening and reform should be the Chinese translation of globalisation. Foreign money builds the factories where the world's shoes, underwear and wristwatches are made. It builds factories to make cars—not just runabouts but BMWs and Audis—for the country's middle classes. And we can all see why. Not only is labour cheap—which is true across much of the world, after all—but it is well-ordered and, above all, not organised. Free trade unions are banned. There is an official, party-run union, but it is not known for its activism. One of the most amusing spats this year was between the state media and Wal-Mart which, rightly seeing China as the new Midwest, has started opening stores here. Wal-Mart does not recognise trade unions, and this policy was applied to China. The media professed outrage that the workers' voice was being silenced, which goes to show that just as left-wing unions have ideological disputes, so do their opponents. The

general assumption of a politician like Tony Blair, who wants to see more and more British investment, is that it is a win-win process—good for Western business, good for China. There is still a feeling that although Tiananmen Square put things back for a bit, all that economic liberalisation must be followed by political freedom somewhere down the line, and that constructive engagement, as it used to be known when South Africa was the issue, will help it along.

If so, there is not much sign of it at the moment: newspaper editors are arrested and dissidents who might say something controversial during the Tiananmen anniversary are put under house arrest. An interesting series of papers by a Harvard University academic had a different interpretation of foreign investment in China: for the most part, it was not just another aspect of a burgeoning local private economy, but a replacement for it. Private entrepreneurs, though theoretically protected in China, get very little access to financing, as bank loans are directed to state firms. The cash, instead, came from foreign firms.

The political advantage for the government is this: the economy is left in the hands of two sets of people with a vested interest in the status quo—state firms run by party appointees, and foreigners, who know they have to keep on the good side of party officials to extricate their cash as profits. The other thing the party has realised in the last 15 years, apart from the fact that ambitious young graduates can soon be bought off with company cars, is that people still have to feel they believe in something. Hence the fuss about the Diaoyutai islands. The what? Well, they are small, uninhabited and entirely unimportant. Nevertheless, the fact that they are owned by the Japanese when the Chinese think they belong to them is enough for regular protests at the Japanese embassy and symbolic raiding parties on the islands themselves, which are a couple of hundred miles away from the Chinese coast. Nationalism is the new ideology of state. Critics of government policy are no longer revisionists or counter-revolutionaries, words which have disappeared from the official lexicon. Instead, when a Hong Kong democrat is kicking up a fuss about elections, some official will turn purple, splutter and in full Colonel Blimp mode accuse him of being a traitor. A young punk rocker was asked by a (foreign) interviewer what made him really angry. 'Japanese occupation of the Diaoyutais,' he replied.

Sometimes this nationalism is turned against the Americans, but mostly the Chinese love the Americans and the American way. While Chinese leaders make up to Europe in the eventual hope of a 'multipolar world', there is no doubt that they prefer George Bush to Romano Prodi when it comes to diplomatic horse-trading. And if you ask people what developmental model China is following, few mention Europe, with its long tradition of monarchs and autarchs giving way to modern economies. China is a Wild West, bringing railways, industry (and prostitution) to its inner vastness as it seeks to subdue those rebellious and primitive natives: Tibetans and Muslims, for example.

One senior diplomat in Beijing expresses his support of his government's engagement with China, and regards the EU arms embargo imposed after Tiananmen as a slightly meaningless formality—as, apparently, does Mr Blair, who is said to want it lifted. In private, should you ask how open China will be in 20 years' time, the diplomat ventures the suggestion that it will be just like now, but richer and more powerful. That is a right-wing prospect, but not necessarily a pleasant one. It is hard to imagine businesses doing anything other than investing in China, but I do wonder how we will all feel when it happens.

Last Christmas was another significant anniversary, of Mao Tse-tung's 110th birthday. The Daily Telegraph interviewed a teenage boy queuing up outside the Great Helmsman's mausoleum on Tiananmen Square. He was a Young Pioneer, and about to be inducted into the Communist Youth League. What was Mao's relevance to him today? 'We learn from Mao that you can always achieve your goal if you struggle and work hard enough,' he said, in all seriousness. And what was his goal? 'To go to America,' he said.

He probably will, and if he comes back it will be to work for Ford or IBM. Fifteen years ago, the tanks were rolling where he stood, and he is a child of that day.

Richard Spencer is Beijing correspondent of the Daily Telegraph.

Squeezing Profits From Propaganda

A long-expected reform of the media has kicked off with newspapers that black-mail subscribers as the first targets. The party, however, is determined to keep editorial control even as it opens the sector to market forces and investment

By Susan V. Lawrence/BEIJING

THE ANNOUNCEMENT on the June 20 evening news was brief. Even on the front pages of official newspapers the next day, it was only a sentence. The Communist Party's Publicity Department (the former Propaganda Department), the State Press and Publications Administration, and the Post Office were barring all newspapers and periodicals, except scientific publications, from taking any subscriptions for 2004 until the end of September. It said the move was part of an effort to stop publications from engaging in coercion to secure subscriptions.

The significance of this move could easily be missed. Scholars and media executives say, however, that it is the first concrete step in a long-anticipated shake-up of the media in China. The immediate target: thousands of small party- and government department-affiliated publications that, by flaunting their official links, force people to buy subscriptions and take out pricey advertisements. A typical publication doing so is a newspaper run by the local tax, commercial or police bureau. Subscribe or advertise and, by implication at least, you are protected against a tax audit, losing your business licence, or worse.

The party's next step, scholars and media executives say, is a plan, which could be approved by the Publicity Department this month, requiring most party and government publications to cut their links with their official parents before subscriptions can resume. They will then be left to swim, or more likely sink, in the marketplace. This mirrors earlier campaigns to force official departments and the military to give up running businesses, on the grounds that such enterprises consumed official resources and fed corruption. Their official backing also gave them an unfair edge against competitors. The crucial difference this time is that the party aims to retain its control over the publications' content even after they are forced into the market. Exactly how this continued editorial hold will be achieved is unclear.

Equally surprisingly, some scholars say the official reform plan, which was originally drawn up by the State Press and Publications Administration, also provides for minority foreign and domestic private investment—perhaps up to a 40% stake—in previously closed parts of the sensitive media industry.

"This kind of regulation will come out very soon," predicts Yu Guoming, a high-profile media scholar who saw an early draft of the plan and believes it will survive vetting by the Publicity Department largely intact. Referring to the investment provisions, Yu, the vice-dean of the Renmin University School of Journalism in Beijing, says that for Chinese media groups to succeed as companies, as the government wants, "they

REFORMS BUT NOT A REVOLUTION YET

- The state seeks to stem losses from the print media without diluting its propaganda message
- So it plans to stop forced subscriptions and cut links between newspapers and official bodies
- It may even allow foreign investment, while retaining its editorial control

have to have capital from all sorts of sources. Just relying on the state channel [of investment] isn't enough, I'm afraid."

Even as they contemplate tapping new, even foreign, capital, the authorities have made it clear they have no intention of allowing nonstate investors into the editorial operations of major media companies. The corporate model being promoted by China's propaganda tsars is one in which media companies spin off their nonsensitive operations into subsidiaries that could be opened to outside investment. Editorial control, meanwhile, is retained by their corporate parents, which are meant to stay 100% state-controlled.

But some observers say allowing non-state investment in any part of the operations of news organizations will inevitably have an impact on content. Du Gangjian, a professor at China's elite training academy for state officials, the National School of Administration, suggests that distributors, because of their proximity to advertisers and subscribers, can easily slide into a role in which they would begin influencing editorial decisions—by noting, for example, that certain kinds of content are good for advertising, while others aren't.

Renmin University's Yu also sees a more market-oriented media as having real implications for content. In China, he explains, only the market has the power to "bargain with the government on the question of control of the media."

The reforms are part of a campaign by a new generation of party managers to improve the economic performance of the media while retaining control over content, experts say. Officials speak openly of their hunger for China to have media conglomerates capable of taking on the world's media giants. They are impatient with a media landscape cluttered with publications that one senior official described on an October television show as "low level" and "redundant." China currently has 70 media groups, 38 of them newspaper groups, and none remotely able to compete with the global behemoths that China wants to emulate, such as AOL Time Warner and Germany's Bertelsmann. Indeed, the head of China's State Press and Publications Administration, Shi Zongyuan, noted in a book last year that sales revenue for Bertelsmann in the 1999–2000 financial year exceeded sales revenue for the entire Chinese press and publishing industry the year before.

The scale of China's ambitions helps to explain longstanding foreign investor interest in its media industry. For ideological reasons, Chinese officials have been wary of allowing in foreign capital. But Yu says that they are starting to recognize the need for infusions of foreign capital and management know-how to power a rapid expansion of media groups so that the country can hold its own internationally. The editor of one mid-sized official newspaper says he, for one, has no ideological objections to foreign in-

vestment. It could bankroll an increase in circulation, he says, and put the paper on a firmer financial footing.

To be sure, the State Press and Publications Administration blueprint on reform now being considered by the Publicity Department wasn't widely disseminated. Senior editors at three leading publications say they haven't been told about its foreign investment provisions, and are sceptical the party would allow that in a sector it is intent on continuing to control.

Media groups, of course, could also raise money by listing on domestic stock markets. Propaganda officials in the past have blown hot and cold on that idea. An article in March in the Beijing-based *Business Post* reported that six newspaper groups had prepared proposals for listing on China's domestic A-share market, but most for now had put the plans aside.

Until recently, foreign investment in China's media was banned. Foreign companies had to make do with cooperative agreements under which Chinese partners used their titles and content, usually in return for a portion of advertising revenue. Newsstands are full of foreign titles from such arrangements, including Chinese-language editions of *Madame Figaro, Marie Claire, Cosmopolitan* and *National Geographic Traveler*.

In December, China for the first time allowed foreign equity investment in a part of the print media—retail newspaper, magazine, and book distribution. The State Press and Publications Administration began accepting applications for such investments on May 1. The official *China Daily* newspaper reported in April that more than 60 overseas companies had opened offices in China with the aim of applying. According to its commitments to the World Trade Organization, China is scheduled to start allowing foreign capital into the wholesale distribution business from December 2004. The deeper significance of the changes in the investment rules now being mooted is that they could also permit foreign investment in other aspects of print-media operations.

The picture for domestic private investment in the media is more complicated. Although regulations officially

restrict investment to state media companies, a number of state, private, and even listed companies hold stakes in Chinese publications.

If broader foreign investment is allowed, a key precondition for official approval will almost certainly be the Chinese firms successful separation of its ideologically sensitive editorial activities—which will remain off-limits to foreign or domestic private investment—from the rest of its business.

A Publicity Department vice-minister, Li Congjun, spoke about the need for such a division in October in a TV discussion programme, a rare public forum where propaganda officials talked about media reform. Chinese media companies, Li said, needed to be managed as companies and to be mindful of their "ideological characteristics" and "obligations to guide correct opinion." Indeed, State Administration of Radio Film and Television Minister Xu Guangchun listed on the programme four things that would never change: the media's role as the party's "throat and tongue," party management of the media and media personnel, and the media's role in "guiding correct opinion."

THE MODEL FOR THE PROPAGANDA TSARS IS COMPANIES SPINNING OFF NONSENSITIVE OPERATIONS INTO SUBSIDIARIES OPEN TO OUTSIDE INVESTMENT

To manage the tension between party-controlled content and the market, Li said that media companies should "separate propaganda business from operations." He gave the Shandong Dazhong Newspaper Group as an example. From 2001, it created seven subsidiaries, each handling a different aspect of the business, such as advertising, distribution and printing. It kept editorial activities in the parent company, and created an investment-management company to control its assets. Foreign investment would presumably be allowed only in subsidiary companies.

The aim of the party's move to end coerced subscriptions and force official papers

A POPULIST STYLE FOR PARTY, NOT PUBLIC, INFORMATION

Communist Party propaganda officials have little time for anyone who suggests China would be better off with a free media. In the party's book, the central role of the media is to "guide correct opinion"—most importantly, "correct opinion" about all the benefits of Communist Party rule.

Now, as always, publications and individual editors and reporters who stray too far from party reporting guidelines face disciplinary action. Since a new party leadership took office last November, the party has closed one paper—*Xin Bao* in the Worker's Daily Group—and suspended another, the *21st Century World Herald*. It has removed editors at top papers, and may well have had a role in the failure of hard-hitting financial magazine *Caijing* to get its June 20 issue onto newsstands. (Eager to play down the controversy, the magazine is blaming unspecified "technical problems.")

Most notoriously, in the early months of the outbreak of Severe Acute Respiratory Syndrome (Sars), the party ordered a blackout on reporting on the virus, in the name of preventing panic. The blackout only fuelled greater panic, and contributed to the disease's spread. The party ultimately backtracked, and allowed a period of unusually open coverage.

The image-conscious new leadership of President Hu Jintao and Premier Wen Jiabao would prefer that no one dwell on such heavy-handed tactics. They would like to be known instead for a handful of more populist media initiatives. Chinese journalists all cite, for example, Hu's instruction that news outlets should cut back dramatically on their coverage of leaders' activities. They say that he issued the order on his first day on the job as Communist Party general secretary, after watching the endless, typically stultifying coverage of his accession on the television news.

The TV news still runs reports of leaders addressing rows of bored cadres and shaking hands with visiting foreign dignitaries, but the stories are now shorter. Hu reportedly told the media to use the time and space freed up by his order to report on "things that people really care about." He even warned the secretaries of fellow leaders to stop calling news organizations demanding coverage, a grateful top editor at a party paper says.

Among his more substantial initiatives, Hu is credited with giving the go-ahead for news organizations to report when the full Politburo has met, and what subjects it discussed. Such information was previously secret. In addition, Chinese television broadcast live tense press conferences at the height of the Sars crisis.

Yet Du Gangjan, a professor at the National School of Administration in Beijing, says that he hopes for more. The propaganda authorities, he says, still tend to feel that "information is the monopoly of government." Du says that work is continuing on a public information law started under Hu's predecessor, Jiang Zemin. But for now, propaganda officials "have no sense of the right to know, or of what public information is," he says.

Susan V. Lawrence

to operate according to market rules is similarly to improve the financial fundamentals of the media industry. One major goal is to cut the numbers of publications. China has more than 2,000 newspapers and nearly 9,000 periodicals. The National School of Administration's Du estimates that half the newspapers are in financial trouble. They drain funds from the party and government, and from victims of coerced subscriptions and advertising. Du estimates the average village government

spends more than 1,000 renminbi ($120) a year on forced subscriptions—up to three years' income for a farmer in some parts of China.

To relieve burdens at all levels, Renmin University's Yu and others say the party wants to retain just one affiliated newspaper group per province and city. The Beijing-backed Hong Kong newspaper *Ta Kung Pao* said on June 24 that "in principle" all district- and county-level party and government papers will be eliminated.

It's unclear exactly how party and government departments will divest themselves of publications. Some scholars suggest that the government may have to drop its requirement that all publications have official sponsors. This could open the way for publications to be openly run by nonstate players. Du sees a more "pluralistic" market as inevitable. "The ruling party cannot prevent the emergence of non-'throat-and-tongue' media," he says.

Where Are the Patients?

Health care and the environment have been overlooked in China's rush to capitalism.
We look at health care below; at pollution on page 63

RONGRONG was only a few weeks old when her parents noticed swellings. They took her to a village clinic and to a rural hospital. Both failed to spot that the baby was severely malnourished, though they charged the equivalent of two and a half months of her father's income as a brick-factory worker. The delay probably cost Rongrong her life. When her parents eventually took her to a city hospital, where a correct diagnosis was made, she was in a critical condition. After seven days, and another three months' salary-worth of hospital bills, she was dead.

The baby, from a village in the central province of Anhui, was one of at least 200 infants around China who were seriously malnourished, and in some cases killed, by consuming substandard milk powder in the past few months. Parents had no way of knowing that the cheap but well packaged product they were buying was illegal and hardly more nourishing than water. And because of the prohibitive cost of health care, parents often failed to get their children treated in time to save them.

After the scandal came to light in April, the government announced that affected children would be treated free (the Chinese press has since reported that some hospitals resumed charging after the departure of the central government officials sent to investigate). Rongrong's parents, like other bereaved families in the area, received compensation from the local government. The payout of 12,000 yuan ($1,450) was the equivalent of nearly two years' income for Rongrong's father.

But the tragedy had taken its toll on Rongrong's mother, who now needs medical treatment herself. To have more time to look after her, her husband has given up his job at the brick factory and returned to subsistence farming. His wife's hospital bills have already amounted to more than three months' worth of his former salary.

Criticism of the affair has focused on the ubiquitous sale of fake and substandard products. But just as big a failing was that of the health system itself. The families who bought the milk powder were mostly poor country people. Undoubtedly fewer lives would have been lost if they had had better access to basic health care.

Mao Zedong, for all his egregious faults, would have done better at providing it. In his day, nine out of ten country people had access to subsidised health clinics run by the much celebrated "barefoot doctors". But in the course of China's relentless march towards capitalism in the past two decades, this arrangement has collapsed. In the countryside, 90% of the population now has no health insurance. In the cities, nearly 60% are uncovered. Out-of-pocket spending on health care is soaring.

When the World Health Organisation (WHO) ranked the public-health systems of 191 countries four years ago, China was placed at 144, behind some of Africa's poorest. India, which has half China's GDP per head, came in at 112. The criteria included fairness of access to health care and fairness of contributions to the cost.

Better to live on the coast

Judged by life expectancy, infant mortality and child-birth deaths, China's record looks impressive. Twice as many children in India die in the first few months of life, and twice as many mothers die in child-birth. At birth, Chinese girls can be expected to live to 73 and boys to 70—a level comparable to medium-level developed countries. But there are huge disparities between regions. In richer areas, such as around Shanghai on the coast, health indicators are as good as they are in many western countries. In western China, they are those of a basket-case country.

After making strong gains in the first three decades of Communist rule, health indicators have changed little in the past quarter-century, despite the extraordinary economic achievements of China, with annual average GDP growth of 9.7% in the past 20 years. Some health experts believe that in parts of the country—particularly in the west where incomes are half the level of the booming eastern seaboard—life expectancy might even be falling.

According to the World Bank, China has lifted 400m people out of severe poverty in the past two decades. But millions have slipped back into it as a result of health-care costs. Millions of others are dying because they cannot afford health care. A government survey three years ago found that some 60% of rural residents avoid hospitals altogether because of the expense. Diseases once declared tamed, such as tuberculosis, measles and snail fever, have been making a comeback. And amid the disarray of the system, a new infection, HIV, is rapidly taking hold.

Things could get much worse. In the coming years, the customary Chinese way of dealing with expensive medical crises—borrowing from family and friends—will become more difficult as the proportion of elderly citizens in the population rises steeply. The UN predicts that by 2040 China will have only two working-age people for every person over 60, compared with 6.4 in 2000. While ageing populations are common in the developed world, the Centre for Strategic and International Studies said in a recent report that China may be the first major country to grow old before it becomes rich.

Too nervous to spend

Public anxiety over the collapse of affordable health care is reflected in China's high savings rate. Despite the parlous state of China's state-owned banks, bank savings by individuals have grown rapidly in recent years. Worries about the fast-rising costs of health care and education—and the lack of private-insurance schemes and other low-risk investment opportunities that might help offset them—are restraining consumer demand and thereby imperiling China's long-term economic growth.

This makes health-care reform a crucial part of China's development strategy. To ease growing pressures on its fragile financial system, China wants to become less dependent on government investment as an engine of growth. But unless consumers feel confident that they can cope with the risk of a serious health problem (as well as with all the other increasingly costly contingencies), it will be difficult to encourage them to spend.

Last year's outbreak of severe acute respiratory syndrome, or SARS, not only concentrated the government's attention on the problem but aroused the world's attention to the potential health threats from a country with a patchy record on epidemic control. Fortunately, SARS never took hold in the rural areas where facilities are the shabbiest. Had it done so, it would have been hard to stop its spread.

Well into the crisis, the government realised this and tried to reassure people that they would not have to pay for any SARS treatment costs. But given the high possibility that SARS-like symptoms might turn out to be caused by another ailment not covered by free treatment, there was the likelihood that many people would have avoided the hospitals anyway.

The outbreak demonstrated how fragile and threatening the health system had become. Suddenly, officials began to turn their attention to neglected health problems such as HIV. China estimates that it could have around 1M HIV carriers at present. The WHO says this could rise to 10m by the end of the decade. Last year, for the first time, top leaders, including the prime minister Wen Jiabao, were shown on television shaking hands with AIDS patients. Officials, at least in some areas, have begun to overcome their prudishness and to promote condom use and the distribution of clean needles to drug addicts.

For President Hu Jintao and Mr Wen improving health care has also become part of a political strategy aimed at salvaging the Communist Party's badly tarnished image. Mr Wen and Mr Hu now stress the need to address the concerns of the marginalised. In 2002, the party set a goal of turning China into a middle-income country with a "well off" population by 2020. But in recent months the emphasis has shifted from simply increasing GDP per head to achieving broader measures of wealth, such as enjoying good health.

But getting there will be far more difficult than fulfilling GDP growth targets. For the past 20 years, the government's financial commitment to health care has been declining. Urban hospitals, though mostly still state-owned, now receive only about 10% of their operational funds from the state. For the rest they have to generate their own revenues, mostly from selling medicine and medical tests (the cost and wilful over-prescription of which is the biggest grievance of patients).

Even immunisation isn't free

Rural hospitals are in even worse financial shape. The most basic ones are run by governments at the township level, the lowest tier of government hierarchy. For most of these administrations, the only source of funding is the trickle of income they receive from higher-level government, plus taxes and fees that they raise from farmers and businessmen. Even preventive medicine now has to rely on fees. The WHO says that China is the only country in the western Pacific region which relies on patients to finance childhood immunisations. Not surprisingly, many peasants now avoid such treatment.

State-owned enterprises once shouldered much of the responsibility for basic health care, including the running of their own hospitals. But with the collapse of many such businesses, workers have been left to fend for themselves. Private businesses are supposed to pay for medical insurance, but most do not bother.

The chronic underfunding of public health has created a culture of cynicism and corruption in China's hospitals. As well as having to pay up front before they are treated, patients frequently complain that in order to get good treatment they need to pay "red packets" (bribes) to doctors and nurses.

In the past year—goaded by the SARS outbreak—the authorities have stepped up their efforts to ensure that country people and the poor get access to basic health care. In the countryside, they have designated more than 300 counties (about 10% of the total) where a new "co-operative medical system" is being tried, with a plan to make it countrywide by 2010. Funding is shared between voluntary participants, the local authorities and the central government. In addition, a new insurance scheme, paid for by central and local governments, has been introduced for the poorest of urban and rural families to cover the cost of serious illnesses. The scheme is due to be implemented nationwide next year.

But both projects have serious drawbacks. Local governments are often unwilling to make the necessary contributions, especially in poorer areas. And individuals are often unwilling to pay for a service they feel they may not immediately need. For the past two decades, local governments have gouged farmers for contributions to an almost non-existent health-care system, with the money being used mostly to pay staff (many surplus to requirements or simply non-existent, with the money being used to line officials' pockets) rather than to pay for services. Consequently, there is reluctance to join any new scheme.

Can market forces provide the answer?

What other options does China have? Some officials, seeing the market as the panacea, suggest that the government should withdraw entirely and let prices be controlled by competition. In recent years, many township governments have sold off or leased their hospitals to private investors. But while in some cases the investment may have helped improve conditions, there is little evidence that the price of health care has fallen.

In the cities, a few hospitals have been built with private money, and businesses have taken over some of the hospitals that used to be run by state-owned enterprises. By 2005, state enterprises are supposed to cease their support for all hospital facilities. Some hospitals will be closed, others merged with bigger ones or sold off.

Privatisation remains controversial. Two years ago the central government tried to limit the trend by decreeing that every township must retain at least one state-owned hospital. But many cities have pressed ahead with their own reforms without explicit approval from Beijing.

Xinxiang, a city in the central province of Henan, has engaged in the biggest sell-off, and one of the most controversial. Earlier this year it sold majority control of all of its five main hospitals to a single state-owned pharmaceutical company, China Worldbest Group. The government was happy to have reduced its health-care burden, and the Shanghai-based company was happy to have guaranteed outlets for its drugs. But what about competition? Many health-care specialists saw the move as the replacement of one overpriced monopoly by another, and the Shanghai government banned further such deals between drug companies and hospitals.

But there are signs that the central government is at last trying to adopt a more coherent policy for health-care reform. In early April, at its annual national conference on health-care issues, the Health Ministry circulated a secret draft of a policy paper outlining the respective roles of the government and the private sector in urban health care. Experts familiar with the document say it suggests that the government retain control of the main hospitals, but let second-tier hospitals be owned and operated privately. One expert estimated that this could involve the privatisation of 60% of urban hospitals. This would allow the government to increase its spending on the hospitals it keeps, thus reducing their financial dependence on charges for medicine and tests.

On the face of it this sounds a good idea. China's problem is not a shortage of medical facilities. It has a relative abundance of them: 1.6 doctors per 1,000 people compared with 0.4 in India, and 2.4 hospital beds per 1,000 people compared with India's 0.8. But this means that official resources are stretched too thin. Concentrating on key hospitals would enable the government to pay doctors a decent wage (though Chinese sources say that the document does not promise this will happen). At present a hospital director earns about the same as a company sales representative with a couple of years of experience. No wonder doctors are demoralised.

But if large numbers of people are still unable to afford treatment, such reforms will create a better system only for the affluent. In the long run, what China needs most is a health-insurance system that works. This should include insurance for private treatment (now non-existent), giving patients a bigger choice of facilities and stimulating private investment in hospitals. It would also need to ensure that those who contribute little or nothing to the system still get some coverage. In other words, the government needs to spend a lot more, particularly in the countryside and among rural migrants to the cities.

One reason why China's health-care system is in such a mess is that the central government's share of tax revenue has dropped in the past 20 years. But relying on local governments to do more will not work. At the township level, the majority are bankrupt. The central government needs to allocate more of its own money—and to force provincial governments (which like to go their own way fiscally) to make sure the allocations reach their intended targets.

Strong incentives, such as tax breaks, will be needed to encourage private business to run hospitals on a not-for-profit basis. Although there has been considerable interest among foreign and domestic private investors, very few large-scale investments have been made because of regulatory uncertainty and questions over profit and pricing. The establishment of a handful of foreign-run hospitals in the big cities for the rich and well-insured are rare exceptions. In poor areas, including much of the countryside, the government will need to remain the primary provider.

Achieving this will involve changing priorities. Prestige projects may have to be abandoned. And there will have to be a fairer allocation of resources to address the current imbalance by which cities currently enjoy 80% of health resources despite having only 35% of the population. And to make sure it all works, there will need to be an effective system of oversight which China now sorely lacks. The idea of good corporate governance is novel enough in China, but in health care it is non-existent. A sea change is needed in everything from hospital management to the way central and local governments spend their money. Even so, slowly and reluctantly as it may be, China is beginning to discover that market forces alone cannot produce good health care.

Manufaketure

Counterfeiting and pirating (that is, making knockoffs of what developed nations have created) are at the heart of the Chinese economic boom. As unethical or illegal as it might be, the Chinese government is not about to stop it.

By Ted C. Fishman

Most of the pharmacies in China that dispense Western-style medicines have an antiquated, if reassuring, air about them. There are no posters on the walls for brand-name drugs. Candy is not for sale. Photo processing is not available. Druggists work in long white lab coats and surgical hats that could have been salvaged from a World War II hospital ship. Some pharmacies require prescriptions for the most potent drugs, others only an earnest chat with a druggist. Drug orders create paperwork that passes through three or four bureaucratic layers before reaching the solemn cashier, who issues a handwritten receipt.

Such an old-fashioned scene might argue for just how far China trails the United States and other advanced economies, where both science and marketing are seemingly years ahead. Yet these pharmacies also represent a current and urgent battleground in one of the most important struggles between the developed world and China's surging economic power. This is the fight over intellectual property and the related investments essential to the knowledge economy, that amorphously defined new world in which better ideas, not faster, cheaper hands create jobs and wealth. Despite their appearances, Chinese pharmacies are stocked with expertly copied versions of some of the world's most profitable medicines, patented products that generate hundreds of billions of dollars' worth of business in the United States, Europe and Japan. Even the very latest miracle drugs sell in China for a fraction, often one-tenth or less, of what their authorized equivalents in the United States cost.

Foreign companies lose control of their goods in two related ways: to counterfeiters who copy products and then sell them under different or altered brand names, and to pirates who make look-alikes and try to pass them off as the real thing. Using a lost-sales calculus, which measures the losses to foreign companies by determining the value of the dubious goods sold, the U.S. Department of Commerce estimates that American companies, as a result of counterfeiting and piracy, lose between $20 billion and $24 billion annually. The Japanese sacrifice even more: $34 billion. Throw in the sales lost by the

European Union, and the cumulative losses for the three economic blocs approach $80 billion.

While losses to American and other advanced economies are high, China's appropriation and dissemination of the world's most valuable products and technologies, if they continue unabated, will ultimately mean a lot more than dollars lost. China's pirating and counterfeiting could radically change the way entertainment, fashion, medicine and services are created and sold. The companies, big and small, that Americans work for could be weakened. Chinese practices might reduce the prices of what we buy, by undermining the powerful companies that now control essential but expensive goods like drugs and computer software—or these practices might, should China's unwillingness to accede to American copyright demands ignite trade wars, drive prices up. A U.S. consular official in China who requested anonymity—few American officials are willing to speak openly about sensitive issues relating to China—told me: "Nothing has a higher priority in our trade policy than the fight to protect American intellectual property. It is every bit as important an effort for us as the war against weapons of mass destruction."

The analogy has some merit. As with stolen bombs, the chief worry about losing control over intellectual property is not that American manufacturers will forego sales opportunities; the fear is that its new "owners" will turn our own innovations back on us and inflict much broader economic damage. For the United States, the world's most formidable producer and exporter of invention, entertainment and trademarked brands, the stakes are highest. William H. Lash III, the Commerce Department official who is leading a new initiative to change China's practices, vows that the Bush administration will take "whatever means are necessary" to force a change.

What makes China so troubling for American and other foreign companies is that the country is both a potential rival, with an alternative legal approach to intellectual property that limits their prospects in China and weakens their competitive strength globally, and a haven for pirates and counterfeiters. Start with the damage that fake drugs, for example, can do. Whether well made or poorly, they knock the genuine thing out of the market.

According to the Chinese government-run press, hundreds of thousands of people in China have died from fake drugs that are either toxic or do not contain the active ingredients that users need. Drug companies report an increased threat from counterfeits entering the legitimate supply chains around the world. John Theriault, a 26-year veteran of the F.B.I. who now helps orchestrate anticounterfeiting efforts on behalf of Pfizer, says the company, working with Chinese authorities, has "seized millions of units of counterfeit pharmaceuticals and thousands of kilos of compounds" used to make them. In the worst cases, the fakes are commingled with legitimate products. "You might have 2 bad pills mixed in with 28 good ones," he says. (In May 2003, 200,000 bottles that had been sold in U.S. pharmacies and that contained counterfeits of Lipitor, Pfizer's cholesterol-lowering pill, were recalled.) Fakes "can ruin a brand and ruin a company," Theriault says.

China's

failure to police industry and to protect intellectual property acts, in effect like one of the greatest industrial subsidies in the world.

But if bad imitations are a big problem, good imitations may be a bigger one. Pfizer happens also to be a prime example of what is arguably the most serious threat to U.S. knowledge-based companies in China: its intellectual-property rules. In the case of drugs manufactured before China agreed (in order to join the World Trade Organization) to adopt patent standards closer to the international norm, production continues as before—that is, without any licensing fees paid to Western companies. Even today, however, Chinese companies, many of them government-run, simply continue to "reverse engineer"—that is, take the known ingredients and work backward to figure out a process that produces accurate copies of—the drugs (including recent blockbusters) and pay the foreign patent holders nothing. Increasingly, China's pharmaceutical companies are rushing to claim patents for their copies before foreign patent owners can assert their rights. This is what happened with Pfizer's Viagra, which has multiple Chinese imitators: the Chinese authorities denied patent protection for Pfizer and opened the market to Chinese knockoffs instead. (Pfizer is appealing.)

Press coverage in China of the Viagra decision made a point of noting that one Viagra pill costs 1 yuan to make, or around 12 cents, yet it sells for 98 yuan, or about $12. That sort of difference is sure to pique the attention of margin-squeezing Chinese manufacturers—and perhaps encourage more copycats to rush into the market. Selling Chinese-made Viagra could turn a company into a future pharmaceutical Goliath, which would please China's rulers.

Certainly China also has a public health incentive to see that drug prices are affordable for its people, who earn on average one-fortieth of what Americans do and who rarely have health insurance. China's strategy often works: the fear of knockoffs entering the market drives the price of the patented drug down, and many important drugs cost less there than they do nearly ev-

erywhere else in the world. (Historically, medicines lacking patent protection, either because a time limit has expired or because countries like India or China simply offer no such protection, can experience price drops of more than 90 percent.)

THE THREAT TO American interests is not hard to identify. According to the Milken Institute, Big Pharma employs 400,000 Americans directly, creates another 2.7 million jobs and contributes $172 billion to the U.S. economy. It is one of the most important engines of the knowledge economy; in 2003 the pharmaceutical industry invested $33.2 billion in drug research. That does not include the nearly $30 billion spent on life sciences by the publicly financed National Institutes of Health, which pays for research that leads to commercial drugs. Weaken the drug industry and you weaken one pillar of the U.S. economy. And Pfizer's trouble with Viagra in China demonstrates just how vulnerable the American knowledge economy is in a world where ideas "protected" by our laws trade freely nonetheless. Behind almost every blockbuster drug, killer software application or computer-chip design is a public infrastructure that has steered uncountable sums and the country's best talents toward their creation.

Consider what an advanced economy like ours does best: make movies, produce television shows watched from Helsinki to Cape Town, turn out global pop stars. We design the software and processes that streamline the operations of giant retail chains and global high-tech manufacturers. We engineer advanced engines and the guts of the world's computers. We devise brands, durable corporate identities and fashions. We conjure new ways to move money and put it to work. We turn the most basic tasks into knowledge work. Modern printers, to note one example, rely heavily on the most advanced automated presses, computerized design tools and management and shipping for delivering materials efficiently to consumers and are as dependent on the latest software and technological innovations as a biotech lab. And those 2.8 million American workers who in recent years have lost their factory jobs? They don't learn new ways to use power tools. They are retrained in front of a computer; they learn to run the robots that do the jobs they used to do.

The trouble with this apparently successful state of affairs is that the stuff we do best exists nowhere and everywhere at the same time. Some of our most valuable things—software codes, pharmaceutical processes, car designs, digital movie files—weigh nothing and, as e-mail attachments, can move at the speed of light. To learn American ideas and procedures is all but the same as owning them. (Unless, of course, laws successfully prohibit their co-option.) In contrast, most of what China makes that finds its way into the world market is physical. The Chinese can borrow and steal the designs to our best products all they want. For instance, 90 percent of all software running on Chinese computers has been pirated and bought openly in stores for around $3 a copy. But if Americans wanted to borrow and steal what China makes, we would have to march in with an army and commandeer Chinese factories and workers. Western powers and the Japanese tried that in the mid-19th and -20th centuries, respectively, and will not repeat the experiment. China, however, can in a sense colonize the developed world simply through careful study and a willingness to go its own way on intellectual-property protection.

If China's commitment to wipe out commercial piracy and counterfeiting were judged by the laws that the country has on its books, the Chinese government would seem to be as strict as any. China has made a great show of cracking down in the past few years. Newspapers and television news programs regularly feature stories about government raids on massive counterfeiting operations. Hundreds of thousands of DVD's and dozens of duplicating machines seized here, a warehouse of CD's there and trucks full of sham designer handbags somewhere else. In December, China passed a much-awaited national law that criminalizes piracy and counterfeiting, allowing courts to jail violators for up to seven years; before, only civil penalties applied. The new law is unlikely to spur enforcement, however. And even if it does send people to prison, that may only prove a boon to the copycat economy. For example, just before Christmas, Sony announced the results of an investigation into Chinese operations that were daily turning out 50,000 fake PlayStation 2 game consoles and accessories: a container loaded with fake parts was found to have visited a prison in Shenzhen just long enough for inmates to assemble the parts. The Chinese themselves take it as given that powerful government interests stand behind the trade in counterfeit or pirated goods. What to foreigners may seem to be an aggressive action against a big piracy ring can look to the Chinese like a sort of St. Valentine's Day Massacre, where one powerful manufacturer uses police cover to eliminate a weaker one.

As the legal code grows fatter, so, too, do the supply and sophistication of fake goods. The places they are sold no longer look like back-alley stalls but like Main Street retailers. Near Beijing's diplomatic row, two outdoor markets once famous for knockoff fashions have been combined into a large, bright department store-like building with escalators, tailors on site and merchants with business cards, international shipping accounts and full stocks of fake fashions, designer tableware, brand-name musical instruments and, of course, thousands upon thousands of fake Swiss watches. The most common punishment counterfeiters face is the confiscation of whatever products they have in stock. Sometimes a pitiful fine is levied. China's National Copyright Administration cites with much fanfare 52 raids on video shops in 2003, but the total fines amounted to $6,900, or an average of $132 for each offender.

China's lax policies on copyright protection offer the country the advantages of both bread and circuses. Andrew Mertha, a political scientist at Washington University who has worked with Chinese and American officials on Chinese intellectual-property law, summarizes the circus side of things: "If you're the Chinese leadership, do you want people idling around in the street, complaining about how unhappy they are, or do you want them home watching Hollywood movies?" In other words, the government is slow to crack down on the piracy of entertainment products because these serve its social agenda. But is there

any doubt that if vendors suddenly found a brisk market for DVD's promoting Tibetan independence or the virtues of Falun Gong, the outlawed religious sect, the DVD business would shrivel up overnight and all those anticounterfeiting laws on the books would find ready application? Indeed, when Sega's new online fantasy sports game "Football Manager 2005" had the gall to suggest that imaginary soccer leagues in Hong Kong, Taiwan and Tibet could be governed locally, rather than by the central government, China's Ministry of Culture banned the game on the grounds that it posed "harm to the country's sovereignty and territorial integrity." Fines reached $3,600.

Because the overwhelming majority of products pirated and counterfeited in China are, for now, sold mainly in China, they provide the Chinese people with "bread" insofar as they make all sorts of other goods affordable. Often, as in the case of medicines and medical devices, some foods, school textbooks and clothing, these counterfeit products are essential goods. Thus, any government crackdown is essentially a tax on China's needy consumers.

Counterfeiters and pirates also serve the country by usurping the foreign technology that China needs to meet its ambitious industrial goals. In 2005, China will most likely be the world's third-largest trading nation, and counterfeiters give the country's increasing number of globally competitive companies the means to compete against powerful foreign rivals that pay for their use of proprietary technologies. In a broader geopolitical context, China's counterfeiters deny the world's advanced economies, especially in the U.S. and Japan, the opportunity to sell to China the valuable designs, trademarked goods, advanced technology and popular entertainment that the Chinese urgently desire but cannot yet produce on their own.

Put another way, China's failure to police industry and to protect intellectual-property acts, in effect, like one of the greatest industrial subsidies in the world. Chinese manufacturers and industries freely exploit foreign ideas and technologies. "China helps distribute technology that has already been paid for by the developed world, often by companies, but also by taxpayers who support the government labs where much of the most important industrial technology begins," says Oded Shenkar, a professor of business at Ohio State University and the author of the recent book "The Chinese Century." "And, seen as a subsidy, this one is a particularly good deal for the Chinese government because it doesn't have to pay for it."

For the most part, China fears no repercussions from its actions because the size and potential of its markets give China an undiminished (for now) power to lure the world's most advanced technology to its shores. For example, China for years has tendered the prospect of large, advanced transportation projects to foreign governments as a way to coax largess and technology from outsiders. When the Chinese government announced that is was considering high-speed magnetic levitation ("maglev") cross-country train routes, Germany and Japan each put together government and business alliances to win future contracts there. The German industrial giants ThyssenKrupp and Siemens formed a partnership to build a nearly 20-mile maglev line in Shanghai to prove that they were up to the job. The line began operating last year while China was said to be considering which of several technologies to use. In December, workers for the German operation videotaped Chinese engineers poking around the maglev train's maintenance building in Shanghai at 3 a.m. one Saturday, apparently in search of confidential information. The manager of the Chinese operation that was a partner of the Germans clumsily excused the prowlers by saying they were merely taking part in a "research and development" exercise. Later that month, the government said that to save money it would eschew foreign designs for Chinese trains and, instead, employ newly developed indigenous maglev technology. Soon China could be exporting maglev trains for half the price Germany or Japan demand.

THE GENEROUS AND optimistic view of China's behavior is that it is a passing phase, and one not all that unusual for countries on the make. European powers once struggled to steal (and even transplant) one another's prime proprietary assets, like Mesoamerican gold, Brazilian rubber and Indonesian cloves. Blue-and-white Delftware was a Dutch attempt to copy China's porcelain works. At the dawn of the Industrial Revolution, American companies paid industrial spies to steal the designs of British machines. American theatrical producers routinely staged foreign operas and plays without permission; publishers sold dubious editions of English novels. More recently, Taiwan circumvented foreign patents and copyrights early in its post-World War II industrialization drive. And countries in Southeast Asia, Latin America, Africa and the former Soviet Union still operate well outside the developed world's norm for intellectual-property protection. Yet no other violators, past or present, match China's potential to change the rules of the world economy through piracy and counterfeiting.

Countries like Brazil or Vietnam may be as lax about copying as China is, but they do not have the industrial infrastructure or the ranks of skilled scientists and engineers to pull off the more ambitious copies of, say, drugs and automotive vehicles. China, however, has the expertise and infrastructure to reverse-engineer and produce nearly anything. And it has a market large enough to support the enterprise. The Chinese motorcycle industry provides a good example. Honda entered China in the 1980's and soon captured one-fifth of the motorcycle marketplace. But cheaper Chinese imitations appeared, and Honda's market share quickly halved. The company found that staying in China required that it enter into partnerships with some of the very companies copying its bikes. Now, with as many as 100 motorcycle makers in China, the country is the largest such manufacturer in the world, producing 15 million motorcycles a year (half of all new vehicles sold worldwide). Still the copying persists. The Japanese government estimates that of the 11 million motorcycles made in China in 2002, 9 million were imitations of Japanese products.

Oded Shenkar, who has long studied the Chinese automobile industry, argues that China's current regime is an essential factor in the country's ability to produce goods cheaply and get them quickly to market. In the U.S., about $1,000 of the price of an average car goes to pay for that model's product develop-

ment; that's money the car maker invested over the course of years. Copiers pay none of that and can rush their products to market. "Almost everything you can think of that is made in China has a very low technology investment embedded in it," Shenkar says. "Drugs, DVD's, every trademark, software and whole production lines get copied. Some of the technology is transferred to China by multinational corporations and one way or another finds its way to other producers; others are simply 'borrowed.' The practices feed one another. Why pay for software to run a production line that is itself an unauthorized copy of someone else's technology and processes?"

This sort of technological expropriation allows China to create industries nearly from scratch. Though it costs tens or hundreds of millions of dollars to develop new-model cars and motorcycles, China is home to hundreds of companies that produce the vehicles, many of them small companies with limited sales. "You can't start an automobile company that sells a few thousand cars a year and still pay the $500 million or more it costs to develop a new model," Shenkar says. "Where else in the world could a company that makes 30,000 units compete with one that makes a million units?" The hopeful analogy that compares China with earlier, now-reformed "borrowers" simply ignores the scale of the long-term advantages that both encourage and result from China's copying.

Unless it comes up with a remedy that forces China to change, the United States will have to find its own solutions. Ken DeWoskin, a professor emeritus of Chinese studies at the University of Michigan and a consultant who advises PricewaterhouseCoopers on China, argues that China, as in the Viagra case, will increasingly take on the veneer of an American-style intellectual-property regime while finding ways at every step to assert its interests within that system. "American pharmaceutical companies will be very seriously attacked by China's approach to I.P.," DeWoskin says. "You can already see how China is changing the rules of the game." Americans, he notes, pay higher drug prices than consumers in other economies can sustain, all for products made here at home. A result is that we underwrite both our companies and the rest of the world's consumption. How much American consumers will tolerate other kinds of similarly expensive economic nationalism is hard to predict, but DeWoskin says he can envision the U.S. economy slowly but surely adopting such measures, much as Japan has to protect its domestic markets and companies. Japan's economy is structured to support national industries over foreign rivals. Japanese consumers, for instance, typically pay more at home for goods manufactured in their own country than consumers outside pay for Japan's products. Without realizing it, Americans have already tilted toward the Japanese arrangement in pharmaceuticals.

One approach that vulnerable companies trumpet is speeding up the pace of innovation and rushing their products to market before Chinese competitors can catch up. But this solution overlooks the extent to which the Chinese themselves are increasingly skilled at hurrying copycat goods to market. Another approach is to sell legitimate goods at lower prices. Already, China's loose intellectual-property protection has done what years of legal and political pressures on the software and pharmaceutical industries in the U.S. have failed to do: forced powerful American companies to rethink, and often reduce, their prices. Chinese and Indian drugs that fall outside Western-style patent protection are drastically cheaper in poorer countries. Microsoft recently introduced less expensive versions of its software in developing countries where patent and copyright protections are weak (though the company has yet to do so in China).

Another approach to the Chinese intellectual-property regime is to leverage its vitality. The Japanese may be showing the way here too. In September, Toyota surprised the world's automobile makers by announcing that it would join with China's government-owned First Auto Works Corporation to start building its Prius hybrid cars in Jilin, a northeast Chinese province. The innovative Prius is one of the world's most sought-after cars—why would Toyota bring its hottest technology to China where it is almost certain to be carefully studied and boldly copied? The company says that it just wants to make more cars to meet demand. But an American management consultant who asked not to be identified told me that Toyota could have a deeper strategy that actually counts on Chinese manufacturers to usurp and adapt some of the car's technology. The car's central and perhaps most expensive component is its battery. China has already taken a sizable piece of the small-battery business away from leading Japanese manufacturers in recent years, thereby pushing battery prices down by 40 percent or more. The country is also a leading producer of electric motors.

China is just the place, in other words, to drive down the price of the Prius's battery and motor, and if that happens it will give Toyota an even bigger jump on the rest of the world's car makers struggling to design and produce their own hybrids. Toyota's move into China could even transform the automotive industry by luring car buyers into hybrids faster. In effect, Toyota may be hoping to ride China's copycat tendencies past American competitors and into the top spot among world car makers—provided, of course, that Chinese manufacturers do not do to Toyota what they did to Japan's motorcycle makers.

It's a dangerous bargain, becoming partner to a system that's a relentless competitor at the same time. The Chinese government recently announced that it would suspend the purchase of large aircraft in 2005, claiming it wants to cool off an overheating domestic aviation industry. It's just as likely that China wants to give its aircraft industry a chance to catch up with foreign manufacturers like Boeing. If so, the American industrial giant, which has pinned much of its future growth on sales in China and has aggressively transferred technology to China in order to secure its place there, may well lose billions in sales—and end up with a competitor that can match its current technology and beat it on cost. Last month, China announced the first international sale of 20 domestically produced midsize passenger planes.

Ted C. Fishman is the author of "China, Inc.: How the Rise of the Next Superpower Challenges America and the World," to be published next month by Scribner and from which this article is adapted.

THE ONCE AND FUTURE CHINA

What of China's past could be a harbinger for its future?

By Jonathan D. Spence

Despite its incredible pace of change, China continues to carry echoes of its past. And yet, the difficulty of drawing any direct links between its past and present is demonstrated by the fact that any topic can shift in perspective depending on where you enter China's vast chronology. What constitutes political stability, for example, has varied dramatically across almost four millennia, and in different periods it has been defined in relation to the greatness of leaders, the peacefulness of imperial successions, the suppression of peasant rebellions, and the handling of foreign incursions—whether religions, technologies, or troops.

Our appreciation of China's economic growth will veer erratically, depending on whether we concentrate on specie and banking, the formation of cities, the creation of trade hubs, or advances in transportation and communication. Our current fascination with high-tech dynamism could be tied to an equally wide range of variants, designed to give China an aura of either preeminence or stagnation. Rarely has China been so weak as when the emperor's ill-equipped army did battle with British forces during the Opium Wars in the mid-19th century. And yet, the sophistication of the Song dynasty's metallurgy or the imposing power of the Ming dynasty's fleets made China a potential global leader long before the competition among states was considered in these terms.

But today, relations among states are discussed very much like a competition or race, and few have run it as well as China in the modern era. Indeed, the prospect of China's rise has become a source of endless speculation and debate. To speak of China's "rise" is to suggest its reemergence. It can also imply a recovery from some kind of slump or period of quietude. But "rise" can also mean that a change is being made at someone else's expense. Must a fall always accompany a rise? If so, then a conflict will occur almost by definition. These are difficult questions made all the more so by the fact that a country as vast and complex as China makes up at least half of the equation.

One arena, however, in which China's past can serve as a useful prologue to the present, can be found in looking at how its territorial extent has evolved over time. This approach can show both how China has come to be the size it is, and perhaps—although this is a more contentious area—how China might change again in the future.

The China of today can be recognizably traced back to the late 16th century and the waning years of the Ming dynasty. One harbinger of what was to come was China's earlier Korean War—in 1592. It was then that the wildly ambitious Japanese military commander Hideyoshi sent a powerful fleet and ground forces to invade Korea, hoping to consume the country and force a passage into China, the greatest prize of all. Despite the ineptitude and factionalism of the Ming court, the Chinese responded powerfully, sending a strong expeditionary force to check the Japanese advance and shore up the Korean king. They ordered major fleets from south China to sail north with reinforcements and supplies, and to interdict the Japanese supply routes. After numerous costly engagements on land and sea, and vast numbers of both civilian and military casualties, the Sino-Korean forces prevailed, and in late 1598, the Japanese withdrew.

So did the Chinese, and that was one important marker for the future: China itself would not try to conquer Korea, but China would react against another power if it interfered in the Korean peninsula, even at great cost. Such interventions by China occurred a second time in the face of renewed Japanese aggression in 1894, and once again in the face of the presumed threat of the U.N. forces sent to check the North Korean invasion of South Korea in 1950. Few probably realize that China's current diplomatic role in the six-party talks regarding the North's nuclear programs has a historical lineage more than 400 years old.

By the same token, a number of China's most complex domestic grievances are rooted in conquests made by Chinese rulers during the 17th and 18th centuries. From 1644 onward, the vast region of Manchuria to China's northeast became part of the country's central concept of its power. In 1683, the Qing emperor ordered naval forces from Fujian province to oust renegade Chinese forces from several islands off the country's southeastern coast.

The emperor's forces dispatched the rebels in a crisp campaign and, in the process, added the fertile island of Taiwan to the growing orbit of the Qing empire. Likewise, unrest on China's frontier led the Qing dynasty to send military forces to Tibet around 1720, and subsequently to incorporate border areas of north and eastern Tibet into the Qing administrative structure, a process that was well underway by the 1750s. It was also in the mid-18th century that Qing expeditionary forces penetrated deep into the Altishahr regions of Central Asia, and to Kashgar, Urumqi, and Ili, leading to Chinese occupation of the vast, mainly Muslim regions of what is now called Xinjiang.

Having gained these territories in the corners of the kingdom, China has been loath to ever let them go. Even when the Qing dynasty fell in 1912, the Republican government, despite its fragility as an administrative entity, sought to hold on to the fullest extent of the empire. After their victory in 1949, the Communists did the same. Today, Muslim unrest and Tibetan nationalism are near-constant sources of tension for China's leadership. And Taiwan, lost first to the Japanese in 1895, and then to the Chinese Nationalists in 1949, is one of Asia's most dangerous potential flash points.

Although relations between China and the United States may be of vital importance to both, from the Chinese perspective, the relationship has been extremely brief. Indeed, there wasn't even a United States for China to have relations with until late into the reign of the Chinese Emperor Qianlong, arguably one of the greatest leaders of the last Chinese dynasty. When relations were established, Americans sometimes behaved admirably. Other times, they were a nuisance, or worse, a menace. Again, it depends who you are and where you settle your gaze. You can see the United States as benevolent in its development of Chinese hospitals and modern medicine. You can see it as destructive in its dissemination of partisan religious tracts by American evangelists to such people as the leader of the Taiping Rebellion. Or, you can see it as thoroughly ambiguous in the 1900s, when U.S. leaders urged the Chinese toward a more republican form of government, which quickly descended into warlordism. To be sure, the Chinese have these images, and many more, in mind when they think about their relations with the United States.

These are the memories and the territorial histories that China has to juggle as it embarks on its myriad new challenges and opportunities: as the defender of an apparently irrelevant revolutionary ideology, as a new kind of regional powerhouse, as the ambiguous heart of a global diaspora, as one of the world's major new competitors for shrinking supplies of fossil fuels, and as the present guardian of an unprecedented amount of foreign exchange and investment. Some of these phenomena can also be tracked through the historian's lens, but some are, I believe, genuinely new. Just why that should be is itself part of the story.

Jonathan D. Spence is Sterling professor of history at Yale University.

DEBATE
CLASH OF THE TITANS

Is China more interested in money than missiles? Will the United States seek to contain China as it once contained the Soviet Union? Zbigniew Brzezinski and John Mearsheimer go head-to-head on whether these two great powers are destined to fight it out.

Make Money, Not War

By Zbigniew Brzezinski

Today in East Asia, China is rising—peacefully so far. For understandable reasons, China harbors resentment and even humiliation about some chapters of its history. Nationalism is an important force, and there are serious grievances regarding external issues, notably Taiwan. But conflict is not inevitable or even likely. China's leadership is not inclined to challenge the United States militarily, and its focus remains on economic development and winning acceptance as a great power.

China is preoccupied, and almost fascinated, with the trajectory of its own ascent. When I met with the top leadership not long ago, what struck me was the frequency with which I was asked for predictions about the next 15 or 20 years. Not long ago, the Chinese Politburo invited two distinguished, Western-trained professors to a special meeting. Their task was to analyze nine major powers since the 15th century to see why they rose and fell. It's an interesting exercise for the top leadership of a massive and complex country.

This focus on the experience of past great powers could lead to the conclusion that the iron laws of political theory and history point to some inevitable collision or conflict. But there are other political realities. In the next five years, China will host several events that will restrain the conduct of its foreign policy. The 2008 Olympic Games is the most important, of course. The scale of the economic and psychological investment in the Beijing games is staggering. My expectation is that they will be magnificently organized. And make no mistake, China intends to win at the Olympics. A second date is 2010, when China will hold the World Expo in Shanghai. Successfully organizing these international gatherings is important to China and suggests that a cautious foreign policy will prevail.

More broadly, China is determined to sustain its economic growth. A confrontational foreign policy could disrupt that growth, harm hundreds of millions of Chinese, and threaten the Communist Party's hold on power. China's leader-

ship appears rational, calculating, and conscious not only of China's rise but also of its continued weakness.

There will be inevitable frictions as China's regional role increases and as a Chinese "sphere of influence" develops. U.S. power may recede gradually in the coming years, and the unavoidable decline in Japan's influence will heighten the sense of China's regional preeminence. But to have a real collision, China needs a military that is capable of going toe-to-toe with the United States. At the strategic level, China maintains a posture of minimum deterrence. Forty years after acquiring nuclear-weapons technology, China has just 24 ballistic missiles capable of hitting the United States. Even beyond the realm of strategic warfare, a country must have the capacity to attain its political objectives before it will engage in limited war. It is hard to envisage how China could promote its objectives when it is acutely vulnerable to a blockade and isolation enforced by the United States. In a conflict, Chinese maritime trade would stop entirely. The flow of oil would cease, and the Chinese economy would be paralyzed.

I have the sense that the Chinese are cautious about Taiwan, their fierce talk

notwithstanding. Last March, a Communist Party magazine noted that "we have basically contained the overt threat of Taiwanese independence since [President] Chen [Shuibian] took office, avoiding a worst-case scenario and maintaining the status of Taiwan as part of China." A public opinion poll taken in Beijing at the same time found that 58 percent thought military action was unnecessary. Only 15 percent supported military action to "liberate" Taiwan.

Of course, stability today does not ensure peace tomorrow. If China were to succumb to internal violence, for example, all bets are off. If sociopolitical tensions or social inequality becomes unmanageable, the leadership might be tempted to exploit nationalist passions. But the small possibility of this type of catastrophe does not weaken my belief that we can avoid the negative consequences that often accompany the rise of new powers. China is clearly assimilating into the international system. Its leadership appears to realize that attempting to dislodge the United States would be futile, and that the cautious spread of Chinese influence is the surest path to global preeminence.

Better to Be Godzilla than Bambi

By John J. Mearsheimer

China cannot rise peacefully, and if it continues its dramatic economic growth over the next few decades, the United States and China are likely to engage in an intense security competition with considerable potential for war. Most of China's neighbors, including India, Japan, Singapore, South Korea, Russia, and Vietnam, will likely join with the United States to contain China's power.

To predict the future in Asia, one needs a theory that explains how rising powers are likely to act and how other states will react to them. My theory of international politics says that the mightiest states attempt to establish hegemony in their own region while making sure that no rival great power dominates an-

other region. The ultimate goal of every great power is to maximize its share of world power and eventually dominate the system.

The international system has several defining characteristics. The main actors are states that operate in anarchy—which simply means that there is no higher authority above them. All great powers have some offensive military capability, which means that they can hurt each other. Finally, no state can know the future intentions of other states with certainty. The best way to survive in such a system is to be as powerful as possible, relative to potential rivals. The mightier a state is, the less likely it is that another state will attack it.

The great powers do not merely strive to be the strongest great power, although that is a welcome outcome. Their ultimate aim is to be the hegemon—the only great power in the system. But it is almost impossible for any state to achieve global hegemony in the modern world, because it is too hard to project and sustain power around the globe. Even the United States is a regional but not a global hegemon. The best outcome that a state can hope for is to dominate its own backyard.

States that gain regional hegemony have a further aim: to prevent other geographical areas from being dominated by other great powers. Regional hegemons, in other words, do not want peer competitors.

Instead, they want to keep other regions divided among several great powers so that these states will compete with each other. In 1991, shortly after the Cold War ended, the first Bush administration boldly stated that the United States was now the most powerful state in the world and planned to remain so. That same message appeared in the famous National Security Strategy issued by the second Bush administration in September 2002. This document's stance on preemptive war generated harsh criticism, but hardly a word of protest greeted the assertion that the United States should check rising powers and maintain its commanding position in the global balance of power.

China is likely to try to dominate Asia the way the United States dominates the Western Hemisphere. Specifically, China will strive to maximize the power gap between itself and its neighbors, especially Japan and Russia, and to ensure that no state in Asia can threaten it. It is unlikely that China will go on a rampage and conquer other Asian countries. Instead, China

> *A* powerful China is likely to try to push the United States out of Asia, much the way the United States pushed the European great powers out of the Western Hemisphere.

will want to dictate the boundaries of acceptable behavior to neighboring countries, much the way the United States does in the Americas. An increasingly powerful China is also likely to try to push the United States out of Asia, much the way the United States pushed the European great powers out of the Western Hemisphere. Not incidentally, gaining regional hegemony is probably the only way that China will get back Taiwan.

Why should we expect China to act differently than the United States? U.S. policymakers, after all, react harshly when other great powers send military forces into the Western Hemisphere. These foreign forces are invariably seen

as a potential threat to American security. Are the Chinese more principled, more ethical, less nationalistic, or less concerned about their survival than Westerners? They are none of these things, which is why China is likely to imitate the United States and attempt to become a regional hegemon. China's leadership and people remember what happened in the last century, when Japan was powerful and China was weak. In the anarchic world of international politics, it is better to be Godzilla than Bambi.

It is clear from the historical record how American policymakers will react if China attempts to dominate Asia. The United States does not tolerate peer competitors. As it demonstrated in the 20th century, it is determined to remain the world's only regional hegemon. Therefore, the United States will seek to contain China and ultimately weaken it to the point where it is no longer capable of dominating Asia. In essence, the United States is likely to behave toward China much the way it behaved toward the Soviet Union during the Cold War.

Nukes Change Everything

Zbigniew Brzezinski responds.

As an occasional scholar, I am impressed by the power of theory. But theory—at least in international relations—is essentially retrospective. When something happens that does not fit the theory, it gets revised. And I suspect that will happen in the U.S.-China relationship.

We live in a very different world than the one in which hegemonic powers could go to war without erasing each other as societies. The nuclear age has altered power politics in a way that was al-

ready evident in the U.S.-Soviet competition. The avoidance of direct conflict in that standoff owed much to weaponry that makes the total elimination of societies part of the escalating dynamic of war. It tells you something that the Chinese are not trying to acquire the military capabilities to take on the United States.

How great powers behave is not predetermined. If the Germans and the Japanese had not conducted themselves the

way they did, their regimes might not have been destroyed. Germany was not required to adopt the policy it did in 1914 (indeed, German Chancellor Otto von Bismarck followed a very different path). The Japanese in 1941 could have directed their expansionism toward Russia rather than Britain and the United States. For its part, the Chinese leadership appears much more flexible and sophisticated than many previous aspirants to great power status.

Showing the United States the Door

John J. Mearsheimer responds.

The dichotomy that you raised between theory and political reality is an important one. The reason that we have to

privilege theory over political reality is that we cannot know what political reality is going to look like in the year

2025. You mentioned that you traveled to China recently and talked to Chinese leaders who appear to be much more prudent about Taiwan than the conventional wisdom has it. That may be true, but it's largely irrelevant. The key issue is, What are the Chinese leaders and people going to think about Taiwan in 2025? We have no way of knowing. So today's political realities get washed out of the equation, and what really matters is the theory that one employs to predict the future.

You also argue that China's desire for continued economic growth makes conflict with the United States unlikely. One of the principal reasons that China has been so successful economically over the past 20 years is that it has not picked a fight with the United States. But that logic should have applied to Germany before World War I and to Germany and Japan before World War II. By 1939, the German economy was growing strongly, yet Hitler started World War II. Japan started conflict in Asia despite its impressive economic growth. Clearly there are factors that sometimes override economic considerations and cause great powers to start wars—even when it hurts them economically.

It is also true that China does not have the military wherewithal to take on the United States. That's absolutely correct—for now. But again, what we are talking about is the situation in 2025 or 2030, when China has the military muscle to take on the United States. What happens then, when China has a much larger gross national product and a much more formidable military than it has today? The history of great powers offers a straightforward answer: China will try to push the Americans out of Asia and dominate the region. And if it succeeds, it will be in an ideal situation to deal with Taiwan.

America's Staying Power

Zbigniew Brzezinski responds.

How can China push the United States out of East Asia? Or, more pointedly, how can China push the United States out of Japan? And if the United States were somehow pushed out of Japan or decided to leave on its own, what would the Japanese do? Japan has an impressive military program and, in a matter of months, it could have a significant nuclear deterrent. Frankly, I doubt that China could push the United States out of Asia. But even if it could, I don't think it would want to live with the consequences: a powerful, nationalistic, and nuclear-armed Japan.

Of course, tensions over Taiwan are the most worrisome strategic danger. But any Chinese military planner has to take into account the likelihood that even if China could overrun Taiwan, the United States would enter the conflict. That prospect vitiates any political calculus justifying a military operation until and unless the United States is out of the picture. And the United States will not be out of the picture for a long, long time.

It's Not a Pretty Picture

John J. Mearsheimer responds.

If the Chinese are smart, they will not pick a fight over Taiwan now. This is not the time. What they should do is concentrate on building their economy to the point where it is bigger than the U.S. economy. Then they can translate that economic strength into military might and create a situation where they are in a position to dictate terms to states in the region and to give the United States all sorts of trouble.

From China's point of view, it would be ideal to dominate Asia, and for Brazil, Argentina, or Mexico to became a great power and force the United States to concentrate on its own region. The great advantage the United States has at the moment is that no state in the Western Hemisphere can threaten its survival or security interests. So the United States is free to roam the world causing trouble in other people's backyards. Other states, including China of course, have a vested interest in causing trouble in the United States' backyard to keep it focused there. The picture I have painted is not a pretty one. I wish I could tell a more optimistic story about the future, but international politics is a nasty and dangerous business. No amount of good will can ameliorate the intense security competition that will set in as an aspiring hegemon appears in Asia.

Zbigniew Brzezinski is a counselor at the Center for Strategic and International Studies. John J. Mearsheimer is the R. Wendell Harrison distinguished service professor of political science at the University of Chicago.

WHY IS CHINA GROWING SO SLOWLY?

For all its success, China is still not living up to its potential

By Martin Wolf

When people discuss China's economic growth, it's usually with awe. But these assessments of the country's economic performance almost always end with one question: How long can China's rapid growth continue? It's an obvious, even banal question. Yet it is not the right one. Because asking it suggests that there has been something extraordinary about the growth of the Asian colossus over the past 25 years. What is remarkable is not how quickly China's economy has grown, but rather how slowly it has done so.

It may seem an odd thing to say of the world's fastest growing economy. But, given where the country stood when economic reforms were introduced in 1978, China should have grown even faster. This, in turn, suggests that, with the right mix of policies, China's economy should not maintain its current rate of growth: It should accelerate.

China's growing gross domestic product (GDP) is Exhibit A for those who laud the country's economic success. Between 1978 and 2003, China's per capita GDP grew at a compound rate of 6.1 percent a year, which amounted to a 337 percent increase a quarter of a century later. It's an impressive performance, but it's not record-breaking. Japan's per capita GDP, for example, topped 490 percent between 1950 and 1973. South Korea outpaced China with 7.6 percent compound growth a year between 1962 and 1990, and Taiwan achieved annual growth of 6.3 percent between 1958 and 1990.

It may not seem fair to compare China to these smaller economies. That's true. China should have outperformed them all. The speed with which a country can grow is a function of how far it is lagging behind the productivity levels of the world's most advanced economies. That is why each generation of catch-up economies has tended to grow faster than the previous one. When China's surge began, its per capita GDP was only a twentieth of that of the United States. Even now, after 25 years of growth, China's per capita output is only 15 percent of that of the United States. Japan's was a fifth of that of the United States in 1950, even before its record-breaking growth surge began.

To be fair, many developing countries haven't caught up to the leaders as quickly as they should have. But most of them lack the fundamental ingredients for success. China can't claim that excuse. It, like Japan, South Korea, and Taiwan before it, possesses a hardworking, cheap labor force; the ability to transfer huge numbers of workers from low-productivity agriculture to higher-productivity manufacturing; political stability; and an effective, development-oriented government.

And China possesses something else few ever do: an extraordinarily high rate of investment. At more than 40 percent of its GDP, the country's fixed investment is probably the highest ever achieved in a large economy. Nor has any country ever been awash in so much capital at this stage of its development. For example, China's per capita GDP (at purchasing power parity, or internationally comparable prices) is roughly the same today as South Korea's was in 1982, Taiwan's in 1976, and Japan's in 1961. But, in those years, Japan's investment rate was just above 30 percent of GDP, and South Korea's and Taiwan's were both below 30 percent. None of those countries invested as much capital at comparable stages in their development as China does today.

So why hasn't Beijing done a better job? Because, China's economy is still highly inefficient. The voracious maw of China's state-owned enterprises accounts for much of this drag. Between 1993 and 2000, more than 60 percent of all loans went to these state-owned behemoths. The country's notoriously high level of bad loans tells you how good an investment they have been: The Standard & Poor's rating agency currently estimates that China's banks have issued about $650 billion in bad loans, or about 40 percent of outstanding loans. If an economy growing at close to 10 percent a year generates bad loans on this scale, the misallocation of capital has to be gigantic. Although countries such as South Korea or Taiwan may not have had as much capital, they obtained considerably more growth for their investment buck. The same was true of Japan in its high-growth phase. The same is true of India today.

Those who object to the idea that China could have grown faster will argue that a country of China's scale required far greater investment than its neighbors to build its infrastructure. And they will suggest that inefficiency should be expected in a country still trying to throw off the trappings of its socialist past. These points, though valid, do not reverse the verdict: Given China's ample opportunities and investment, it should have raised its living standards even faster than it did.

The bottom line is clear. Do not think China's rapid growth is either extraordinary or a flash in the pan. It is neither. The social and political obstacles to China's rapid growth are considerable. But the opportunity remains enormous. China's economic boom could well be in its middle, not its end.

Martin Wolf is associate editor and chief economics commentator at the Financial Times *in London.*

A Grand Chessboard

Beijing seeks to reassure the world that it is a gentle giant.

By Ashley J. Tellis

*I*n 2003, Chinese President Hu Jintao's advisors hatched a new theory. Dubbed China's "peaceful rise," it held that, in contrast to the warlike behavior of ascending great powers in the past, the economic ties between China and its trading partners not only made war unthinkable but would actually allow all sides to rise together. The theory's name didn't survive power struggles within the Communist Party, but the general idea lives on in new and updated formulations such as "peaceful development" and "peaceful coexistence." Regardless of the label Chinese apparatchiks ultimately agree on, one thing is clear: China spends a great deal of time worrying about what other countries think about it.

And for good reason: While China's economic growth over the last 20 years has generated tremendous wealth at home, it has also stirred apprehension abroad. Beijing knows that the United States and countries throughout Asia are casting a wary eye in its direction, worried that China could ultimately become a regional hegemon that threatens their security. It has become obvious to Beijing that a new Chinese grand strategy is required—one that would allow it to continue its economic growth, technological modernization, and military buildup without provoking other countries into a costly rivalry. The China we see striding on the world stage today is cut from the cloth of that new grand strategy.

Beijing began by making nice in its own neighborhood. It has sought to develop friendly relations with the major states on its periphery—Russia, Japan, India, and the Central and Southeast Asian states— that are potential balancing partners in any future U.S.-led, anti-Chinese coalition. This good neighbor approach is dramatically different from its behavior of the 1990s. Instead of invoking Chinese claims in territorial and maritime disputes as it did

during that decade, Beijing today has made a special effort to assure other states that it has the best intentions. China agreed to codes of conduct where territorial disputes have economic consequences, such as the South China Sea. It began to resolve border disputes with important neighbors, such as India. It started to take its nonproliferation obligations much more seriously than before, including efforts to tighten export controls of potentially dangerous dual-use technologies. And it expressed a willingness to shelve political disputes that cannot be reconciled immediately, so long as none of the other parties (such as Taiwan) disrupts the status quo. In 1994, during Washington's nuclear standoff with Pyongyang, Beijing's role was minor. Today, it is the driving force behind the complex six-party talks on North Korea's nuclear arsenal.

> *B*eijing knows that the United States is casting a wary eye in its direction.

No relationship factors more into this diplomatic about-face than China's relationship with the United States. Beijing has gone out of its way to mollify Washington, trying to demonstrate that it has neither the intention nor the capability of challenging U.S. leadership in Asia—even as it seeks to promote a regional environment where a U.S. political-military presence will eventually become unnecessary. Toward this end, Beijing has used the war on terror to position itself as a U.S. partner. Yet, it has also sought to preempt a potential U.S. led coalition by deepening economic ties with American allies such as Japan, South Korea, Taiwan, and Australia. These countries would pay a steep economic price if they were to support any U.S. led, anti-Chinese policies in the future. And China has adroitly exploited every manifestation of regional dissatisfaction with America's obses-

sive and overbearing war on terror, seeking to cast itself as a friendly, non-interfering alternative to U.S. might in the region. It is even proposing new institutional arrangements wherein China can exercise a leadership role that excludes the United States, such as the East Asian Economic Zone.

China has sought to make its presence felt outside of Asia, too. Much of China's diplomatic globetrotting has been driven by the need to secure stable energy sources to fuel its gigantic economic machine. China is now routinely sending trade missions not only to Central Asia and the Persian Gulf but also to Africa and Latin America. And, as if giving notice of its full arrival as a great power on the world stage, China has become a much more robust player in the United Nations, the World Trade Organization, and other international bodies. More interesting, China has become acutely conscious of the need to promote its culture abroad, partly because it recognizes the benefits of "soft power," but also because it believes that a genuine appreciation of Confucian rectitude will go a long way in mitigating suspicions about how Beijing might exercise its future power.

This strategy of emphasizing peaceful ascendancy in word and deed will likely satisfy Chinese interests until it becomes a true rival of the United States. At that point, China will face another strategic crossroads. Whether a turn toward strident assertiveness or deepened accommodation represents the future of China's geopolitical orientation, only time will tell. But Washington should recognize that if it mishandles its relations with its current or prospective partners, it might be faced with an absence of allies precisely when it needs them most. China's current grand strategy is focused on making that scenario a reality.

Ashley J. Tellis is senior associate at the Carnegie Endowment for International Peace.

LIFTING ALL BOATS

Why China's great leap is good for the world's poor

By Homi Kharas

China's economic juggernaut has forced the world to make room. For rich nations, it's just a matter of adjusting their economic strategies. But how is China's rise affecting poorer countries? Governments in Asia, Latin America, and Central and Eastern Europe watch the Chinese export machine and worry about keeping their manufacturing jobs at home. The anxiety is understandable, but a closer look suggests that China's success will help, not hurt, most developing countries. The power of its economy—and the power of its example—will advance the fight against poverty.

Today, China has a lock on a large portion of the export market in North America, Europe, and elsewhere—markets that poor countries covet. This situation would spell trouble for many, but for the fact that China has also become one of the developing world's best customers. Forty-five percent of China's $400 billion in annual imports comes from developing countries, and these imports rose by $55 billion in 2003. Indeed, China runs a trade deficit with the developing world. Chinese demand for basic commodities (produced primarily in poorer countries) is so strong that it has pushed up prices for food staples and industrial raw materials such as aluminum, steel, copper, cotton, and rubber. For the millions of farmers around the world who depend on revenue from these products, the global price boom has come at just the right time, reversing decades of slumping prices.

China has also become the center of a virtuous regional trade cycle that benefits Asia's developing countries. True, China sucks up vast quantities of raw materials, but four fifths of its imports are now manufactured goods, including office machines, telecom equipment, and electrical machinery. Neighboring countries are feeding the trade boom by exporting components and machine parts to China for final assembly. Korea and Taiwan have benefited the most, but the Philippines, Thailand, and Indonesia saw their annual exports to China shoot up by roughly 30 percent last year. Other re-

gional production networks are developing, notably in automobiles and garments, so the gains from this trade will probably endure even if one sector lags.

China's economic impact is powerful. But so too is its example. The country has become a showcase of what open markets can achieve. It is reinvigorating the debate on how trade can reduce global poverty. China already has a large agricultural sector relatively undistorted by the types of subsidies and tariffs found in the United States or Europe. Its free trade credentials will only grow as it complies with increasingly stringent World Trade Organization (WTO) commitments. Global quotas on textiles and clothing, for example, disappeared on Jan. 1, 2005. If economic liberalization allowed China to post 9 percent growth over three decades and lift 300 million people from poverty during that time, then surely other countries can make significant gains by knocking down barriers.

The standard excuses for poor development performance—an uneven global playing field and exploitation by foreign investors—lose credibility when set against China's record. And there are signs that the lesson is taking hold. China's example was likely an important catalyst in India's reforms and growth surge during the last decade. Latin American countries are starting to take notice: Chile and China are contemplating a free trade agreement, and Mexico and Brazil are sending high-level trade and investment missions to China. As China engages in free trade agreements with neighboring countries, Southeast Asia is likely to benefit even more.

During its communist heyday, Beijing often championed the cause of the developing world, at least in its rhetoric. Now, as a large, successful trader, China is in a far better position to put meaning behind its message, shaping the rules in the WTO and other international bodies to address development concerns. China is already active in the Group of 20, a forum in which rich countries and the largest developing coun-

tries exchange views. In many cases, China's interests coincide with those of other developing countries, many of which look to China for support. For example, China wants to promote freer global trade in agriculture, a key concern of poorer countries. China might also add its voice to the chorus of developing countries that seek safeguards for their service sectors.

> China has become the center of a virtuous regional trade cycle.

Of course, China's rise does come at a cost for some. Those poor countries that rely on commodity imports take a hit as China's demand pushes up prices. China is such an efficient producer of garments that it will likely dominate textile markets, now that the global system of quotas has disappeared. That scenario will hurt garment workers in such countries as Bangladesh and Cambodia, whose jobs and wages depend on protected markets. Maquiladora industries in Central America that export to the United States under preferential agreements are already exiting the market, fearing the coming competition with China. Similarly, the advantages conveyed to some of the world's poorest countries through free-market access agreements with the United States and Europe will decline, as global trade barriers come down and efficient producers such as China begin to compete. Still, the benefits of China's economic rise, and of a more liberal and fairer global trading system, outweigh the costs.

Decades ago, Japan, Germany, and South Korea showed the world how to develop with a strategy based on exporting manufactured goods. China's rise may offer an equally compelling example of how open economies can spur rapid growth. For the developing world, it's something to emulate, not fear.

Homi Kharas is chief economist for East Asia and the Pacific region at the World Bank.

DANGEROUS DENIALS

China's economy is blinding the world to its political risks.

By Minxin Pei

There is a Chinese proverb that says "one spot of beauty can conceal a hundred spots of ugliness." Today, in China, there are few things as beautiful as the country's economic growth. But it is premature to dismiss the inherent instability in China's authoritarian politics. The country's rapid economic growth may be blinding us to systemic risks in Chinese domestic politics that, if poorly managed, could explode, threatening the survival of the regime. There is no question that China's economy is on the rise—but so are the risks of political crisis.

To be fair, some of the dangers China is facing simply come with the challenge of being a developing country racing toward a market economy. Shaking off socialism isn't easy for any nation. When you are the world's most populous country, the chances for socioeconomic disaster are enormous. Income inequality, for example, is to be expected. The period from 1980 to 1997 saw a 50-percent rise in inequality in China. Labor migration is natural. But China is experiencing the largest movement of rural labor in history. In recent years, Chinese cities have absorbed at least 114 million rural workers, and they are expected to see an influx of another 250 to 300 million in the next few decades. Under the circumstances, it's hardly surprising that China's effort to establish a new social safety net has fallen short, especially given its socialist roots. It would be a Herculean task for any government.

But China's isn't just any government. It is one that rests on fragile political foundations, little rule of law, and corrupt governance. Worse, it has consistently placed the highest value on economic growth and viewed all demands for curbing its discretion and power as threats to its goal of rapid modernization. The result? Social deficits in education, public health, and environmental protection. But it is hardly surprising, since promoting high growth advances the careers of government officials. Thus, China's elites devote most of their resources to building glitzy shopping malls, factories, and even Formula One racing tracks, while neglecting social investments with long-term returns. So for those who wonder how, if China's political system is so rotten, it can deliver robust growth year after year, the answer is that it delivers robust growth year after year, in part, because it is so rotten.

But the Chinese Communist Party knows that the people will tolerate only so much rot. Corruption is a rising concern. The party's inability to police its own officials, many of whom are now engaged in unrestrained looting of public assets, is one of Beijing's greatest worries. These regime insiders have effectively privatized the power of the state and use it to advance personal interests. Their loyalty to the party is questionable, if it exists at all. The accelerating effect on the party's demise resembles that of a bank run; more and more insiders cannot wait to cash in their investment in the party.

Of course, the Chinese government, like other authoritarian regimes, is constantly threatened by internal power struggles. Again, Beijing has not only bucked the naysayers, but its ability to weather internecine strife appears to have improved markedly since the 1989 Tiananmen tragedy. The recent transfer of power from Presidents Jiang Zemin to Hu Jintao turned out more smoothly than expected, perhaps signifying that the party has acquired a higher degree of institutional maturity. But it may still be too little, too late for an increasingly pluralistic and assertive population. Although the government managed to build an elitist ruling alliance of party officials, bureaucrats, intellectuals, and businessmen, the durability of this alliance is questionable. And, as in other countries, exclusionary politics inevitably breeds alienation, resentment, and anger. This does not mean that a social revolution is imminent in China. But should a crisis hit, all bets are off.

Thankfully, all of these risks are manageable if China confronts them with bold political reforms rather than denial and delay. But this may be wishful thinking. Beijing has thus far preferred these risks to the gamble of democratic reforms. The only thing certain about China's political risks is that they are on the rise. And that reality is hardly a thing of beauty.

Minxin Pei is director of the China Program at the Carnegie Endowment for International Peace.

The Emperor Is Far Away

Understanding the Challenges Faced By the New Leadership

A Conversation with Ezra Vogel

HARVARD INTERNATIONAL REVIEW:

China enjoyed remarkable growth rates in the last 20 years. What accounts for this accelerated rate of development, and how has China's economic liberalization affected its attitude toward the West?

China's opening to the West began in 1970, when the Chinese leadership decided that the biggest risk was no longer the United States but the Soviet Union. Practically no Westerners had had direct contact with China, and initial progress was slow. China remained a relatively closed country until 1978, when its leadership decided to adopt an official policy of foreign opening. By the mid-1980s, China began to grow quite rapidly and continued at an ever greater rate in the early 1990s. The average per capita annual income is now about US$1,000, so China remains relatively undeveloped, though many countries are much poorer. This process of opening and growth has allowed the Chinese government to prepare its people to take their place in world affairs.

Historically, China has been the dominant power in the region, but it was never a global power. China began to take part in world affairs for the first time in the 19th century. For the subsequent 150 years, it was much weaker than other world powers and suffered oppression from the outside world. But once China opened in the early 1980s, it was able to take part in the world system. One of my favorite Chinese expressions is "linking tracks." In the 1930s, some of the Chinese warlords had no railroads because they had a narrower gauge than the national railway, leaving a wider distance between the rails. The warlords had to design a way to make the tracks compatible in order to form a national railway system. Now, China uses "linking tracks" to describe the process of adjusting various traditional practices so that they can interface with the global system.

This is an enormous change, and on the whole, China has done remarkably well. China today has become one of the world's leading trading states with a substantial trade surplus. This development marks extraordinary progress over a short period of time from the backward, isolated country that China once was. Furthermore, this is a much more exciting period within China than the outside world realizes. The 1989 Tiananmen protest and response received so much attention that even today China is often thought of as one big jail. But some of the most exciting cultural and intellectual growth in the world is taking place in China right now. There are a lot of smart people thinking about how to combine diverse Western influences with Chinese history. There are extraordinarily creative programs being conducted in every field on the grand scale—it is truly a renaissance. China can be criticized for being too repressive, but the change for the average person in China has been overwhelming.

Do you ascribe the opening of China to structural change, individual action, or a combination of factors?

I think there were three reasons for the opening. The first is that in its basic foreign policy strategy, China first identifies its greatest foreign adversary and then seeks allies against it. After the late 1940s, China saw the United States as the major enemy. By 1970, when it decided that the Soviet Union was the greatest enemy, China sought cooperation with the West. That decision had a serious impact on the re-organization that began China's entrance to the world system. Second, the Cultural Revolution ruined the country and was extremely painful to many people. There was real chaos because so many of the leaders had been in jail for an extended period of time. Then, after Mao Tsetung died in 1976, the new leadership began to think, "What should we do?" The timing was right, and the West was receptive to an opening from China. The third factor is individual leadership, and China has been very lucky in this regard. There were a number of circumstances that made it possible for Deng Xiaoping to become the paramount leader after Mao. He guided the country through an extremely difficult time and achieved rapid growth and openness in the midst of extreme poverty, corruption, and chaos. To be able to manage that process took extraordinary skill and good sense.

Do you think that subsequent leaders like Jiang Zemin have followed in Deng's success?

Mao united the country and became such an icon that it was almost impossible to say anything critical about him. Deng had a kind of revolutionary background that made him an especially strong figure. The situation with Jiang is quite different. The pattern of liberalization under his rule reminds me of a company that is tightly controlled by its owner and that later gets taken over by a corporation with an administrative system that makes it impossible for a single individual to control the business.

The key variable is not the type of person in charge. Instead, it is the whole situation surrounding the decision to change. Jiang Zemin was brought in as a compromise after Tiananmen. He was acceptable to those who wanted a crackdown as well as to more radical interests. He did not have much experience in central party politics. His rise in government is analogous in the United States to a governor who is elected president and installed in the White House with no prior experience in Washington. Although Jiang was at one time a minister, that post was very different from being in a position of genuine power. Heading a ministry is a very technical job compared with running the Chinese Communist Party, and I think Jiang proved to have greater strength and sounder judgment about international affairs and many local interests than people would have expected. Jiang did not know the details of running the Party headquarters, but I think he managed the process quite well. China flourished under him. He provided a very

different kind of leadership—much less dramatic than Deng or Mao, who, it can be argued, had problems in their later years. Jiang was a good leader for his time.

A great deal of Hu Jintao's experience is from his time in the western provinces, particularly in Tibet. What do you think of China's future regarding ethnic minorities like the Tibetans or the Uighurs of Xianjiang?

Hu had a very tough problem when he was in Tibet. People in the United States think "the more freedom the better." In the 1980s, some top Beijing leaders felt the same way and took measures that increased popular expectations and hopes. This led to chaos and demonstrations. Hu took part in leading a crackdown in response, but Hu did it without alienating the local community. In his attention to root problems, he demonstrated concern for the issues of the local population. I think minorities feel that Hu is a person they can deal with, someone who understands them.

Minorities compose roughly eight percent of China's population. When Jiang visited Harvard University in 1997, some responsible for his security were less concerned with the thousands of demonstrators for the Tibetan movement and more worried about the small number of Uighurs, who have demonstrated a willingness to resort to violence. In contrast, Tibetan Buddhists have generally avoided violence since their uprising in 1959. The United States itself no longer considers the Uighurs "freedom fighters;" after September 11, they became Islamic terrorists. The current US attitude is now closer to the Chinese view that some Uighurs present a problem to social order. Xinjiang is officially led by a Uighur government and Uighur officials, and they use their traditional language. The Uighurs have not assimilated as much as the Hui, another Muslim minority in Xingiang.

The minority populations of China include very different kinds of people, and the diversity among the

groups extends to their grievances. On one hand, there are the Koreans, who have a higher level of education and income than the average Chinese and, as a result, form a kind of privileged minority. Then there are a number of groups, particularly in southwest China and mountainous regions, who have been pushed back by the advancing Chinese in a way similar to the experience of Native Americans in the United States.

The Chinese government will make an effort to bring aid to backward areas, but there will still be far more investment in the area near the eastern coast because capital flows are now governed by a free market, and investors will make more money in the east. The amount of money the government will put in the west will not be nearly enough to counterbalance that lure. Nonetheless, the Chinese government will do quite a bit to build infrastructure and transport links to the western part of the country. They will also try to build schools and universities in the region. You can argue about whether this is enough or not, but at least there will be some national program of assistance.

But even this approach can be complicated when minority issues are at stake. In the 1990s, for instance, as money began to flow into Tibet and the government began its construction projects, many merchants, such as the Hui, moved from nearby areas in search of profit. Many outsiders came to Lhasa, the Tibetan capital, because the government gave financial aid. These people are working to respond to market opportunities, not because of official assignments.

When discussing economic modernization, what do you think of the internal changes in the way the domestic Chinese economy is run? How does this impact China's foreign trade?

China was a planned economy when it began to open in 1978. The big factories were all entirely state owned. The government dreaded having private enterprises, which were only allowed on a small scale. After 1978, private enterprises were gradually allowed to exist and grow. In the last decade, the government has allowed foreign companies and their subsidiaries to develop infrastructure in the country. During its liberalization period, the former Soviet Union was under the influence of the World Bank, so it tried to privatize right away. In contrast, China felt—I think rightly—that this would not be a good policy. In China, there were no private investors and no real experience with the free market. The unemployment situation was so severe in China that privatization could have led to massive social unrest. What China wanted was to build up enterprises first and then gradually put more pressure on the state enterprises to become competitive, a process that continues today.

But in the last decade, foreign investment has become the cutting edge of the export market. There were few foreign companies in China in the early years of reform, and generally they were required to have Chinese partners. As more foreign companies were allowed in, the vast portion of Chinese exports are now produced by more competitive foreign companies. Other countries at comparable stages of development, like Japan, were far more restrictive in allowing foreign enterprises to enter. For China, that policy means ceding a certain portion of profits to foreign countries, but it also means that China grows and acquires modern technology. China is now modernizing fairly rapidly, moving up the chain from the handicraft industry of the 1970s to a labor-intensive light industry and toward higher technology production. China cannot yet compete with Japan or Korea in terms of high technology products, but there has been notable progress. One example of this progress is that Chinese businesses are beginning to make memory chips and computer parts.

This industrial evolution will only accelerate in coming years as China's World Trade Organization (WTO)

membership forces it to allow more foreign firms into the country. China has learned from the Japanese Ministry of International Trade and Industry how to manage the process of slowing the entry of foreign firms while simultaneously telling local companies that they must get ready for the transition. The officials supervising the Chinese auto industry have decided that the only way to manage the coming competition is to allow many foreign companies to have joint ventures in China. Modern foreign industries are coming in, and the Chinese who are operating their plants are learning very quickly. So the Chinese auto industry is rapidly developing labor and management skills as well as higher levels of technology. After all, if they do not adapt, these businesses will be pushed out. On the other hand, the pressure from WTO membership will inevitably lead to problems with copyrights and patents. There will also be other internal issues with admission to the WTO , but they will not stop China from continuing to catch up with the industrialized world.

Is this a rosy picture of Chinese industrialization? Does the Chinese government still face problems with industrialization, urbanization, and a migrant population of workers?

To say that the picture is rosy is too simple and ignores a host of issues. Certainly the picture for China is rosier than for some, but there are many problems. In the 1980s, when farmers began to produce more because individual households received profits for the first time, farm incomes increased rapidly. Then agricultural productivity improved, so agricultural prices did not keep pace with the rise in price of industrial goods. As a result, agricultural wages stagnated. As Chinese agricultural production becomes more efficient, the country will need fewer agricultural laborers. There are people in need of employment leaving rural areas, and there are urban factories that require work-

ers. So it follows that there will continue to be massive migration to urban centers.

Migrants face great problems in the cities. The 700 million people currently involved in agriculture must undergo huge adjustments. Consider that in Japan in 1947, agriculture workers represented around 50 percent of the workforce; today, they are less than three percent. China will go through a similar process. The question is whether cities can expand fast enough to provide sufficient opportunities to accommodate this influx of workers. Factories are not going to be enough. There also has to be a large service sector. Nonetheless, factories can do a great deal to generate wealth and provide income for local governments to build infrastructure.

Migration poses many problems. If you were an employer and were getting almost unlimited workers coming into a place, how high a wage would you pay? Probably what the market would bear, with labor standards high enough for the workers to be willing to stay. That is typical of the free market. The important companies in China, because of foreign pressures and because of their knowledge, provide marginally better wages and circumstances for their workers. But there is a wide range of wages and working conditions. There are some local entrepreneurs, from places like Taiwan and Southeast Asia, who believe that cheaper is better. They tolerate minimal working conditions and incur many injuries among their employees. This system promotes exploitation, but it is how the market functions.

On the other hand, life in some of the factories I visited in the coastal areas reminds me of small college life in the United States. Many of the workers in these factories are young men and women who live together in a dormitory setting. They have clean, modern products, and the factory atmosphere even functions like a campus. These workers have a new kind of life, including compensation in the form of a salary, a good part of which they often send back home. In some isolated rural communities, the average income from the migrants is higher than the total locally generated income. Many of these workers return to their homes after a few years of factory work.

Do you think that Hu Jintao will continue the positive trends initiated by his predecessors? Is there reason to have confidence in his ability to address the challenges facing China?

Hu Jintao is a product of his generation. Like Jiang, he graduated in engineering around the time of the revolution. But Hu combined a high-technology background with very high positions in the western provinces of Gansu, Guizhou, and Tibet. Because

Beijing thought he handled the situations there quite well, he was brought into the premier leadership group. China's Standing Committee of the Politburo functions like a corporation of overseers. There are now nine members, but previously there were seven. Hu has been a member of that select group for the past decade, the only member of his age group to have that experience, and is therefore thoroughly familiar with all major decisions for the past decade. Hu is like a Washington insider, and his ascent in government is analogous to the elevation of a corporation's vice-president to the top post.

Hu is a bright, cosmopolitan person with strong natural talents and

10 years of friendship with top officials. This latter characteristic is truly remarkable because it demonstrates that he can get along with all kinds of different people. This is the first time since Mao that China has successfully prepared for a transfer of power. Mao selected all kinds of successors who failed, but now China has someone who has been intimately involved in the highest levels of Chinese politics for a decade. In my opinion, Hu is not in a big hurry. He knows that he has a lot of talent and support. I think the chances are that Hu will have a very promising future.

The Defense of Xinjiang

Politics, Economics, and Security in Central Asia

CHIEN-PENG CHUNG

As a state long noted for its potentially destabilizing ethnic heterogeneity, China has been extremely mindful of the northwestern region of Xinjiang, which is often viewed as one beset by what the Chinese have termed the "three evils" of separatism, fundamentalism, and terrorism. However, this mindfulness extends far beyond domestic policy alone. Indeed, China's role in Central Asia is inextricably tied to its desire to strengthen its political control over, economic links with, and security posture in the adjacent Xinjiang region.

The principal mechanism for achieving these intertwining aims is the Shanghai Cooperation Organization (SCO). Founded on June 15, 2001 by Russia, China, and the Central Asian republics of Kazakhstan, Kyrgyzstan, Tajikistan, and Uzbekistan, the SCO calls for closer political and economic cooperation and coordinated action among the member states to fight the "three evils," whether in Xinjiang or in the neighboring states themselves. In fact, at its inaugural meeting the SCO decided that China may be allowed to intervene militarily in Central Asia to combat terrorist threats at the request of regional governments. The organization's purpose was strengthened at the SCO's latest annual meeting on June 7, 2002, in St. Petersburg as the presidents of the organization's member states legally created an SCO Anti-Terrorism Center.

At the same time, the grouping is of great interest to China, not only to reinforce its hold over Xinjiang, but also to curb the rising influence of the United States in the region, in tandem with Russia if possible. The 2,060-mile border between China and the three Central Asian republics of Kazakhstan, Kyrgyzstan, and Tajikistan is in many places mountainous and difficult to patrol.

Since China's major concern for the region is to ensure its peace and stability as a means of guaranteeing the security of its own restive western frontier, the Chinese leadership considers the presence of US military bases in Kazakhstan and Kyrgyzstan an attempt by Washington to bolster its own influence in Central Asia at the expense of China, Russia, and Iran. The US presence in Central Asia after the terrorist attacks of September 11, 2001, and the subsequent war on Afghanistan threaten China's defense of Xinjiang and challenge Beijing's nascent but conspicuous political, economic, and strategic roles in neighboring Central Asia.

The SCO and Domestic Stability

China spearheaded the establishment of the SCO largely as part of a security strategy to prevent Uighur separatists from using Central Asian states as a base for separatist activities in Xinjiang. The joint fight by SCO states against terrorism and threats to national sovereignty has won China assurances from Muslim Central Asian states that they will not provide assistance to their religious and ethnic brethren engaged in militant separatist activities in Xinjiang. Indeed, Islam is a salient characteristic of Xinjiang's eight-million strong Turkic-speaking Uighur ethnic group, which constitutes around 45 percent of the region's population. Thus, Beijing's commitment to regional security through the SCO to fight separatism and terrorism is based not only on its fear of violent Islamic rebellion in Central Asia affecting the Uighurs, but also its concern for being branded an anti-Muslim country by Central Asian republics or Middle Eastern states if it suppresses major uprisings in Xinjiang.

This concern is valid, for Central Asia is host to a Uighur diaspora of about 500,000 individuals, with 200,000 of them in Kazakhstan alone. According to a statement released by the People's Republic of China (PRC) State Council Information Office on January 21, 2002, Uighur separatists were responsible for 200 terrorist incidents in Xinjiang from 1990 to 2001, killing 162 people and injuring more than 440 in their agitation for an independent state of "Eastern Turkestan" or "Uighuristan," their preferred names for Xinjiang.

These incidents increased in frequency through the 1990s, generating concern in the PRC government about ethnic rioting, assassinations, bombings, and oil production sabotage in Xinjiang. Beijing has been pressing Central Asian governments to cut off Uighur groups like the Nozugum Foundation and the Kazakhstan Regional Uighur Organization that reside among their diaspora communities and may be aiding Uigher separatists in Xinjiang. The Chinese government also believes that several well-known terrorist groups operating in Central Asia, such as the Hizb-ut-Tahrir and the Islamic Movement of Uzbekistan, have recruited and trained Uighur separatists in the past. Under strong pressure from Beijing, the governments of Kazakhstan and Kyrgyzstan have shut down Uighur political parties and newspapers operating in their countries—a development seemingly abetted by the SCO.

China hopes the SCO will bolster the territorial integrity, economic revival, and secular character of the authoritarian, impoverished, and ethnically diverse regimes in Central Asia that are struggling to curb rising sentiments of Pan-Turkic nationalism, Islamic extremism, and terrorist activities in the region. A peaceful, stable Central Asia would secure China's western borders against separatists, preventing manpower, funds, arms, and propaganda materials from crossing into Xinjiang and adding to an already volatile crisis of stability.

Despite the desire of Central Asia's political leadership to maintain beneficial ties with China and discourage the spread of instability, many people in the region have a place in their hearts for the secessionist struggles of their ethnic and religious kin in Xinjiang. It is perhaps not surprising that, according to a state-wide poll conducted in Kazakhstan by a Russian newspaper in 1998, only 9.4 percent of Kazakhs supported the development of good relations with China. Beijing has also expressed concern over the apparent failure of the authorities in Kyrgyzstan to act effectively against Uighur activists on Kyrgyz territory. In 1998, Uighurs in Kyrgyzstan staged a protest demonstration outside the Chinese embassy in Bishkek. On July 1, 2002, a Chinese diplomat and a businessman were both assassinated in the Kyrgyz capital. According to the Kyrgyz Interior Ministry, the murders were linked to local members of a Uighur separatist movement called the Eastern Turkestan Liberation Front.

China's Roles in Central Asia

China has been looking at the oil and gas resources of Central Asia to satisfy its energy requirements for rapid economic growth, especially since hydrocarbon development in Xinjiang, touted as the next oil and gas bonanza for China, has so far failed to live up to its promises. China's oil consumption rose

from 2.1 million barrels per day (bpd) in 1990 to about 4.6 million a decade later, with domestic production remaining relatively stagnant. Foreign experts generally predict that China will import three to four million bpd in 2010 and five to eight million bpd in 2020.

China's hopes to open new oil fields in Xinjiang have met with only modest success. Though the petroleum resources are estimated to be substantial, their remote and forbidding location in the bone-dry Taklamakan Desert render extraction commercially unattractive, especially given the low price of crude oil in recent years that has only shown a gradual upward trend lately. Nonetheless, Beijing has plans to build a 2,500-mile pipeline costing US$4.8 billion and capable of carrying 7,500 to 12,000 gallons of gas per day to link the Tarim Basin gas reserves of Xinjiang with the port of Shanghai on China's east coast.

The central government is leaving the funding of the project up to state corporations, local governments, and foreign partners, which include bidders from US, Dutch, Hong Kong, and Russian oil and gas interests. However, like other joint-ventures for infrastructure development in Xinjiang, progress has been slow, as potential domestic and foreign investors are not confident that China's legal system is transparent or effective enough to resolve potential disputes arising from the construction and operation of such a large trans-regional energy project.

By exploring the possibilities for energy cooperation with Central Asian states, China hopes to diversify its sources of crude oil imports beyond the volatile Middle East supply and reduce its reliance on the US goodwill that ensures the flow of China's imported oil supplies from the Persian Gulf. In 1997, Beijing's state-owned China National Petroleum Corporation (CNPC) acquired shares and development rights to Kazakhstan's Atyubinsk oil field and successfully outbid US petroleum interests for the country's largest oil field at Uzen. China had hoped that securing significant and long-term crude oil supplies through a pipeline from CNPC's Kazakh oil fields to Xinjiang would render economically attractive the planned east-

bound pipelines that would integrate Xinjiang with the industrial and political heartland of eastern China. Unfortunately for China, the project to construct the US$9.5 billion, 1,860-mile Kazakh-Xinjiang stretch of the pipeline system has since been put on hold, as the Kazakh government has been unwilling to guarantee the annual minimum volume of 20 million tons of crude oil that would make the pipeline profitable for Beijing. Instability in Central Asian countries and possible troubles in Xinjiang from Uighur separatists, who have already mounted attacks on oil installations and convoys, have also made potential investors reluctant to commit to building the infrastructure for transporting energy resources to China and to markets further east in Japan or Korea.

Indeed, a major purpose of the SCO is to guard against external disturbances to the domestic tranquility of Xinjiang, which would dampen the confidence of investors and flow of tourists to the region. However, Beijing also realizes that only by improving the material welfare of Uighurs will their attention be truly deflected from the persistent calls of Xinjiang separatists to engage in acts to detach the region from China. China's strategy for promoting Xinjiang's economic development calls for integrating Xinjiang into an emerging Central Asian market. As early as 1994, aware that Central Asia constituted a potentially important export market for products from Xinjiang and other parts of China, former PRC Premier Li Peng espoused the need to revive the ancient commercial "Silk Road" running through Central Asia from Europe to China through a proposed network of roads and railways in a project that will rival a similar proposal backed by Western countries to link Central Asia with Europe called the Transport Corridor Europe Caucasus Asia (TRACECA).

The SCO offers an effective method to achieve this same goal, opening doors for China's general domestic economic strategy of "Great Western Development" by promoting trade and investment between Central Asia and western China, where Beijing has pumped more than US$55 billion into infrastructure building, ecological projects, and education programs, with US$12 billion to be spent on railway construction alone from 2001 to 2005. Such domestic economic development can stabilize Xinjiang and, as PRC leadership hopes, maintain the legitimacy of communist rule by maximizing China's global stature, economic gains, and border security.

> **With a strong US ally in their midst, and Russia now a full partner of US-led NATO, Central Asian states may have less need than before to accommodate Chinese demands to crackdown on Uighur separatists.**

However, the economic benefits of such regional trading seem to have accrued more favorably to China than to the Central Asian countries considering the terms of trade between China's exports (including textiles, consumer electronics, food, fertilizers, and machinery) and its imports (consisting mainly of mineral, agricultural, and animal products). Kazakhstan and Kyrgyzstan have so far failed to prevent the smuggling of consumer goods from Xinjiang, which has led to a thriving black market across Central Asia, and the Chinese authorities have offered little tangible help.

Although trade between China and the Central Asian countries has quadrupled over the last 10 years, it is still less than US$2 billion, which, when compared to China's current annual trade levels of almost US$500 billion, leaves room for expansion. During the St. Petersburg summit, Kazakhstan's President Nursultan Nazarbayev moved to hold consultations on a common tariff policy for all SCO states, and Kyrgyz President Askar Akayev called for the removal of trade barriers and discriminatory trade policies among the group's members. To expand Xinjiang's more than US$1.3 billion border trade with Central Asia, which constitutes some 60 percent of its total trade value, the Chinese government may wish to support these measures, standardize foreign investment codes and export quality among SCO countries, and actively interdict cross-border smuggling.

Eyeing US Interests

Facing a severe crackdown on terrorist activities by the Chinese government in the last two years, many separatists in Xinjiang who escaped arrest have adopted non-violent means to continue their struggle. Others, however, have fled to Afghanistan to join what is left of Al Qaeda, sought refuge in Muslim religious schools in Pakistan, or surfaced in Central Asia to train with the extremist anti-government groups. With a strong US ally in their midst, and Russia now a full partner of US-led NATO, Central Asian states may have less need than before to accommodate Chinese demands to crackdown on Uighur separatists.

Indeed, tensions between separatists operating from Central Asia and PRC military presence in Xinjiang are expected to rise. One of the most organized and radical separatist groups, the Home of East Turkestan Youth, also known as Xinjiang's Hamas, reportedly now has 2,000 fighters operating along Xinjiang's Tajik and Kazakh borders. Separatist groups may even come to expect assistance from Central Asian governments or the United States in orchestrating further anti-Chinese disturbances in Xinjiang. At the same time, since the US invasion of Afghanistan in October 2001, China has not withdrawn its military forces along the border with Afghanistan or those adjacent to Tajikistan and Pakistan-held Kashmir.

In October 2002, units from China's Xinjiang Military District and Kyrgyzstan conducted a two-day joint military exercise involving several hundred troops and dozens of armored personnel carriers and helicopters aimed at combating cross-border terrorist activities. This exercise was supposedly directed against the East Turkestan Islamic Movement (ETIM), which is classified as a terrorist organization by the United Nations. According to Beijing, ETIM was behind numerous terrorist activities in Xinjiang, instigating 166 violent incidents and maintaining 44 strongholds and arsenals that have been smashed by PRC authorities over the past year. However, China's first military maneuver with a Central Asian state may also be interpreted as an attempt to strengthen

Xinjiang's defensive shield against any unwanted incursions from Central Asia.

The US troop deployments in Central Asia and the destruction of the Taliban regime in Afghanistan have led to the decimation of Pan-Turkic and Islamic insurgent groups operating in the region and forced their remnants to lie low for the time being, greatly reducing the threat of religious terrorism emanating from Central Asia and inciting Uighur separatists in Xinjiang. US military strength in Uzbekistan and Kyrgyzstan is likely to rise to 6,000 in the near future. The United States is also expected to expand Manas airbase in Kyrgyzstan, only 200 miles from the Chinese border, into a major base of support for air operations and the hub of a regional electronic intelligence network which covers western China.

But this strong US presence poses an uneasy threat to PRC interests. Despite the current attention to Islamic terrorists and Iraq, some analysts in US military and security establishments still believe that China will remain its long-term competitor for global power and influence. Should the United States perceive China as a threat to its strategic interests—for example, if the United States fears a Chinese attack on Taiwan—it is now much more possible for Washington to put direct military pressure on Beijing from Central Asia. At the very minimum, the United States will have enough clout with the governments of Central Asia to make sure that they do not take China's side in the event of conflict between that country and the United States. China's push for a nuclear weapons-free zone in Central Asia during the second SCO summit reflects its anxiety that the United States, which has yet to endorse this vision, or a third country unfriendly to China may introduce tactical nuclear weapons into the region. Hence, Beijing is unwilling to dismantle its nuclear weapons test site at Xinjiang's Lop Nor, nor will it allow Xinjiang to be included in such a proposed zone.

Equally pressing, however, is that China has reason to worry that its economic influence in Central Asia will be compromised by this renewed US interest in the region. Beijing wants to make sure it retains the welcome mat put out by Kazakhstan and other Central Asian governments in negotiating future concessions to sink wells or lay oil and gas pipelines from the region, at the expense of US oil interests, which are now very active there.

This threat of encirclement by the United States is perhaps China's greatest concern. US troops were already stationed in Japan, South Korea, and parts of Southeast Asia. Now they are also in Central Asia, while Washington has improved ties with Moscow and New Delhi. Still, by discrediting religious-based violence and giving the Chinese government a pretext to designate Xinjiang separatist groups as terrorist organizations, the events of September 11 and the US presence in Central Asia have led to a more quiet and predictable security environment in Xinjiang's surroundings, and that alone is a small, but definite, blessing for Beijing.

CHIEN-PENG CHUNG is Assistant Professor at Nanyang Technological University's Institute of Defence and Stategic Studies, Singapore.

From *Harvard International Review*, Summer 2003, pp. 58-62. Copyright © 2003 by Harvard International Review. Reprinted by permission.

A Big Awakening for Chinese Rivals

Hong Kong and Shanghi Look Afar

By Kai-Yin Lo

HONG KONG Over the past decade, China's premier business centers, Hong Kong and Shanghai, have engaged in a friendly rivalry, touting their relative strengths in hopes of attracting investment. Now they are waking up to the theories of a growing number of experts that cities must nurture their "creative capital" in order to entice capital of a more conventional sort.

Both are now taking a cue from experts in urban regeneration like John Howkins and Richard Florida, who argue that cities succeed by establishing themselves as fun places to live that attract the creative classes—writers, musicians, publishers, architects—and provide an environment for this elite to change society.

The cities' efforts are paying off with the return of people like the Hong Kong-born architect Edwin Chan, a co-planner of the Guggenheim Museum in Bilbao, Spain, and now a partner in the Los Angeles-based architectural firm of his boss, Frank Gehry. Chan fled his home city at an early age to pursue a career in the arts abroad. So did the composer Bright Sheng, a Shanghai native who is Leonard Bernstein Professor of Music at the University of Michigan in Ann Arbor and who received the MacArthur Foundation Fellowship in music in 2001.

When he left China, Chan complained that Hong Kong was "not nurturing." Sheng found Shanghai "suffocating." Skip forward 20 or 30 years and both men are regular commuters, lured by China's buzz and the exploding opportunities now on offer.

Hong Kong and Shanghai are taking as reference points cities like London, where long-term planning starting in the Thatcher era succeeded in revitalizing the city, resulting in a surge in property prices, multinational investment, tourist income and a flowering of the arts.

As the Hong Kong-Shanghai competition goes beyond business to enter new spheres, from the arts and design to sports, lifestyle, entertainment, tourism and the convention business, each city is taking steps to manipulate its image for greater emphasis on creativity.

Shanghai's determination to enter the world league of cities is reflected in the breakneck pace of building, as well as the high priority given by the city's leaders to "cultural enterprises," backed by national leaders in Beijing who cherish their Shanghai affiliations. Since the early 1990s, the Shanghai government has invested $230 million in cultural complexes in the city center, more than any other Chinese city, although Beijing is quickly catching up.

Shanghai, which will hold the World Expo in 2010, is using this as a target date for reaching world-class status. The World Expo plans were on the agenda in October at a conference called by Mayor Han Zheng, with an international advisory committee that enlisted world business leaders, and the mayors of Paris and Seoul, to provide suggestions on how Shanghai might boost its international competitiveness.

By comparison, Hong Kong may seem a bit tired, despite its famous skyscrapers, efficient infrastructure and legacy of British institutions.

The difference between the two represents a difference in stage of life, with Hong Kong taking a more relaxed approach to events that Shanghai is experiencing for the first time.

Hong Kong leaders envision a spectacular 40-hectare, or 100-acre, cultural center in the West Kowloon district that will increase the city's appeal, but they have run into snags of a type that Shanghai has yet to encounter: Citizens' groups have questioned not only the finances of the plan, but whether the development, no matter how iconic, will provide a basis for organic and sustainable growth in arts and culture.

The competition is at its sharpest when it comes to positioning for business. Shanghai is the logical choice for multinationals that need to keep a close watch on the vast Chinese consumer market they desire. Hong Kong, a traders' city, is friendly to exporters, small and medium-size enterprises and financial services.

Hong Kong remains significantly more attractive than Shanghai as an operational base because of its mature business culture, its bilingualism in English and Chinese, its efficiency and its rule of law, which includes vigilance over intellectual property rights, especially protection against rampant Chinese counterfeits.

Somewhat belatedly, the Hong Kong government has recognized the power of the creative industries to energize its economy and to add value to the advanced manufacturing industries in the adjacent Pearl River Delta on the mainland, where Hong Kong business controls 70 percent of the work force. This recognition has resulted in a $32 million innovation and design fund for helping small creative industries and a renewed pledge for support from the city chief executive, Tung Chee-hwa, in his policy address last week.

Under way is an aggressive program to establish Hong Kong as the hub of design in the region. Lifestyle Asia, an annual event organized by the forward-looking Hong Kong Design Center and the Hong Kong Trade Development Council, has become Asia's foremost design conference. Last autumn, three major design and branding conferences took place, attracting a stellar list of speakers, from the architects Tadao Ando and Frank Gehry to the designer Philippe Starck, the Guggenheim Museum's Thomas Krens, the innovation strategist Larry Keeley, and the luxury brand supremos Bernard Arnault of LVMH and Santo Versace.

The conventional wisdom about Shanghai is that it dazzles with its speedy development of "hardware," from commercial buildings to cultural and entertainment venues, while Hong Kong is strong on institutional and legal "software." The truth is more complex, and requires a look at the cities' cultural background.

Historically, both cities achieved prominence through their foreign connections, Shanghai as a treaty port, Hong Kong Island as a British imperial possession, together with its leased territory on the adjacent mainland. Still, both absorbed and assimilated the dominant regional cultures of their Chinese hinterlands.

In Shanghai's case, that meant drawing upon one of the preeminent centers of China's literary culture in the Song and Ming dynasties (the 10th to 17th centuries), the so-called Jiangnan region centered in nearby Suzhou and Hangzhou. These cities were renowned for their creativity, taste and individuality, in contrast to the formal style associated with government-dominated Beijing. "Suzhou style" was reinterpreted as Shanghai style as that city prospered. By the 1930s, together with Berlin and Chicago, Shanghai attracted global attention with its mix of sophisticated opulence and the demi-monde.

Both Shanghai and Hong Kong are indelibly associated with their colonial pasts. However, after 56 years of Communist rule, Shanghai is currently reveling in its economic growth and new-

ness. Colonial residues remain in the stately edifices on the Bund, the city's famous waterfront, and the hybrid Chinese-European residences still standing in the old French concession, largely retained for their tourist value.

In contrast, Hong Kong has kept many virtues of its British colonial masters, if little of their architecture. Its legal institutions, individual liberties, deep-rooted ethical standards, vigilance against corruption and efficient civil service are the cornerstone of its business, political and civic culture.

Visiting the city recently, Bruno Marquet, executive director of the Pompidou Center in Paris, paid Hong Kong a backhanded compliment by describing it as "Chinese, but not very Chinese." This perception reflects Hong Kong's attractiveness to Westerners, but does not adequately measure the connection between Hong Kong's culture and its hinterland, historically looked down upon by northerners as provincial.

Hong Kong remains an extreme example of southern China's linguistic, ethnic and cultural diversity. If Shanghai's strength is the strategic vision of its leadership and the renewal of the colorful culture of its heyday from the 1920s to 1940s, Hong Kong is spurred by a kind of creative combustion and resourcefulness.

An emblem is the late James Wong Jim, an icon of Cantopop music, whose lyrics were the defining expression of Hong Kong's golden era of popular culture in the late 1970s and 1980s. His composition "Under the Lion Rock," the title song of a television series, speaks to the sentiments of many in Hong Kong, who came by chance or as refugees after the Communist revolution in 1949, and thrived despite adversity.

In a seminar at the University of Hong Kong last year, Wong said: "We were not consciously creating anything. Hong Kong culture is just ordinary living." His definition encompasses the mix of commercial instincts, hard work and entrepreneurship that translates into a Hong Kong lifestyle based on business success, materialism, flexibility and an international outlook.

The recent clamor in Hong Kong for a greater political say signifies the maturing civic consciousness of the city's populace, who are determined to defend their liberties, rights and way of life. The Basic Law, the 1997 charter by which China governs Hong Kong, stipulates that this lifestyle—and the capitalist system that makes it possible—will remain for 50 years.

Shanghai, menawhile, held a Hong Kong Culture Week for the first time last autumn. The theme of an exhibition at the Grand Theater was Hong Kong's colorful lifestyle, which Shanghai, and indeed the rest of China, are striving to emulate.

Over time, differences between Shanghai and Hong Kong are likely to persist, although in some respects the cities may come closer together. They may not be rivals in the true sense but, like other major brands, they thrive on competition with each other.

Kai-Yin Lo is a designer, consultant and writer based in Hong Kong.

ONE COUNTRY, TWO SYSTEMS:
Getting Beyond Boundaries

By William H. Overholt

Many commentators around the world are reading the big July 1 demonstration and its aftermath as a victory for human rights and as one more signpost along the road of Hong Kong's deterioration. They are right about the first, but hopefully dead wrong about the second.

Here are a few thoughts of an old expat.

To friends who despair: Article 23 issues are important. It is appropriate that people are excited about them. But while you worry about your rights and demonstrate for them, it is useful to keep some perspective.

In any arrangement like "one country, two systems", there are important boundary issues that cannot be resolved in the short space of a constitutional document like the Basic Law. The US Supreme Court has spent two centuries trying to resolve boundary issues between, for instance, the First and Second Amendments to the Constitution.

Most big boundary issues have been resolved with minimum confrontation. The capitalist economic system is intact without significant skirmishes. The court system is intact following a couple of border skirmishes that were minor in scale but crucial in substance—one over the Chinese government's initial commitment on the structure of the Court of Final Appeal, the other over the scope of the initial finding on immigration. Both worked out satisfactorily, if not perfectly. The free press is intact. The right to demonstrate is far better than in pre-1990 British days, as demonstrated on July 1 and a thousand other occasions.

The immigration issue was messy but has been resolved in a manageable way. Article 23 has become the big border skirmish. Resolve this one satisfactorily for Hong Kong people and the basic boundaries of the two systems have been delineated. Victory is within grasp for both Hong Kong and the mainland.

To my friends who fear Article 23: It is right to fight so that homes cannot be invaded without a court order. It is right to insist on a law that cannot be used, in principle, any way, ever, against a peaceful organisation operating in Hong Kong and not using its Hong Kong organisation to undermine the Chinese government. It is right not to depend on the goodwill of future officials in implementing the laws. The rule of law means depending on the law, not on the individuals implementing it.

At the same time, if you believe in the rule of law then there must be legal expression of "one country, two systems". The most basic rule of the system is, you can't subvert me and I can't subvert you. Hong Kong is the big winner from this basic rule. There is no way that Hong Kong can allow active undermining of the Chinese government without it responding, and anyone with a map can see who will win. There must be Article 23 legislation with enough teeth so that Hong Kong can halt any substantial activity that would undermine the basic rule of mutual non-subversion. Anybody who denies that, however noble the principles cited as excuses, is selling out Hong Kong's autonomy and freedom.

But for "one country, two systems" to be viable, decisions must come from Hong Kong, not the mainland. They must be limited and subject to judicial review. The point is to do it in a way that secures Hong Kong's freedom, not in a way that abandons key freedoms.

To my friends in Beijing: Your response to July 1 so far has been wise and restrained and the whole world has noticed. Don't get too nervous. Everything that has happened since July 1, 1997, shows that Hong Kong people do not want to cause trouble for the mainland. Polls consistently showed that Hong Kong people admired Zhu Rongji more than any local leaders. Those Hong Kong politicians who ran on obsessively anti-China platforms have lost support. Pre-1997 fears of Hong Kong subverting the mainland have proved false.

Positive, even patriotic, feelings for the mainland in Hong Kong have been remarkably high. The demonstrations are not directed at the national leadership—they are defensive, not offensive. People feel they are defending their fundamental rights, and their anger is more focused on local than national leaders.

If you support them rather than hinder them, you will consolidate their comfort with "one country, two systems". If you make them feel suppressed, you will create wealthy, well-educated, well-organised enemies for generations. Please understand this moment as an opportunity to consolidate a success, not a threat to Chinese stability.

After the Article 23 crisis is resolved, you need to tackle the problem of Hong Kong's economic stagnation and its connection to the political structure. Without reforms ranging from education to competition policy, Hong Kong's economy will weaken and political discontent will grow. It is falling behind Shanghai in such areas. Reforms stagnate because the chief executive, the legislature and the civil service block each other.

There are theoretically two ways to resolve this. First, one can revert to a more dictatorial system like the old British governorship. Second, one can allow more open political organisation so that leaders gain a broad, organised popular mandate to implement reforms, as in South Korea. Reverting to the British system will cause upheaval. Trying to remain the same, and blaming problems on the personalities of Tung Chee-hwa or the opposition, will ensure continued stagnation and discontent.

It is time to start moving forward, carefully, with protection for national interests, but forward nevertheless. If you use the Article 23 debate to ally with the Hong Kong people, they will trust you and you can trust them.

To my democratic friends: This is a time for showing how strongly you feel about basic rights, such as the sanctity of the home. It is a time for resolving the Article 23 issue on terms that a democratic spirit can live with for decades to come. It is not, however, the time to press for drastic changes in the way Hong Kong is governed. To resolve Article 23, you must fight on particular issues with all appropriate passion, but you must not create fear on the mainland that a revolutionary movement might seize control of Hong Kong. The Chinese government gets paranoid about that quite easily, and tragedy could ensue. Doing one important thing at a time, using your head as well as your heart, is no betrayal of principle.

Assuming that Article 23 is satisfactorily resolved, it is essential for the democracy movement to take this (hopefully) auspicious occasion to distance the democracy movement from gratuitous anti-mainland rhetoric. We have come a long way from one prominent democracy leader's pre-1997 insistence that after July 1, 1997, Chinese soldiers would be arresting people on the streets of Hong Kong and that he personally was likely to be killed or jailed. We need to move farther away; there will be no democracy without some degree of comfort in Beijing. It is possible to be passionate about particular issues, to be furious at particular situations and to be principled at all times, while always avoiding exaggeration and always phrasing issues in the most constructive way.

So far, the Chinese government's response to the crisis has been favourable to your side. Express appreciation. If it mishandles the situation, there will be plenty of time for retribution, but this is a chance to show the leadership that they must take you seriously, and that they benefit from taking you seriously.

To my business community friends: On issues like Article 23 you need to be conspicuously part of the solution, not part of the problem. Business community leaders have admirably defused the immediate crisis, but vital lessons must be learned from earlier complacency. Hong Kong has prob-

lems that cannot be solved through the unquestioning acceptance of old ways and easy accession to whims from the north. Some of the business leadership was at risk of being seen as part of the problem with Article 23, which risked hellish future confrontations. In my experience, mainland leaders do not mind being counselled to change policy, as long as the message clearly has constructive intent.

Business leaders often argue that Hong Kong is not ready for democracy, because the citizenry doesn't understand business. But, equally, business leaders often fail to understand the depth of feeling in the community. Moreover, much of the opposition to sound business policy is a direct result of non-business politicians feeling excluded from real influence and therefore having no stake in doing anything other than articulating the demands, however excessive, of their constituents. Give them a real stake, and the best will position themselves as responsible managers in the hope of getting the big jobs.

Beyond that, the business leadership has failed to press Hong Kong forward into the reforms needed to keep Hong Kong competitive. From education reform to competition policy reform, the business community needs the organised support of the broader community, and it needs to acknowledge that sometimes its support of narrow interests is making Hong Kong fall behind Shanghai. These weaknesses precisely mirror the weaknesses of the opposition. The business community is married to the rest of the community, but somewhat estranged; this is a good time to seduce and a bad time to scratch old wounds.

July 1, 1997 remains an historic victory for human decency, political restraint and diplomatic compromise. We must never forget that the alternative was some local version of the Indian army's march into Goa or the Indonesian army's march into East Timor. Given the ideological distances and distrust of the early 1980s, the wisdom and restraint of Britain and the mainland were amazing. Today the task is much easier, but we of the current generation have yet to show whether we are equally mature as we move beyond that early compromise.

It is not surprising that implementing "one country, two systems" involves important ambiguities, powerful clashes of interests and powerful emotions. The key to the future is whether both sides recognise that they are an inch from victory in delineating the basic boundary of the system, and therefore conduct themselves with confidence and restraint and focus on vital issues. Both sides can passionately support their vital interests and win, because ultimately their vital interests do not conflict. If, on the other hand, either side reacts with unfocused emotion, then it risks snatching defeat from the jaws of victory.

William Overholt, who was an investment banker in Hong Kong for 16 years, holds the Center for Asia Pacific Policy chair at RAND, a think-tank based in Santa Monica, California.

The Hong Kong Legislative Election of September 12, 2004

Assessment and Implications

WILLIAM OVERHOLT

Statement of William H. Overholt[*] Asia Policy Chair, RAND Corporation

Before the Congressional-Executive Commission on China

September 23, 2004

Summary

The recent Hong Kong election was noteworthy for:

- Very gradual democratization;
- Recent new restrictions on the pace of future democratization that clearly frustrate a majority of Hong Kong people;
- Chinese central government fear of the democracy movement leading to repressive tactics that are largely legal but ultimately contrary to its own interests;
- Some unsettling incidents of legal and illegal intimidation prior to the election;
- A high turnout election in a calm atmosphere with an outcome that was not affected by the incidents;
- A voting majority above 60% for pro-democracy candidates;
- An electoral system that nonetheless translated the pro-democracy majority vote into a majority of seats (35/60) for pro-government conservatives;
- A clear mandate for a strategy of democratization and moderation;
- Weak, semi-competent, scandal-ridden political parties poorly representing their social bases;

- A democracy movement caught between a rising, frustrated consensus on the necessity of more rapid democratization and a deepening consensus against direct confrontation with Beijing;
- Deep division in China over proper policy toward Hong Kong;
- Considerable hope in Hong Kong for an understanding that accommodates both Hong Kong's democracy aspirations and China's security concerns;
- Policy proposals in the U.S. that expressed understandable frustration but risked undermining the democracy movement.

*The opinions and conclusions expressed in this testimony are the author's alone and should not be interpreted as representing those of RAND or any of the sponsors of its research. This product is part of the RAND Corporation testimony series. RAND testimonies record testimony presented by RAND associates to federal, state, or local legislative committees; government-appointed commissions and panels; and private review and oversight bodies. The RAND Corporation is a nonprofit research organization providing objective analysis and effective solutions that address the challenges facing the public and private sectors around the world. RAND's publications do not necessarily reflect the opinions of its research clients and sponsors.

Gradual democratization/Absence of democracy/ Rising frustration

Hong Kong has been experiencing very gradual democratization. Up to the time when China demanded Hong Kong back from the British, 100% of legislators were appointed by the British Governor. Effective with this election, 0% of legislators are appointed.

Notwithstanding this gradual improvement, the system has not progressed to the point where even very popular views can effect structural change or ensure policy change. China's central government handpicks the Hong Kong Chief Executive through a carefully chosen small committee that has no autonomy. The central government has less control over the legislature, but the elitist functional constituencies constituting half of the legislature (30/60) heavily weight electoral outcomes in favor of candidates who follow the Chief Executive's wishes; that gives the Chief Executive effective control over most policy issues.

Dirty events/Clean election

A number of intimidating incidents and violations of people's freedoms occurred prior to the election. Beijing efforts to contain the democracy movement have been directed primarily not at this 2004 election but at staunching pressures for universal suffrage elections in 2007-8. Chinese officials and media announced in late 2003 and early 2004 that Hong Kong could only be ruled by patriots and put a newly restrictive interpretation on "patriots." The Politburo Standing Committee issued a quasi-constitutional "interpretation" of Hong Kong's Basic Law that barred universal suffrage elections in 2007-8. (China has the unambiguous legal right to make that decision; the issue is not whether it is legal but whether it is sensible policy.) A Chinese fleet sailed through Hong Kong harbor for the first time since 1997, and the Peoples Liberation Army held its first-ever military parade in Hong Kong. Equally prominent were carrots designed to win favor from Hong Kong people, most notably measures that successfully reflated the Hong Kong economy, visits by Olympic athletes and a finger of Buddha, conciliatory albeit uncompromising visits from Beijing dignitaries, and gradually increasing willingness to consult quietly with pro-democracy figures.

Second, and quite separately, there was also a series of human rights and democracy violations affecting the current election whose origin and intent were more obscure. There were isolated reports of attempts from people on the mainland side of the border to influence votes, including demands for cell phone photographs of their completed ballots. Three radio station hosts resigned after alleged intimidation. A Democrat Party candidate was imprisoned for soliciting a prostitute. Office fronts belonging to three pro-democracy figures were vandalized. Some commentaries lumped such incidents together as part of a concerted campaign by Beijing to influence the election.

The reality behind these violations was more complex. Some were unambiguous violations of ambiguous origin. Some may or may not have been actual violations. The head of a movement opposing further landfills in Hong Kong's harbor was threatened, resigned his position, and left Hong Kong. The vandalism definitely occurred. In all probability there were some actual cases of people in China trying to impose voting choices on Hong Kong people.

However, unlike the clear effort to repress demands for universal suffrage in 2007–8, the origins and intents of these violations related to the 2004 election remained unclear. It is difficult to imagine Beijing taking a serious interest in the Save the Harbour movement, easier to imagine action by enraged local business interests, and successor Christine Loh seems not to have been intimidated. Radio host Albert Cheng, who had been physically attacked in the past after for publicly denouncing triad criminals, said he resigned because of threats, but he then ran for election, giving his abrasive views a much bigger megaphone, and won. Apparently he felt intimidated about one job but not the other; he certainly did not moderate his views. Radio host and former

conservative politician Allen Lee resigned following what he believed was an intimidating phone call that referred to his virtuous wife and beautiful daughter; it transpired that the phone call came from a retired Chinese official, Cheng Sousan, who had made such calls to quite a number of people, who apparently didn't feel threatened, and Beijing immediately identified the person in question. Was this intimidation, or an elderly gentleman seeking news?

Democrat Party candidate Albert Ho was arrested for soliciting a prostitute. Fearful democrats could reasonably infer malice when a single Democrat was arrested at this particular time although numerous other politicians, officials and executives were vulnerable to arrest for the same offense over the years and few or no others have been arrested. On the other hand, despite the scandal, the Hong Kong government certified Ho as a candidate even though it might have been able to interpret the law restrictively. If the goal was to hurt the Democrats in the election, Albert Ho was a strange target, since nobody gave him any chance of election. Was such an arrest part of a grand Beijing intimidation plan or some local prosecutor trying to impress his boss?

I do not know conclusively whether Beijing strategy or local political entrepreneurship or business vengeance was behind any of these cases. Anyone who claims to know must elucidate details and show evidence. It is difficult not to notice that Beijing's repressive posture regarding 2007-8 exhibited a very clear strategy, with sticks and carrots clearly proportionate to the (regrettable) goal it sought to achieve, whereas the incidents affecting the 2004 election made no strategic sense either individually or as a group. To put it another way, Beijing has so far taken a clear repressive stand on the issue of structural changes in the electoral system, but there is as yet no persuasive evidence that it is interfering with the election process itself.

Third, there were occasions of election day incompetence. Long lines formed at some polling booths and some ballot boxes were not big enough to accommodate the consequences of larger turnout, larger ballots, and crumpled ballot sheets. There is an argument that pro-democratic voters tend to vote later and therefore may have suffered more discouragement from late-day delays. Conversely, there are reports of more votes than eligible voters in some of the functional constituencies won by democratic groups.

Through the fog of conflicting evidence on such incidents, five things stand out.

- The functional constituency structure is designed to allocate seats disproportionately to conservative forces and did so.

- No commentator of standing, including the most partisan, has argued that any of these instances of intimidation, rights violations or incompetence significantly affected the basic shape of the election outcome. Exit polls and election results tallied to the degree expected in a proper election. The balloting process was basically clean and calm despite the problems.

- In longer perspective the main consequence of the anti-democratic incidents has probably been to broaden and deepen the appeal of the democracy movement.
- There has been a permissive atmosphere in which threatening incidents have become more common than in the past. The Hong Kong government has an indisputable responsibility for ensuring an atmosphere of rigorous observance of people's rights, and it will at some point have to provide a thorough account of how vigorously it protected rights, what scale of investigative resources it devoted to identifying potential malefactors, and most importantly whether the permissive atmosphere disappears.
- The body of Hong Kong's freedoms of speech, press, religion, assembly, rule of law and so forth, remains intact, but has accumulated dents and scratches at a rate that raises concerns.

The real issue for Hong Kong democracy is not the detail of this legislative election but whether there will be substantial, early progress toward a system that would give Hong Kong people more direct leverage over the officials and decisions that affect them or whether, on the contrary, democratization will be indefinitely stalled.

The election outcome

The election itself enjoyed a record turnout of 55.6% and a calm atmosphere. Clearly a majority of Hong Kong people felt that their votes mattered and that they were comfortable voting.

Pro-democratic groups got over 60% of the vote but only 25 of 60 seats. Beijing takes heart from conservatives' continued numerical control of the legislature, while democrats demonstrated, and slightly increased, their dominance of the popular vote. Among the conservatives, the Liberal Party gained substantially and won its first ever popularly elected seats. Much of its popularity was due to the fact that it has not been a conservative rubber stamp. Liberal Party leader James Tien resigned from the government last year to oppose the controversial anti-subversion law, and the Liberal Party platform calls for universal suffrage elections in 2012. Hence the Liberal Party's gains demonstrate simultaneous support for wider suffrage and for moderate strategies.

While the results send a strong message to Beijing that Hong Kong's majority wants wider suffrage, they also demonstrate a continued embrace of moderation by a large center of gravity of the electorate. There have been huge controversies over the antisubversion bill of 2003 and over suffrage for the 2007-8 elections, but and the Hong Kong majority is standing firm about these issues but is equally firm about avoiding gratuitous confrontation.

An important caveat to the electorate's embrace of moderation comes from the elections of abrasive former radio commentator Albert Cheng and disruptive Trotskyist "Long Hair" Leung, which constitute a warning that segments of public opinion can take a different turn if aspirations are frustrated too long. Cheng is the Ralph Nader of Hong Kong

and Leung is analogous to a leader of the old 1960s "Weatherman" faction of Students for a Democratic Society. Conservative groups associate opposition to democracy with "stability," but the election of "Long Hair" indicates that rigidity and social frustration could cause future instability.

Collectors of historical ironies should note that the single most unsettling aspect of this election for Beijing was Hong Kong's first-ever election of a disruptive Marxist, and the most upsetting thing for Hong Kong's democrats was Beijing's insistence on further entrenching rules that give special advantages to Hong Kong's leading capitalist interest groups.

An immature party system

It would be a mistake for either Washington or Beijing to view the election results as a clear image of the electorate's sentiments. Not only are the rules such that democratic groups' majority of the popular vote translates into a minority of seats, but also immature political parties only partially translate the breadth and intensity of democratic sentiment.

Democratic political parties are split and much weaker than the social forces they represent. There are several distinct parties among the democracy advocates. The Democratic Party of Hong Kong has a total of 638 members (according to its website on September 15, which cites July 2004 figures) and negligible ability to raise funds from Hong Kong citizens. It is deeply divided between an elitist leadership and a populist base, and between older leaders who are confrontational toward China and younger supporters who are far less so. It lacks distinctive policies on the principal social and economic issues facing Hong Kong.[1]

For some years new leadership, under Yeung Sum, has run the Democratic Party of Hong Kong, with Martin Lee continuing to serve as a primary spokesman toward foreigners because of his exceptional command of the English language. In addition, other democratic groups have arisen. Audrey Eu is now the most popular figure in the democratic movement, running first in popularity among legislators compared to Martin Lee's seventh, and her Article 45 Concern Group has, according to HKU POP polls, slightly exceeded the Democratic Party in name recognition among the electorate.[2] Political figures like Audrey Eu, Ronnie Tong, Alan Leong, and Margaret Ng are coalescing into what may become a formal political party.

The conservative DAB, which won the most seats, is better organized than any other party. Its links to its constituents are based on detailed study and emulation of the major U.S. parties. DAB events are well funded due to the contributions of the local subsidiaries of Chinese state enterprises—a large advantage in any polity. It receives loyal support from the trade union leadership. (Over 90% of the union functional constituency vote went to conservative groups.) But it has lost credibility from support of last year's government-proposed anti-subversion law, from abandonment of past promises to advocate democratization, and from some deeply ideological leadership. In the previous election, it was severely set back by leadership scandals, and its improved position this time is largely a bounce-back from those scandals.

The issue of outside influence over Hong Kong campaigns continues to have great salience. Many in China charge that the democratic movement is manipulated by the United States and support their charges by citing Martin Lee's long reliance on an American strategy advisor, his vigorous solicitation of foreign support, and his pre-1997 characterization of laws restricting foreign political party donations as a human rights abuse. Grants from American NGOs, his warm welcome in Washington in March of this year, and the National Endowment for Democracy's presentation to him of a democracy award modeled on the statue of freedom in the 1989 Tiananmen Square demonstrations have been emotionally gratifying for some Americans, but their main consequence has been to bolster the hardliners in Beijing and to fuel controversy inside Hong Kong's democracy movement. In recent years, Lee's foreign support has undoubtedly hurt his party more than it has helped. Every conversation I have about Hong Kong in China, even with the most sympathetically liberal figures, quickly homes on this issue of U.S. manipulation.

Having said that, anyone who has lived in Hong Kong, as I have, knows that those long lines of middle class families demonstrating against tough anti-subversion laws and in favor of greater democratization come from the heart and could not imaginably be mobilized by foreigners. U.S. favoritism toward Lee may in fact have weakened the ascent of stronger leaders in his own party and also slowed the competitive rise of parties more likely to be able to consolidate the democratic movement. A lesson from the business world: any party that depends for long periods on foreign NGO donations is never going to learn to raise money itself. The rising stars of the democracy movement are not those with particularly strong foreign connections. The charge of U.S. domination of the democracy movement is false, but our own actions make it difficult to convince a skeptical observer.

Mainland Chinese influence on the other hand is everywhere manifest. Mainland officials authoritatively exhort members of the Chief Executive Selection Committee to back Tung Chee-hwa. While the subsidiaries of mainland firms operating in Hong Kong are local entities, the extent to which they finance the DAB by funding its events certainly gives Beijing great leverage. DAB leaders reverse their policy positions, including on democratization, when Beijing demands it.

Where does Hong Kong go from here?

Hong Kong's future path will depend on the wisdom of leaders in Beijing and Hong Kong. Success, even if defined narrowly in classic Hong Kong terms as stability and prosperity, will require compromise on both sides. Instability

and decline will result from rigidity or confrontation on either side.

Hong Kong immediately after the election is quiescent. Conservatives among the leaders in China may see this as confirming their view that a combination of prosperity and firmness will squelch the democratic movement. Many Chinese as well as foreign experts recognize that as an illusion. There was a time when Hong Kong people were apolitical and obsessed with economic growth to the exclusion of political concerns. Two things have changed that. First, there is a pervasive sense among political aware groups that Beijing chose an ineffective leader for Hong Kong, then insisted on re-selecting him, and that Hong Kong's future therefore depends on Hong Kong people being given a chance to choose their leadership. Second, the Tung government's handling of the Article 23 controversy of 2003 created for the first time very focused popular fears about their freedoms. A Chinese policy of trying to push back the tide will not bring stability, whereas a policy of gradually channeling the tide will benefit all parties.

The center of gravity of Hong Kong opinion wants both moderation and democratization. It recognizes that confrontation with Beijing in the service of democratization is self-defeating, and hence it seeks to reassure. The most important democratic leaders in Hong Kong, including Martin Lee, have for instance recently been emphasizing their consensus acceptance of Chinese sovereignty over Hong Kong and also Taiwan. Some reached out to China by re-labeling the July 1, 2004, demonstration for democracy as a "celebration of civic society." From personal experience I can testify that most people in the democratic movement celebrate China's successes. But a clear majority also demands improvement of the current system and, if the policy of democratic reassurance fails to find partners in Beijing, political pressure will build up like steam in a covered kettle. When and how that steam will vent I cannot predict, but eventually it will.

While the strategy of reassuring Beijing while pressing hard for greater democracy provides the only strategy that has any chance at all of success for Hong Kong's democracy movement, there is no assurance whatever that it will succeed. That depends on politics in Beijing, and I cannot predict the outcome of that process. In pure policy terms, there is a great divide between the top leaders' current choice of a hard line and the view of large numbers of officials and scholars with expert knowledge of Hong Kong that the hard line is self-defeating. Policy analysis has suffered from what I call the Three Confusions: confusion of Hong Kong, where there is virtually no separatist sentiment, with Taiwan; confusion of the meaning of traditional lawful demonstrations in Hong Kong with disruptive demonstrations in the mainland; and confusion of the anti-China tactics of a few older democratic leaders with the moderate loyal sentiments of the overwhelming majority of the democratic movement. There is reason to hope that, with greater experience on the part of the new leaders, such confusions will dissipate.

Purely political considerations, however, dim the prospects for such intellectual clarity in the short run. Perceptions of Hong Kong have become tied to a crisis atmosphere regarding Taiwan. Moreover, any leaders who might wish to pursue a more generous approach to Hong Kong are exquisitely vulnerable to the charge that they are insufficiently attentive to the security of the nation. In China as in our own country, there is no more serious charge.

Such overwrought charges have been magnified during a transitional period of divided leadership in 2003-2004, as they have been during own election. With the retirement of all the top leaders of the pre-2003 era transitional stresses should decline. In addition, Beijing leaders are exhibiting more willingness to talk with leaders of the democracy movement. In the past they have largely limited senior Chinese consultations to Hong Kong groups that have strong business interests to oppose democratization, but now they are broadening their contacts and possibly their vision. That is a good start. But the prosperity and stability they seek will eventually require substantial steps toward the democratization that is enshrined as the ultimate goal in the Basic Law, a document that Chinese leaders wrote themselves.

The key strategic considerations for the democracy movement are two. First, democratization will never happen unless the central government is comfortable with it. (The Basic Law shows that in principle they can get comfortable with it.) Second, in an executive-led government, the key to giving the people some influence over policy is to give them traction over the choice of Chief Executive. Short of direct universal suffrage election of the Chief Executive, which China banned for 2007-8, there is an infinitely divisible range of possibilities from the present near-zero traction up to broad popular election of the Selection Committee, which would then function like the U.S. Electoral College.

The key strategic consideration for China should be straightforward. Because of recent demonstrations, the central government fears instability in Hong Kong. But repression of popular desires for wider suffrage will cause instability whereas satisfying them will ensure stability and continued loyalty. The argument to the contrary is based on what I have called the Three Confusions. The argument that Hong Kong can be stabilized by purely economic means is obsolete. The argument that democratization in Hong Kong will destabilize the rest of China is wrong; ever since Deng Xiaoping invented one country, two systems, there has been broad acknowledgment that the Hong Kong system is different. While the argument that the central government can't make political concessions as a result of demonstrations in Hong Kong without encouraging demonstrations in the mainland has some validity, any capable mainland politician of good will should be able to overcome this by making the case that broader suffrage was encouraged by the Basic Law and negotiated with parties that are emphasizing a policy of reassurance.

U.S. interests and policy

The U.S. has large interests in Hong Kong. Tens of thousands of Americans live there, and tens of billions of dollars of American money are invested there. We enjoy the ability of our Navy to visit Hong Kong. But economic and strategic interests are mostly not at stake in the debate over Hong Kong democracy. When Americans and American businesses leave Hong Kong, they predominantly move to Shanghai, which is less democratic. Militarily the Hong Kong port calls are a convenience, not a necessity, and anyway they are not at stake unless we have a larger confrontation.

For the purpose of this hearing, therefore, the American interests at stake are our fellow feeling for the Hong Kong people, our sympathy for the democratic movement, and our hope that China under its new leaders can become as comfortable with democracy in Hong Kong as they have become with the rule of law in Hong Kong.

U.S. policy has a frustrating dilemma. Americans love democracy and would like to support it in Hong Kong, but we have limited positive leverage and great negative leverage. Stating our views emphatically and reasoning with Chinese officials can help; most are in fact open to dialogue. Ultimately, no matter what we do, there is no assurance that China's central government will move in the direction we prefer. The best we can do is to argue our case and to avoid actions that would impair chances for a broader suffrage.

There have been proposals to express our concern over China's recent hard line by removing Hong Kong's status as a separate customs territory or removing its exemption from export controls. Changing Hong Kong's separate trade status would cause grievous harm to precisely those Hong Kong people they purport to help. Removing its exemption from export controls would destroy the ability of banks, including our own banks based there, to upgrade their computers; that would destroy Hong Kong as Asia's and America's regional banking center and cause grievous harm to the people we wish to help. Turning to political strategy, confrontational policies would defeat the moderate strategy of the democratic forces in Hong Kong and the desire of Hong Kong people for a strategy of moderation as clearly expressed in this month's balloting. Nothing serves China's hardliners better than an ability to portray the Hong Kong problem as a confrontation with the United States rather than a negotiation with some of their own people. Times may change, but for now the American posture most supportive of Hong Kong's democratic forces combines a clear voice with avoidance of confrontation.

Put another way: We Americans have every right to press China to show some respect for the clear mandate the Hong Kong people gave for a policy of democratization and moderation. When we make that case, we incur our own obligation to show respect for the second part of the mandate as well as the first.

There are also clear implications of this analysis for the roles of U.S. government-related NGOs. Teaching all political parties in Hong Kong how to organize and raise funds from the electorate provides an unexceptionable service. The parties advocating democratization benefit disproportionately from such a service, because they don't have Chinese enterprises funding their events, but the service itself does not discriminate between the DAB and the Democratic Party, and, equally important, it does not favor one democrat over another. On the other hand, with anti-democratic conservatives basing their influence on an argument that democratization in Hong Kong equates to instability, a policy of systematic American favoritism toward one particularly anti-Chinese figure, and awarding him a statue that associates Hong Kong's democracy movement with Tiananmen Square 1989, seriously damages the prospects of democratization. The ancient rule of the medical profession is valid here: When you seek to help a patient, first do no harm.

Notes

1. See for instance its website statement of education policy, a subject where major reform is a vital issue for Hong Kong's future: `http://www.dphk.org/e_site/index_e.htm`
2. See the Hong Kong University Public Opinion Poll surveys of August 9-16, 2004, and September 14, 2004. `http://hkupop.hku.hk/english/release/release241.html`

Hong Kong: "One Country, Two Systems" in Troubled Waters

"Chinese central government officials are reluctant to allow political reform in Hong Kong to proceed too rapidly or to be driven primarily by public demonstrations and aggressive pro-democracy activists."

CRAIG N. CANNING

China's unique "one country, two systems" experiment with Hong Kong began its seventh year on July 1, 2003. That day also saw an estimated 500,000 Hong Kong citizens take to the streets in a massive public demonstration opposing antisubversion legislation sponsored by Hong Kong's government.

The July 1 demonstration precipitated a series of significant developments: the government's temporary withdrawal of the antisubversion legislation, the resignations of two top Hong Kong officials, new initiatives for political reform spearheaded by pro-democracy advocates, and verbal attacks in the local and mainland media characterizing democrats and other reformers as unpatriotic. In April 2004, Beijing issued historic new interpretations of Hong Kong's constitution that effectively closed the door on efforts to hold direct elections for the territory's next chief executive in 2007 and the Legislative Council in 2008.

Some political reform activists pronounced the central government's actions the death knell for one country, two systems, while pro-Beijing leaders and government officials in Hong Kong as well as central government representatives minimized their impact. Despite these widely disparate interpretations, it is clear that public protests and political activism in Hong Kong over the past year and a half reached new heights. Clearly, the activism also produced concerns that prompted central government leaders to seize the initiative in defining their authority in the struggle over the timetable for democratic reform in Hong Kong. In doing so, China's leaders did not end the one country, two systems experiment, but they did open an important new chapter in it.

"A HIGH DEGREE OF AUTONOMY"

"One country, two systems" originated in the early 1980s with Deng Xiaoping, the principal architect of China's post-Mao reforms, who introduced the concept during Sino-British negotiations for Hong Kong's return to Chinese sovereignty after 156 years of British colonial rule. Deng's proposal was designed to bridge the gap separating China's socialist system from the system of governance that had evolved in Hong Kong under the British. In theory, the one country, two systems framework sought to preserve and protect Hong Kong's distinctive political, social, and economic system—its rule of law, self-governance, and freedoms of assembly, speech and religion—for 50 years after the territory's reversion to Chinese sovereignty. The details of the one country, two systems model were ironed out in the late 1980s and enshrined in the final version of the Basic Law, Hong Kong's mini-constitution, in 1990.

The Basic Law codifies the relationship between the People's Republic of China and the Hong Kong Special Administrative Region, the official name for Hong Kong upon its reunification with China. It promises the territory a "high degree of autonomy" while specifying the fundamental rights and duties of Hong Kong residents and stipulating the structure and functions of the executive, legislative, and judicial branches of government. It gives Beijing responsibility for Hong Kong's defense and foreign policy, but allows for independent participation by the Hong Kong government or international organizations in a wide range of fields, including trade (membership in the World Trade Organization, for example), finance, communications, culture, and sports. The ultimate authority for interpreting and amending the Basic Law is vested in the

Chinese central government, specifically the Standing Committee of the National People's Congress.

Hong Kong government under the Basic Law is designed to limit popular participation in politics and to concentrate decision-making authority in the hands of a chief executive, his hand-picked advisers, and a cabinet whose members head Hong Kong's civil service bureaucracy. The chief executive is selected by an 800-person Election Committee and then formally appointed by Beijing. The Basic Law also calls for a 60-member Legislative Council. Popularly elected representatives from geographical districts within Hong Kong will occupy 30 seats after a September 2004 election—an increase from 20 seats in 1997 and 24 seats in 2000. The other 30 seats are chosen by "functional constituencies" representing major trade and occupational sectors in Hong Kong such as business, finance, labor, law, and education.

Article 68 of the Basic Law states that the "ultimate aim" is to achieve direct popular election of all Legislative Council representatives sometime in the future but only "in accordance with the principle of gradual and orderly progress." In sum, negotiated autonomy came with an expectation that Hong Kong would be governed, at least for its first decade, by a powerful executive branch, a relatively weak legislature, an independent judiciary (appointed by the chief executive but on the recommendation of an independent commission), and limited popular participation. That participation can be changed only after elections in 2007 and 2008 with two-thirds majority support in the legislature, the chief executive's agreement, and Beijing's ultimate approval.

The chief executive is of crucial importance to the one country, two systems experiment. Conservative businessman Tung Chee-hwa, son of shipping magnate C. Y. Tung, took office on July 1, 1997. Born in Shanghai in 1937, Tung moved to Hong Kong with his family 10 years later. He attended university in England, worked for General Electric in the United States, and eventually assumed control of the family business in Hong Kong on his father's death in 1982. Known for his integrity and commitment to China and its heritage, Tung Chee-hwa seemed the ideal person to serve as Hong Kong's first chief executive. He was also Beijing's top choice.

Despite his qualifications, Tung Chee-hwa has had mixed success in his role as Hong Kong's top official. A series of events beyond Tung's control—especially the 1997 Asian financial crisis—combined with the blunders and missteps of a fledgling politician have kept his public opinion-poll ratings consistently low, except during the first months of his tenure. He was reelected to another five-year term by the Election Committee in February 2002 and reappointed by Beijing the next month.

Perhaps the most serious challenge to one country, two systems during Tung's first term was the "right of abode" controversy. The problem arose when permanent residency seekers mounted a legal challenge to the Hong Kong government after it introduced a new ordinance in July 1997 tightening regulations for proving "right of abode" status.

The new rules required proof that at least one parent was a Chinese citizen holding permanent residency in Hong Kong at the time of the claimant's birth. After Hong Kong's Court of Final Appeal upheld the claimants' legal challenges to the new ordinance in January 1999, the Hong Kong government asked the Standing Committee of the National People's Congress to review the case.

The Standing Committee's review produced a new interpretation of immigration provisions in the Basic Law that supported the Hong Kong government's position. But the Standing Committee's decision in turn generated a massive new legal challenge that again ended up before the Court of Final Appeal in Hong Kong. In January 2002 the Court of Final Appeal ruled against the claimants and for the Hong Kong government, acknowledging the ultimate authority of the Standing Committee of the National People's Congress to re-interpret Basic Law provisions. The court, in effect, reversed its own 1999 decision.

THE SECURITY-LEGISLATION TEMPEST

A greater test of the one country, two systems model came just a few months into Tung's second term. In September 2002, the Hong Kong government released a public "consultation paper" outlining the main provisions of antisedition and antisubversion legislation it planned to introduce in the Legislative Council. Having carefully avoided this potentially divisive issue during Tung's first term, government leaders explained that the legislation was needed to revise or eliminate outmoded laws left over from the British colonial administration. The principal motive, however, was the obligation under Article 23 of the Basic Law, as Beijing authorities occasionally pointed out, that Hong Kong "enact laws on its own to prohibit any act of treason, secession, sedition, subversion against the Central People's Government, or theft of state secrets, to prohibit foreign political organizations or bodies from conducting political activities in the Region, and to prohibit political organizations … of the Region from establishing ties with foreign political organizations." In brief, Beijing wanted Article 23 security legislation passed as soon as the Hong Kong government deemed it feasible in order to prevent Hong Kong from becoming a base, as it had been at times during the revolutionary tumult of the early twentieth century, for Chinese or foreign groups actively opposing or seeking to overthrow the Chinese government.

The government's consultation paper drew immediate attention from Hong Kong citizens. A central concern was a provision in the proposed legislation granting the government authority to ban for national security reasons organizations in Hong Kong that were illegal in mainland China—groups such as Falun Gong, the spiritual organization branded an "evil cult" and suppressed in China since 1999 but still active in Hong Kong. Would the proposed legislation undermine Hong Kong's autonomy by extending mainland China's laws into Hong Kong despite fundamental differences between the two legal systems? Moreover, what legal rights would targeted groups or individuals in Hong Kong possess after implementation of Article 23 anti-subversion legislation? Could Hong Kong journalists who criticized the Chinese Communist Party or Chinese central government face charges of sedition, thereby undermining the territory's freedom of the press?

> Beijing authorities have long hoped that the one country, two systems formula would eventually enable not only Hong Kong's but also Taiwan's reunification with mainland China.

These and other questions generated controversy and revealed sharp divisions in Hong Kong society over the following months. Large demonstrations both for and against the proposal were held. In early 2003 the Hong Kong government announced that it would formally introduce the bill in the Legislative Council in the spring, and a vote was eventually set for July 9, 2003.

In March an unanticipated health crisis arose as some Hong Kong residents became ill with the mysterious, highly contagious, and potentially lethal form of atypical pneumonia called Severe Acute Respiratory Syndrome (SARS), and this crisis added to the controversy and anti-government criticism over Article 23. Although the World Health Organization (WHO) pronounced the disease under control in June, a total of 1,755 Hong Kong citizens had been stricken, of whom 299 died. According to WHO statistics, SARS infected 8,098 people worldwide between November 2002 and September 2003, killing 774. The vast majority of victims, more than 7,000, resided in mainland China and Hong Kong.

SARS' toll in Hong Kong extended far beyond the personal tragedies of individual illnesses and deaths. The territory's economy, still struggling from a series of downturns in previous years, took an even harder hit because of SARS. The deflation plaguing the economy continued, and economists began to revise downward by as much as 2 percentage points their projections for Hong Kong's GDP growth in 2003. The government became the target of widespread criticism for failing to quickly recognize the seriousness of SARS as well as for its tardiness in taking preventive measures to combat the disease.

Tung's government forged ahead with its plan to pass a security bill, apparently unaware that Article 23 legislation was rapidly becoming a lightning rod for public anxiety, merging broad dissatisfaction with the economy, SARS fears, sagging confidence in government leadership, and worries about the future of Hong Kong's freedoms under one country, two systems. In the spring of 2003 the government introduced the security bill to the Legislative Council, which held three public consultation sessions while its Bills Committee scrutinized the legislation's provisions in preparation for a vote on July 9. But squabbles, finger-pointing, and walkouts detracted from the consultation and legislative-review processes. Although the government modified a few aspects of the bill, its efforts failed to mollify the legislation's opponents. As July 1, 2003, and the sixth anniversary of Hong Kong's reversion to Chinese sovereignty approached, leaders of the Civil Human Rights Front—a loose coalition of 45 nonprofit groups—and other opponents of the security bill formulated plans for a major public demonstration against Article 23 legislation.

JULY 1, 2003

From the Hong Kong government's perspective, July 1, 2003, should have been a grand occasion. Wen Jiabao, the popular Chinese prime minister appointed in March by the National People's Congress, was scheduled to pay his first visit to Hong Kong. He planned to participate in ceremonies marking the sixth anniversary of Hong Kong's reversion, witness the signing of an important new free trade agreement between China and Hong Kong, and consult with Tung Chee-hwa and other government leaders.

On June 29, the first day of his three-day visit, Wen joined Chief Executive Tung in observing the signing ceremony of the Closer Economic Partnership Arrangement (CEPA), the free trade agreement negotiated over the previous 18 months. By lifting tariffs on many Hong Kong goods exported to China, reducing restrictions on mainland tourists, allowing mainland Chinese to purchase property in Hong Kong, and constructing a new bridge linking Hong Kong with the Pearl River Delta, CEPA was expected to provide a much-needed boost to the city's economic recovery efforts after it took effect on January 1, 2004.

At the conclusion of the CEPA signing ceremony, Tung, looking ahead to the anniversary celebrations on July 1, reportedly reassured Wen about the public demonstration planned for that day, commenting that no more than 50,000 to 60,000 people were expected to participate. Instead, an estimated 500,000 Hong Kong residents took to the streets on a hot summer afternoon to stage a peaceful and orderly protest. This was not only the largest public demonstration in Hong Kong since reversion in 1997, but also the largest on Chinese soil since May and June 1989, when a million Chinese participated in pro-democracy demonstrations in Tiananmen Square and about a million Hong Kong citizens marched spontaneously to condemn Beijing immediately after the bloody June 4 crackdown on the pro-democracy movement.

Despite the enormous public protest, Tung continued his campaign to advance the security legislation, offering to remove the provision giving the executive branch authority to ban Hong Kong organizations outlawed by the Chinese central government but remaining insistent on a July 9 vote. Opponents of the bill were outraged and called for Tung's resignation. Tung eventually deferred the vote after James Tien, chairman of the pro-Beijing Liberal Party and a member of Tung's Executive Council, resigned from the council on July 6, following a quick trip to Beijing. The loss of Tien's support and presumably the votes of many in his party had dimmed prospects for the bill's passage.

Tung's decision to postpone a Legislative Council vote on the security bill did not end the furor. Pro-democracy reform activists and opponents of the legislation now called not only for Tung's resignation but also for constitutional changes allowing the direct election of the next chief executive in 2007 as well as the direct popular election of Hong Kong's fourth Legislative Council in 2008.

On July 16, 2003, Tung sacked two members of his cabinet: Secretary for Security Regina Ip, the government's main champion of Article 23 legislation; and Finance Secretary Antony Leung, who had attracted public criticism for purchasing a luxury automobile a few weeks before introducing a new auto tax policy, thus dodging the tax. Tung held a press conference the following day, promising to do a better job as chief executive, but he did not apologize for the security legislation uproar. Two days later Tung flew to Beijing for high-level meetings with the central government's top leaders—including new President and Party General Secretary Hu Jintao, Prime Minister Wen Jiabao, and Vice President Zeng Qinghong, a close associate of former President and Party Secretary Jiang Zemin, one of Tung's key supporters. The Beijing trip and its media coverage made clear that, despite his difficulties at home, Tung retained support in Beijing and would not resign in the face of massive protests.

In early August 2003, Tung appointed Henry Tang as the new financial secretary. Ambrose Lee was designated secretary for security. Shortly after his appointment, Lee in mid-August announced his intention to launch a fresh round of public consultations on the security bill, thereby signaling the government's determination to persevere with Article 23 legislation. Predictably, the bill's opponents objected again. Facing strong pressure from opponents of the bill and from his own supporters, Tung finally withdrew the legislation on September 5, 2003.

THE REFORMERS' MOMENT

What appeared to the pro-democracy camp as a string of victories since July 1 fueled further efforts to achieve political reform. Political activists began to focus their attention on the 2004 Legislative Council elections. Their strategy was straightforward: if the Democratic Party and other pro-democracy parties could capture all 30 seats open to direct popular election in 2004—riding the wave of anti-security legislation and pro-constitutional reform sentiment—and if they could also cultivate support from a few legislators in the 30-seat functional-constituency section of the Legislative Council, pro-democracy lawmakers would command a majority in the legislature and might be able to push through constitutional reform measures. And if Chief Executive Tung could then be persuaded to back such reforms, Beijing would face an unpalatable choice of either approving the reforms or defying both the executive and legislative branches of the Hong Kong government as well as popular opinion.

Despite the many "ifs" in this strategy, two developments in the second half of 2003 provided some evidence that it might be feasible. First, the number of registered voters increased sharply by 150,000 in July, following the protest demonstration. Second, many Democratic Party and other pro-democracy candidates won election in Hong Kong's district council races on November 23, 2003, while members of a pro-government party and other candidates associated with the government's security bill suffered defeat. The significance of these developments was not lost on attentive observers in Beijing.

Central government leaders could not ignore the implications of the massive public protest of July 1, 2003.

Chinese central government leaders revealed their worries in several ways in late 2003 and early 2004. On December 3, President Hu Jintao signaled Tung in a meeting in Beijing that he was concerned about recent political developments in Hong Kong. Hu's statement was underscored the next day when China's official New China News Agency published a treatise by four well-known legal scholars that laid out several principles for Hong Kong's political development. The bottom line in their argument was that Hong Kong lacked the right under the Basic Law to implement political reform on its own; the Chinese central government would make the final decisions. These and other subtle actions by Beijing made the central government's message clear: whatever steps Hong Kong might take toward political reform would require close consultation, careful review, and ultimate approval by the Chinese central government.

THE STANDING COMMITTEE RESPONDS

Despite these signals from Beijing, demands for political reform focusing on future elections persisted into 2004. As the Hong Kong government continued to face calls for political reform, Beijing took firm action. On April 6, 2004, the Standing Committee of the National People's Congress issued an interpretation of the Basic Law stating that the Hong Kong chief executive must obtain approval from the Standing Committee before introducing any electoral reform bills to the Legislative Council. In effect, the Standing Committee asserted its right to approve not only the final product at the end of the reform process but its initiation as well. The Standing Committee also ruled that Hong Kong's Legislative Council lacked the right to introduce electoral reform legislation on its own, thereby reserving that privilege for the executive branch.

On April 26, just three weeks later, the Standing Committee issued another set of rulings in response to a Hong Kong government request for guidance in regard to the 2007 and 2008 elections. The Standing Committee explicitly forbade the direct election of the chief executive or the Legislative Council because it would contravene the voting rights and procedures spelled out in the Basic Law. The rulings also upheld procedures stipulated in the Basic Law allowing bills from the executive branch to pass more easily than those introduced by the legislative branch. And finally, the Standing Committee refused to approve an increase in the number of Legislative Council seats open to direct popular election in 2008, stating that the precisely balanced ratio with constituency appointments—currently 30/30—must remain fixed.

Anticipating that these decisions would produce political fallout, the central government dispatched several officials to Hong Kong to explain the Basic Law interpretations in person. As expected, public protest demonstrations took place in Hong Kong in April 2004. But the largest of these, according to estimates of protest organizers, attracted about 15,000 individuals—a figure well below those for other public demonstrations since July 1, 2003. In early May 2004, Democratic Party member Martin Lee, one of the best known pro-democracy advocates in Hong Kong, attempted to introduce an amendment in the Legislative Council charging the Standing Committee with "an abuse of power" under the Basic Law. However, the Hong Kong government quickly advised the Legislative Council that such a motion was "out of order." Council President Rita Fan agreed, and Lee's amendment was blocked.

Notwithstanding the demonstrations and other efforts by pro-democracy advocates, the Standing Committee rulings of April 2004 and the actions of Chinese central government officials and pro-Beijing leaders in Hong Kong made clear that in the near term little if any hope remained for political reform—including direct elections in 2007 and 2008—beyond the increase in directly elected Legislative Council seats already written into the Basic Law.

A May 2004 public opinion poll revealed that support for universal suffrage in the 2007 and 2008 elections had dropped sharply—from more than 80 percent of those polled in July 2003 favoring universal suffrage for both the Legislative Council and chief executive elections down to 66 percent in May 2004 for the legislative balloting and 55 percent for the election of the chief executive. Among the factors helping to explain these changing poll results were long-awaited indications in early 2004 of an improving economy, as well as Hong Kong residents' growing recognition that ultimate authority for constitutional change rests with the Chinese central government and that the reform timetable would not be dictated by street protests and legislative activism.

Despite these indications that Hong Kong citizens had registered Beijing's message, another large public protest occurred on July 1, 2004. Approximately 350,000 people joined a peaceful demonstration against the Standing Committee's recent rulings and in support of democratic reforms in Hong Kong. Although Beijing had declared its preeminent authority over the territory's political reform, the July 2004 protest illustrated that Hong Kong citizens are still willing to openly express their discontent.

BEIJING TAKES OFF THE GLOVES

Why did worried leaders of the Chinese central government adopt a more aggressive and interventionist stance toward Hong Kong in the first half of 2004? While the answers lie primarily in problems that the leadership perceived in Hong Kong itself, Beijing's actions were also prompted in varying degrees by China's leadership transition, the country's rapid economic growth and its potentially destabilizing consequences, developments in Beijing's relations with Taiwan and the United States, and China's evolving role in the international community and global economy.

Central government leaders could not ignore the implications of the massive public protest of July 1, 2003. Even more alarming to Beijing was the subsequent initiative by pro-democracy activists to accelerate the pace of political reform in a drive to achieve universal suffrage in the elections of 2007 and 2008.

In this context the leadership of Chief Executive Tung posed several dilemmas for Beijing. First, Tung's sustained unpopularity helped fuel the discontent manifesting itself in public demonstrations and political reform initiatives. In addition, Tung's political ineptitude reduced his administration's ability to cope with "people power" and other political reform challenges in Hong Kong, virtually compelling central government authorities to become more directly involved but also inviting accusations that Beijing was infringing on the "high degree of autonomy" promised Hong Kong in the Basic Law. Yet to remove Tung from office before the end of his second term would reflect poorly on the central government's leaders who had endorsed him twice. It also could encourage more "people power" and legislative activism, and besmirch the one country, two systems experiment in its early years.

Retaining and supporting Tung, as Beijing officials opted to do in July 2003, meant that they would have to grapple more directly with the political reform initiatives emerging in Hong Kong. Consequently, Hu Jintao, Wen Jiabao, and other high-ranking central government leaders sought to influence the situation in Hong Kong in late 2003 and early 2004. They did so through their comments to Hong Kong government officials and their public statements urging the people of Hong Kong to concentrate on economic development and to consider the overall national interest. It was also in this context that the Standing Committee in Beijing issued its interpretations of the Basic Law in April 2004 foreclosing the possibility of direct elections in 2007 and 2008.

Central government decisions may also have been influenced by recent leadership changes in China. Some analysts suggest that Beijing's hard line on political reform in Hong Kong reflects political maneuvering within the Communist Party associated with the transition from third- to fourth-generation leaders. They argue that the continued influence of former President and Party Secretary Jiang Zemin, currently chairman of the Party's Central Military Commission, and his allies is mainly responsible for Beijing's hard-line decisions. But there is no consensus on this assessment. Other experts suggest Hu has already begun to neutralize Jiang's influence.

A more compelling argument points to China's sustained high-level economic growth, which is generating not only rapid social and economic change but also wide-ranging problems such as high unemployment and underemployment, sharp regional disparities, mass labor migration, insufficient health care and health insurance, rapidly widening economic extremes, and a growing gap in living standards between urban and rural China. Any single one or combination of these problems could swiftly trigger substantial popular opposition to the Party or government in China. Consequently, Chinese central government officials are reluctant to allow political reform in Hong Kong to proceed too rapidly or to be driven primarily by public demonstrations and aggressive pro-democracy activists. Little wonder that Prime Minister Wen and other top leaders frequently remind Hong Kong of the value of—and the symbiotic relationship they see between—"prosperity and stability."

THE TAIWAN FACTOR

Two aspects of the "Taiwan factor" must also be considered. First, Beijing authorities have long hoped that the one country, two systems formula would eventually enable not only Hong Kong's but also Taiwan's reunification with mainland China. Taiwan's leadership, it should be noted, has invariably criticized one country, two systems and any hint of Beijing's intrusion into Hong Kong's autonomy, stressing that the Hong Kong model is inappropriate for Taiwan. The recent struggle over political reform in Hong Kong has only strengthened Taiwan's hand in this argument.

Second, from Beijing's perspective the democratization movement in Taiwan has precipitated a variety of unwelcome developments since the Nationalist-dominated government first legalized opposition parties in 1986. In addition, democratization in Taiwan has coincided with a steadily growing sense of Taiwanese identity—one political manifestation of which is the pursuit of independence and international acceptance of Taiwan as a sovereign nation separate from China.

In Hong Kong's case, independence is not an issue. The practical impossibility of achieving an independent Hong Kong, given the territory's extreme dependence on mainland China for everything from its water supply to its economic wellbeing, is recognized and accepted on both sides of the border. What the "Taiwan factor" contributes to central government leaders' attitudes toward Hong Kong is recognition of the inherent volatility and unpredictability of the democratic process, a lesson recently reinforced for them by the narrow and acrimonious reelection of Taiwan President Chen Shuibian in March 2004.

Sino-American relations also have been a factor in the Chinese leaders' calculations. The improved bilateral relationship since 9–11, along with China's growing importance to the United States in its efforts to resolve the dispute over North Korea's nuclear weapons program and to stimulate growth of the American economy, has reduced the likelihood that the Bush administration would mount a serious challenge to Beijing over its Hong Kong policies.

China's entry into the World Trade Organization in November 2001, its deepening engagement with the global economy, and its growing involvement and stature in the international community are also influencing Hong Kong-mainland relations. Hong Kong no longer appears indispensable to Beijing for China's trade and economic development. Today many observers believe the situation has reversed itself: Hong Kong's economic fortunes seem highly dependent on China and on Beijing's policies. The effect has been to increase Beijing's and reduce the reformers' political leverage.

THE FATE OF "ONE COUNTRY, TWO SYSTEMS"

Does all of this mean that one country, two systems is dead? Far from it. Both Beijing and Hong Kong have much to gain from seeing this unique experiment through to a successful conclusion. The Basic Law is a "living document"—its provisions provide the framework for life and governance in the Hong Kong Special Administrative Region, but they must be revisited and reinterpreted as conditions and circumstances change.

The recent interpretations of the Basic Law by the Standing Committee of the National People's Congress underscore Beijing's determination to ensure that political reform in Hong Kong takes place gradually and that it not be dictated by street protests or what Beijing may perceive as opportunistic political activism. Since the Standing Committee's rulings in April 2004, some voices in Hong Kong government, media, and other circles have begun calling for less confrontational "new approaches" involving improved dialogue and a process of quiet, measured steps. These steps may hold the promise of achieving future political reforms on a mutually acceptable timetable.

The central government's recent decisions appear to tie Hong Kong firmly to Beijing's interpretations of the Basic Law regarding political reform. But if the struggles between Hong Kong and Beijing amount to a new chapter in one country, two systems, it seems far from the final chapter. The nature of Hong Kong's self-governance and of its general relationship to mainland China is yet to be determined.

CRAIG N. CANNING is an associate professor of history at the College of William and Mary.

See Me, Hear Me, Touch Me, Heal Me— The Rise of Alternative Medicine

Teng Sue-feng

In the park, people both young or old who pass by the "foot massage path" take off their shoes and walk along it a few times; nearby, a blind masseur offers his services. Gyms and beauty parlors are constantly pushing all kinds of new alternative therapies to attract nine-to-fivers, from hydrotherapy and essential oils to navel therapy and ear candling. Spurred on by liberal doses of media hype, many people are willing to give them a try.

Whether it be to treat disease or to relieve stress, these various home-grown and imported therapies have long been in vogue among the public at large, and have carved out their own territory in the healthcare landscape. In the medically advanced 21st century, how is it that alternative therapies are not only able to survive, but even enjoy increasing popularity?

Case 1: Mr. Chiang's left shoulder is paralyzed after a road accident. Although the bones have knitted together again after surgery, his muscles have gradually withered and lost strength. So he is visiting a traditional Chinese medicine clinic on Taipei's Chunghsiao East Road, where the doctor, Ma Chih-hsiang, inserts fine needles into Mr. Chiang's acupuncture points and passes an electric current through them to stimulate his muscles. In the next bed is a Mrs. Liu, whose arms are swollen because of diabetes. After six sessions of acupuncture and electrical stimulation, the swelling in her arms—which were as thick as her thighs—has gradually gone down again.

Case 2: Bank worker Mr. Hung was last year diagnosed with liver cancer. His doctor suggested he undergo chemotherapy, but Mr. Hung felt that chemotherapy would not get to the root of the disease, so instead he decided to tackle it by changing his dietary habits. He went to southern Taiwan to find apples that were not treated with pesticides, and went over to a macrobiotic diet to restore his health. A year later when Mr. Hung returned to hospital for another examination, his two-centimeter tumor had disappeared.

When patients go their own way

The quickening pace of life, environmental degradation, and stress bearing down from all sides impact us physically, emotionally and spiritually. Although people today are living longer, they are inescapably plagued by chronic ailments and stress. The standard procedures of mainstream medicine, such as drugs and surgery, cannot effectively relieve patients of the fear they experience when battling with disease. Faced with these difficulties, more and more people are willing to try alternative treatments, which have become popular worldwide.

In the cold of winter, people in Taiwan have long been in the habit of buying some Chinese herbal ingredients to add to their food, to make dishes that are both nourishing and health-promoting

In November 1991 the US news magazine *Time* ran a cover story that examined the phenomenon of patients turning their backs on mainstream Western medicine to seek alternative therapies, including acupuncture, reflexology, hypnosis, Ayurvedic medicine and many more.

An opinion poll commissioned by *Time* showed that 30% of the US residents surveyed had tried some form of unconventional therapies. The magazine estimated annual national expenditure on alternative medicine at US$27 billion. The growth of alternative medicine reflected "a gnawing dissatisfaction with conventional, or 'allopathic' medicine." For all Western medicine's brilliant achievements, such as vaccines, penicillin and organ transplants, it also had some severe failings, and the one most complained about was the endless waiting for doctors who viewed one as a sore back, an inoperable tumor or a cardiac case, but never as a person.

In view of the rising popularity of alternative medicine, in 1992 the US Congress set up the Office of Alternative Medicine under the National Institutes of Health, and in 1998 the Office was expanded into the National Center for Complementary and Alternative Medicine. The center categorizes over 40 types of treatment as alternative medicine. They range from familiar staples such as acupuncture, moxibustion, Chinese herbal medicine, *qigong* and shiatsu, to more exotic-sounding varieties such as homeopathy, magnetic healing and energetic medicine. (Homeopathy is a traditional European treatment method in which the practitioner first diagnoses the patient's symptoms, then prescribes tiny doses of drugs that in larger quantities would elicit the same symptoms. The drug's affinity for the disease boosts the body's own defense mechanisms. This is similar to the Chinese principle of "fighting poison with poison.")

Mass movement

If the rising popularity of alternative medicine in the US is a reaction against the rapid development of Western scientific culture, in Taiwan it is more a case of people taking a fresh look at some "old stuff" that is deeply rooted in local traditional culture.

Associate Professor Ting Chih-yin of National Taiwan University's school of public health has been researching alternative therapies for many years. She divides them into four major categories, as seen from the user's perspective. The first category comprises those therapies that involve the injection, ingestion or inhalation of substances; these include traditional Chinese medicine (TCM), herbalism, macrobiotic foods, health foods, and aromatherapy. The second comprises physical therapeutic methods applied externally by a practitioner, such as acupuncture, massage, chiropractic, *guasha*, hydrotherapy, and cupping. The third comprises physical, mental or spiritual exercises to be practiced by users themselves, such as *qigong, taijiquan*, yoga and meditation. The fourth comprises methods involving the redirection of supernatural forces, such as feng shui and *shoujing* (a Taiwanese folk ritual to calm a child after a frightening experience).

In order to understand the Taiwanese public's attitudes towards alternative medicine, Ting conducted a questionnaire-based survey in which she asked: "In the past year, have you used any of the following methods to deal with illness?" The 16 choices included Chinese medicine, other herbal medicines, and macrobiotic foods. As many of 75% of those polled had used some form of alternative medicine.

Among those who had used therapies of the first type, involving the intake of substances, the highest proportion had taken Chinese herbal medicines (44%), followed by health foods (24%), macrobiotic foods, other herbal medicines, and aromatherapy. Frequently used physical therapies were *tuina* massage, *guasha*, and other types of massage (all over 20%). For supernatural methods,

shoujing (10%) and shamanism scored the highest, while only 2.8% of those polled had turned to feng shui or fortune telling.

Ting Chih-yin notes that research in countries such as the UK, the US and Australia shows that demographically speaking, people there who accept alternative medicine tend to be white-collar workers, middle-class, relatively well educated, and with a high degree of self-control, such as cultural and creative workers, environmentalists, and feminists. But in Taiwan, there is almost no differentiation according to level of education-alternative therapies are used by people throughout society. Looking more closely, this is related to the fact that what is referred to as alternative medicine also includes many therapies from traditional Chinese medicine, and these are currently enjoying a market revival.

Ancient wisdom

From the point of view of TCM practitioners, some of the alternative treatments currently popular in Taiwan, such as massage, cupping and *guasha*, are in fact traditional external treatment methods used in TCM, and are not "alternative" treatments at all.

Thousands of years ago, before the advent of the science of anatomy, our ancient Chinese ancestors developed the theory of meridians, based on their experience and understanding of the structure of the human body. According to this theory, running through the body are 12 channels or meridians, along which are distributed various acupoints (acupuncture points) corresponding to various organs. Within these meridians flows the "qi" or vital energy, composed of yin and yang. Qi is regarded as the source of the body's life, and if a meridian is injured, or if the balance of the qi is upset or its flow is obstructed, the function of the corresponding organs will be disturbed and disease will result.

The TCM treatment methods of cupping, *guasha*, and various types of massage and manipulation such as *tuina*, are all based on the theory of meridians. In cupping, for example, heat is used to expel the air from small suction cups, so that they attach themselves to the skin. The cups are placed over certain acupoints and over the afflicted part of the body, and the low pressure within them causes the tissues beneath to become engorged with blood. As explained by TCM dictionaries, the mild congestion of blood under the skin allows the qi to flow freely through the meridians, thus invigorating the qi and the blood (in TCM theory, the qi controls the flow of blood). Cupping is said to invigorate blood circulation, promote and normalize the flow of qi, relieve pain, dispel "wind," "cold" and "dampness" (pathogenic factors in TCM) and eliminate toxins. The ailments it is considered appropriate for are colds and flu, chronic rheumatoid arthritis, asthma, lower back pain, and upper back and shoulder pain.

Massage and manipulation techniques such as *tuina*, meanwhile, attempt to deal with localized injuries and pain. The Qing-dynasty work *Yi Zong Jin Jian* ("*The Golden Mirror of Medicine,*" published 1749) avers that "by pressing on the meridians to free the flow of stagnant *qi*, and by rubbing the swollen parts to dispel the swelling, the ailment may be cured."

A renaissance of tradition

The ancient art of TCM and various folk methods of maintaining health have few side effects, and are almost an element of everyday life. For instance, in the cold of autumn and winter, many ordinary people go to herbal apothecaries to buy some dried herbs which they directly infuse or boil into teas, or else add to their food as a tonic. If they put their back out or twist an ankle, they will still go to a dilapidated neighborhood martial arts studio to ask the bonesetter to manipulate the affected part.

With ingenious packaging by beauty centers, ear candling, a procedure originally devised by Native American peoples, has become a fashionable therapy. It is claimed to promote lymph purification and relieve migraine and dizziness.

We have now arrived in the 21st century and the age of genetic medicine, yet some tried and tested folk remedies with gentle and predictable curative effects have not receded before the advance of science. On the contrary, some traditional treatments that had been abandoned or were rarely used by TCM practitioners have been commercially repackaged and are enjoying a "renaissance" amid the fashion for alternative treatments.

Recently there have been reports of members of the public suffering burns while undergoing "navel therapy" for slimming or to treat ailments. Such accidents have focused renewed attention on this therapy, which combines traditional medicine and modern commercial innovation.

Navel therapy is a form of moxibustion, in which a container holding a burning moxa stick is placed over the patient's navel. It is mentioned as long ago as in the 4th-century TCM classic *Zhou Hou Bei Ji Fang* ("*Handbook of Prescriptions for Emergencies*"). The treatment involves applying medication to the umbilical area (the *shenque* acupoint) as a paste, patch, ointment, liquid, smoke or steam, to prevent or treat disease. In TCM clinical practice it is traditionally used for acute diarrhea, dysentery, coldness of the limbs, and prostration or exhaustion in chronic disease.

As for the effectiveness of navel therapy for weight loss, Cheng Sui-tsung, president of the Chinese Internal Medicine Society of the ROC and director of Liu Fu Tang TCM clinic, has his reservations.

"If there is no organic problem, a woman with *qi* deficiency and 'cold' of the uterus can be treated by navel therapy to improve her constitution and improve the flow of *qi* and blood, allowing her to conceive more easily." Cheng further explains that the danger in navel therapy is that "if you don't control the temperature correctly, or if burning moxa leaves fall onto the skin, or if you use the wrong herbs, the skin around the navel may be burned, stained black or infected."

New keep-fit fashions

Apart from the high level of acceptance in Taiwan of rejuvenated versions of native treatment methods, Ting Chih-yin's research also shows that the public here are also very willing to try imported therapies such as aromatherapy and macrobiotic diets, and even entirely new therapies.

For instance, ingenious packaging by beauty centers has transformed "ear candling," which is derived from a native American remedy, into a fashionable "aromatic therapy for cranial purification." An ear candle is a hollow beeswax candle something over ten centimeters long, encased in cotton cloth, with a wick at one end and a small hole at the other through which the smoke can emerge. The candle is placed into the patient's ear and the smoke slowly flows into the ear canal, giving a sensation of heat. Beauty centers claim that ear candling accompanied by a lymph drainage massage using essential oils can relax the nerves and also effectively improve problems such as giddiness, a stuffy nose, ringing in the ears, poor memory, and unstable moods.

Lighting a burner spiked with essential oils to fill the room with flower fragrance can be an effective way of relieving stress.

The colorful imported bottles of fresh-smelling essential oils of rose and lavender offered for sale at department-store cosmetics counters have reawakened modern people's sense of smell and made them willing to have close encounters with flower fragrances. Essential oils derived from different parts of plants, such as flowers, leaves and seeds, contain many chemical compounds. After the tiny molecules of these compounds enter the body through inhalation, rubbing, massage or fragrant baths, they are said to be transported to the areas of the brain that control moods, where they promote the secretion of serotonin and endorphins. This has the effect of stabilizing moods and relieving stress. In countries such as Germany and France, essential oils are used in medical treatment.

"Wellness centers" tout their ability to stave off aging, and woo customers with scented hot spring baths, quartz sound wave therapy, and whole-body pummeling with contoured mallets. They stress that these therapies can adjust slight lopsidedness or other deformations of the spine caused by modern people's poor posture, thus restoring mental and physical balance.

No elixir of life

In fact, whether the alternative therapies on the market are home-grown or imported, their curative effects are often deliberately hyped up and exaggerated. There have also been many reports of incidents such as people's neck vertebrae being injured by powerful water jets in spas, or patients suffering strokes due to chiropractic procedures.

The Chinese believe that the system of meridians holds the key to health or disease. Acupuncture is one of the treatment methods that reveals the core spirit of traditional Chinese medicine.

As for the idea that ear candling can cure ringing in the ears or improve the memory, one ENT specialist points out that the cervical nerves, vagus nerve, glossopharyngeal nerve and facial nerves all pass through the outer ear canal, so it is possible that the smoke and heat of an ear candle may stimulate these nerves to some degree. But it is very doubtful whether this can produce any curative effects. If some people feel more relaxed after ear candling, it may have more to do with the soft lighting and music in the beauty center—in other words, the effect is entirely psychological.

In fact, the terms "alternative medicine" and "folk remedy" do not appear anywhere in the healthcare legislation administered by the Department of Health. There only references to "actions not governed by the healthcare regulatory system," which means that as long as treatments do not involve such things as bonesetting, prescribing medicines to be taken internally, using medical instruments, or other invasive procedures, then they are not subject to regulation. Treatments that fall outside the regulatory net in this way include the external use of herbal preparations, massage techniques such as *tuina* and shiatsu, *guasha*, reflexology, *shoujing*, amulets, incense ash, cupping, *qigong* and the like.

Lin Yi-hsin, chairman of the DOH's Committee on Chinese Medicine and Pharmacy, states that the reason the DOH does not regulate folk remedies is because—in keeping with the principle that the government should not interfere unnecessarily in citizens' lives—it believes there is nothing wrong with ordinary members of the public practicing therapies such as *guasha* or *tuina* in their own homes. But if businesses exaggerate their curative effects, then there is a need for review.

However, in the absence of a regulatory framework, and with very variable levels of skill among persons engaged in alternative medicine businesses, the quality of alternative therapies is a real cause for concern. Even people who are very open to alternative medicine have begun to be less than satisfied with the healthcare environment it offers.

Mr. Lin, a writer who generally speaking is not averse to *tuina* massage, once went to a TCM traumatology clinic near his home for some skeletal manipulation. The hundred-square-meter clinic was packed full of patients, and in the middle were three reclining chairs attended a young *tuina* masseur. The masseur worked on Lin's neck and back and rotated his shoulders and arms for 20 minutes. Lin heard the sound of his own joints cracking, and felt that his muscles had relaxed quite a lot. But the next day as soon as he woke up he felt that his neck was even more painful than it had been before his massage session.

Just as we frequently hear reports of medical disputes involving conventional hospitals, there are also many problems associated with alternative medicine. But the degree of acceptance of some alternative treatments among the public means that the regulatory authorities cannot ignore their value.

To cite reflexology as an example, during 20 years of energetic promotion in Taiwan, Swiss priest Father Josef Eugster taught students throughout the island, and thus trained many later practitioners. But these people never had a legal status. In 1991, Father Josef and over ten reflexologists registered with the Ministry of the Interior as the "ROC Foot Reflexology Health Association" (now the Chinese Foot Reflexology Association, CFRA); but the MOI categorized the association as a sports club. After persistent efforts, in 1993 the Department of Health finally deemed the association's activities to be a traditional treatment that is not subject to the healthcare regulatory system, and the practitioners were at last freed from the fear of being prosecuted as "underground doctors."

Cupping has a similar effect to a hot compress—it can accelerate local blood circulation and promote metabolism.

At present, countries such as the PRC, Japan, South Korea and the US already have compulsory licensing systems for professional medical masseurs. In Taiwan, the CFRA and Father Josef are currently working to define a training, assessment, certification and licensing system for reflexologists.

Cross-sectoral dialogue

Apart from uneven levels of quality, the main reason why alternative medicine is not welcomed by the regulatory authorities or mainstream medical practitioners is that it is difficult or impossible to test the effectiveness of alternative treatments by scientific methods.

Dr. Julia Tsuei, an internationally respected gynecologist with a background in Western medicine, is director of the Clinic for East-West Medicine in Taipei. Twenty years ago she began researching TCM and various kinds of alternative medicine. Now aged 76, she says she is devoting her life to finding scientific methods of testing. "TCM simply treats disease by methods that are unfamiliar to Western medicine. An acupuncture needle inserted into the same place can reinforce the *qi* in one person but reduce it in another. It's very mysterious, but the real issue is, which methods are the most effective? Today, everyone is confounded by the lack of standardized test methods," she laments.

Tsuei explains that the biggest difference between Chinese and Western medicine is their different understanding and explanations of "*qi*" (energy). When diagnosing and treating imbalances in the body, the great majority of alternative medicine practitioners lay great emphasis on adjusting energy, which TCM practitioners speak of as "nourishing the *qi*."

Fortunately, in 1975 the German physician Rheinhold Voll invented a method of measuring biological energy. His instrument, called an electro-dermal screening device (EDSD), makes use of the property of the meridian system in guiding the *qi*, and for the first time converts the *qi* energy of the meridians into electrical energy that can be measured in order to diagnose disease. Western medicine previously did not acknowledge the existence of channels for "*qi*" in the human body. But since the development and clinical application of the EDSD, measurements on the surface of the skin can be used to detect the movements and distribution of energy within the body, from which one can deduce whether the structure and function of the various organs within the body are in balance, and also whether the patient is in a harmonious psychological state. These methods are called "bioenergetic medicine" or "information medicine."

The method of treatment currently used at the Clinic for East-West Medicine is to first use the EDSD to discover which bodily system has a problem, and then use pathology tests with reagents to make a detailed diagnosis. For instance, from the pancreatic meridian one can determine whether the levels of urea, insulin, cholesterol and triacylglycerols are too high or too low; while from the liver meridian system one can determine whether the patient has hepatitis or cirrhosis, or has been affected by pesticide pollutants.

Body, mind and soul

Proponents say that as well as measuring changes in bodily organs, bioenergetic medicine can also detect moods and psychological problems.

Ten years ago Julia Tsuei began to introduce the Bach flower remedies, which originated in Britain in the early 20th century. Different flower essences are directed toward different moods or psychological states, and at present practitioners use 300 different flower essences to address over 70 moods. Tsuei compares the process of selection to the way an office worker has to spend time in front of her wardrobe every morning choosing what color

clothes to wear that day. Many people may pick them according to whim, but when selecting remedies one has to use scientific methods to weigh this "intuition."

Flower remedies are not the same as aromatherapy, which is popular in Taiwan. As with other similar European remedies, the method used is to add a few drops of various flower essences to pure water and sip the mixture from time to time. Electromagnetic fields emitted by these concoctions are said to resonate with the imbalanced portions of the body and mind, to cause electrochemical changes in the brain and glands. This produces a self-healing effect that brings the emotions into balance.

Reflexology has found a widespread following among the Taiwanese public. In parks large and small one can find foot massage paths surfaced with small round pebbles.

Julia Tsuei believes that flower remedies are simply a way of reacquainting modern people with nature. "The greatest threat to people today, apart from pollution, is all kinds of stress. Physical illness can lead to psychological depression, but on the other hand a deeply depressed mood can also be the cause of disease," says Tsuei. Bioenergetic medicine aims not only to detect imbalances in organs and systems in the body, but also to find the causes of the disease, determine whether the causes of imbalance are physical or psychological, and eliminate toxins from the body. Tsuei says that only if it achieves all these things can it be called a holistic medicine that takes care of body, mind and spirit.

But there are still very few people with a background in mainstream medicine who, like Julia Tsuei, have developed an interest in alternative medicine and gone on to research it.

"I don't believe everything can be verified scientifically, because the concept of yin and yang and the five elements in TCM, and biology in Western medicine, are two completely different theories with a different vocabulary, structure and concepts," says Professor Ting Chihyin. TCM's and Western medicine's ways of understanding human health are still poles apart, but US research shows that the driving force behind the development of alternative medicine is not a desire for it to "replace" mainstream medicine, but rather for it to play a "complementary" role. The same is true in Taiwan: surveys show that 48% of people seeking alternative therapies have first visited an institution offering Western medical treatment.

In other words, although most patients seek alternative therapies because of the limitations of mainstream medicine, most of them do not regard alternative medicine as the only answer.

Joint guardians of health

It will still take a long time to establish just what specific curative effects alternative therapies have. But since the former Office of Alternative Medicine at the US National Institutes of Health was expanded into a "national center" in 1998, its annual research budget has grown hugely, from US$2 million then to US$150 million today. At present, 76 US medical schools offer courses in alternative medicine. In the international academic community, bioenergetic medicine organizations are holding frequent symposia, to take the first steps toward putting the field on a sound scientific footing. They have also begun active dialogue with practitioners of Chinese and Western medicine. It looks as if the health authorities in Taiwan ought not to completely ignore alternative medicine.

At swimming pools and health centers one can see powerful downward-pointing jets of water that are said to help relax neck and shoulder muscles. But to avoid neck injuries, they should be used with caution.

As for patients, whatever kind of therapy they are looking for, surely they also hope that the practitioners can look to their deeper psychological and spiritual needs, for only when body, mind and spirit all receive proper care can one achieve true health.

Translated by Robert Taylor

US-CHINA: QUEST FOR PEACE

Taiwan a deal-breaker for US security

Opinion by Henry C K Liu

The United States argues that the terms and validity of the 1982 communique—one of three documents setting forth the terms of the US-China-Taiwan relationship—depend upon assurances from the People's Republic of China (PRC) to resolve "the Taiwan question" by peaceful means only.

On July 14, 1982, the US gave Taiwan the Six Assurances that it:

- Had not agreed to a date for ending arms sales to Taiwan.
- Had not agreed to hold prior consultations with the PRC regarding arms sales to the Republic of China (ROC).
- Would not play any mediation role between the PRC and the ROC.
- Would not revise the Taiwan Relations Act.
- Had not altered its position regarding sovereignty over Taiwan.
- Would not exert pressure on the ROC to enter into negotiations with the PRC.

It has further been revealed in recent years that US president Ronald Reagan also secretly assured Taipei that if Beijing ceased its commitment to peaceful resolution of the Taiwan question, the August 17, 1982, US-PRC communique would become null and void.

As is well known, the PRC has never rejected the use of force as an option in resolving the Taiwan question, calling into dispute US good faith in signing the 1982 communique while secretly assuring Taiwan of a precondition. Moreover, the current US administration of President George W Bush has declared that there should be no unilateral change in the status quo by either party.

This policy entails three elements:

- Taiwan should not declare independence.
- Neither side should use force.
- Taiwan's future should be resolved in a manner mutually agreeable to the people on both sides of the Taiwan Strait.

In addition, the US has said it does not "support" Taiwanese independence. This means that Washington does not support Taiwanese independence unless all the people on both sides of the Strait agree to it, which in reality could mean never.

Bush flouts accord on Taiwan non-recognition

The Bush administration, in defiance of the basic precondition of non-recognition of Taiwan for normalization of US-China relations, also believes that Washington should maintain robust, albeit unofficial, diplomatic relations with Taipei, on the grounds that peace across the Taiwan Strait is an important US interest and Taiwanese actions, especially provocative ones, fundamentally affect US interests. Regular dialogue and contact with Taiwanese officials are rationalized as necessary to improve communications and limit political surprises. This is a flimsy excuse, since there are already adequate unofficial channels of communication between Washington and Taipei without ostentatious visits between government officials designed merely to boost Taiwan's official status.

The Bush administration has been clear that it expects the parties on both sides of the Strait to act responsibly in support of regional stability, as if they were equal parties. Furthermore, Washington continues to encourage dialogue between Beijing and Taipei on political as well as security issues. This is in contradiction even to Reagan's Six Assurances to Taipei, which maintained that the US would not play any mediation role between the PRC and the ROC and would not exert pressure on the ROC to enter into negotiations with the PRC.

The Bush administration also believes that the United States should assist in finding opportunities for greater international representation for Taiwan in such organizations as the World Health Organization. Its argument for this belief is that it is the right thing to do for the 22 million people of Taiwan, who deserve representation in the international community, especially on issues affecting their health, their economic welfare and the security of their planes and ships. Another reason is that the less Taiwan feels diplomatically isolated and the more it feels part of the international community, the less likely it will be dissatisfied with the status quo and the less likely to undertake provocative actions that could undermine peace and stability across the Strait.

But China only opposes Taiwan's participation in international organizations as an independent entity. Taiwan can participate in international organizations the same way Hong Kong does, as a highly autonomous part of China. But Taipei not only refuses to participate as Taiwan, China, it is even beginning to decline the appellation of Taiwan, Republic of China, and wants to be known only as Taiwan, thus turning the issue to one of independence.

More US moral imperialism directed at China

Finally, US policy encourages political liberalization on the mainland as the best hope for a peaceful resolution of the cross-Strait relationship. This of course is mere moral imperialism in the form of peaceful evolution. Political developments in China will respond only to internal Chinese needs, and are not undertaken to enhance US geopolitical interests. This policy of moral imperialism unwittingly has the counter-effect of deterring political liberalization in China by casting it as a movement against Chinese national interests.

Last June 1, Bush discussed Taiwan with Chinese President Hu Jintao on the sidelines of the Group of Eight summit of industrialized nations in France. In a press briefing later that day, a senior Bush administration official described Bush's comments, triggering speculation on whether US policy toward Taiwan had changed. Within a few days both Taiwan's cabinet spokesman Lin Chia-lung and American Institute in Taiwan chair Therese Shaheen said there had been no harmful change in US policy toward Taiwan.

On the other hand, the People's Daily noted on June 13 that after deviating from the policy of the previous six US administrations, Bush's Taiwan policy had now moved back to the mainstream. It described Bush's "non-support" as "opposition" to Taiwan independence and aligned the United States' "one-China policy" more closely with Beijing's "one China principle". Without directly correcting the Chinese interpretation, the same US senior official said: "On Taiwan, the president repeated our policy of a 'one China' policy based on the three communiques, the Taiwan Relations Act [TRA], no support for Taiwan independence … The president also said … if necessary, we will help Taiwan to the extent possible to defend itself."

Bush's remarks were widely interpreted as a softening in US support for Taiwan. This was the first time "no support for Taiwan's independence" had been elevated to the level of the TRA and the three communiques, and Bush did not mention peaceful resolution with the assent of the people on Taiwan. The April 2001 Bush promise of "whatever it took" to help defend Taiwan, committing the United States to maximum effort in Taiwan's defense, was replaced with the new wording, "If necessary, we will help Taiwan to the extent possible to defend itself" with two qualifications. The phrase "if necessary" is superfluous if it simply means "if China were to attack Taiwan" because US intervention presupposes Chinese military aggression. There is little question that if the People's Liberation Army (PLA) were to launch a blitzkrieg against Taiwan, the island could not alone repel the invading forces.

Taiwan would clearly need US assistance. China has always insisted on its right to the force option on the independence issue. Beijing in fact has put a still-unspecified time limit on peaceful resolution by warnings of military action if Taiwan refuses to acknowledge Chinese sovereignty indefinitely. The qualifying phrase "if necessary" can be interpreted as a US hedging against Taipei's opting unilaterally to make accommodations with Beijing. Similarly, the phrase "to the extent possible" could be a hedge against the eventuality of Taiwan's leadership caving in to PLA actions so quickly that US intervention would make no difference.

Would Taiwan really fight for independence?

Although not publicly acknowledged, Washington is reported to be steadily losing confidence in Taiwan's resolve to fight for its survival as a de facto independent nation. With economic integration with the mainland, Taiwan's economy is increasingly dependent on the motherland. Because of massive outflow of capital, technology and people from Taiwan to the mainland, the Taiwanese economy is being hollowed out of manufacturing. And yet all the island's political parties are intent on early implementation of direct links, further weakening Taiwan's desire for political independence and US-backed security.

After a US B-2 (mistakenly, the United States said) bombed the Chinese Embassy in Belgrade in May 1999 during the North Atlantic Treaty Organization military action in Kosovo, the administration of president Bill Clinton worked to rebuild its relationship with the PRC. Taiwan "president" Lee Teng-hui's July 9 statement that Taiwan and China should deal with each other on a "special state-to-state" basis upset the US State Department. While Washington and Beijing have restored some momentum to their bilateral ties with a landmark agreement on China's entry into the World Trade Organization, the relationship between Washington and Taipei continues to drift.

Lee's "state-to-state" announcement created serious strains between Taiwan and the United States. His provocative statement, so soon after the embassy bombing, could not have come at a more sensitive time for Washington, which viewed Lee's behavior as reckless. The Clinton administration delayed a trip to Taiwan by a Pentagon delegation and leaked to the media the possible scaling back of deliveries of F-16 spare parts, sending a clear signal that Taiwan could not go too far down the "state-to-state" road without risking its security ties to the US. Even some of Taiwan's longtime supporters in Congress criticized Lee's remarks as being unhelpful.

In June 1998, Clinton publicly stated the "Three Nos":

• No support for Taiwan independence.
• No support for "two Chinas".
• No support for Taiwan's participation in state-based international organizations.

Taiwan also sensed US pressure for an interim arrangement whereby Taipei would agree not to declare independence in exchange for Beijing's pledge not to use force. Faced with these developments, Lee chose to assert that Taiwan was already an independent, sovereign state and Beijing and Washington needed to deal with this reality. Nevertheless, it is a de facto sovereignty that depends directly on the US skirting official non-recognition of Taiwan.

Cross-Strait status quo unacceptable to Beijing

The interim agreement proposals from US academics and officials actually fanned misunderstanding. Assistant secretary of state Stanley Roth and other US diplomats suggested that an interim agreement might be useful for improving cross-Strait relations, in the spirit of buying time and stabilizing the status quo. But the status quo was a continuing state fundamentally unacceptable to Beijing.

Lee's "state-to-state" announcement was coached by Republican China hand James Lilley, a former Central Intelligence Agency director with close ties to Taiwan and former ambassador to China in the administration of president George H W Bush, in order to give the Clinton administration a diplomatic headache—and it went further than a declaration of "independence". In alluding to the German model of reunification, Lee identified Taiwan with West Germany, reuniting with a poorer and less developed mainland, as East Germany.

Taiwan realizes that it will need to enter into political discussion with the Beijing government sooner or later and the "state-to-state" formula is a strategy to reject Beijing's idea that it is merely a province. Meanwhile, as Beijing insists on Taiwan being a Chinese province, Taipei insists on being an independent state.

Taiwan's adoption of a "state-to-state" formula was triggered by a perceived softening of US support as the Clinton administration in its second term became more receptive to China's "one country, two systems" formula, as applied to Hong Kong, as a possible solution to the Taiwan issue. Taiwan's state-to-state formulation, in turn, increased US annoyance at Taipei acting as an unwelcome obstacle to US global geopolitical strategy that requires Chinese cooperation. Washington's desire to restrain Taipei's provocative behavior fueled Taiwan's anxieties, and caused Taipei to exert increasing independence even from Washington with the support of the right wing in US domestic politics. "Selling out Taiwan" became a campaign issue in the 2000 presidential election.

The Bush administration's robust support of Taiwan has been diluted primarily because of Taipei's failure to commit unequivocally to non-provocative acceptance of the status quo and secondarily because of America's need to elicit China's help in fighting global terror after September 11, 2001, in resolving the North Korean nuclear standoff, and in supporting (or not opposing) the United States on Iraq in the United Nations Security Council. This shift in the US stance is tactical, however, rather than strategic. The US still perceives geopolitical interests in maintaining the status quo on Taiwan. The recent appointment of Princeton professor and prominent sinologist Aaron Friedberg to the post of deputy national security adviser and director of policy planning on Vice President Dick Cheney's staff tends to support this interpretation (see The Struggle for Harmony Part 1: Myths and realities about China, June 13, 2003, and Part 2: Imagined danger, June 14).

Some see inevitable US-China confrontation

Friedberg sees potential long-term dangers that a modernized PLA may pose to US security and has written about the inevitability of a US-China confrontation. After the September 11, 2001. attacks, world geopolitics has become highly volatile. The US-Taiwan relationship is not exempt from the dynamics of changing international relations. After a private meeting in the White House with Chinese Premier Wen Jiabao last December 11, during Wen's state visit, Bush told the press: "We oppose any unilateral decision by either China or Taiwan to change the status quo. The comments and actions made by the leader of Taiwan indicate that he may be willing to make decisions unilaterally to change the status quo, which we oppose." Thus "oppose" has officially replaced "non-support" for Taiwan independence.

The area of converging US-Taiwan interests is shrinking because of changing geopolitics. The single most important consideration remaining is US concern that allowing Taiwan to fall back into the arms of the PRC by force or coercion could prove detrimental to US leadership in East Asia, particularly in terms of security arrangements with US allies, especially Japan and South Korea. Yet the test of maintaining security is in US diplomatic skill in avoiding war, not its war-winning capability. US allies in East Asia know that while a victory in war may protect US prestige, it would nevertheless leave their respective countries in ruins.

Such war-deterring diplomatic skill requires yielding to China on issues of its fundamental and vital national interest, such as the issue of Taiwan. Taiwan is a deal-breaking issue, as the American saying goes. Pushing the Taiwan issue to a military solution would represent a massive failure of US diplomacy in Asia. Even if the US 7th Fleet with its two carrier groups and the overwhelming force-projection capabilities from US bases in Japan should manage to thwart a PLA invasion of Taiwan, the Taiwan that was left after the bomb smoke clears would not resemble anything worth defending. On the other hand, even if China should succeed in regaining Taiwan through military conquest, the gain of a war-torn economy would be a Pyrrhic victory.

The Taiwan issue is a political issue, and all parties agree that it needs to be solved with political accommodation. Yet political options exist only within limits. Failure to exercise political options will lead to war. Clausewitzian concepts of war notwithstanding, war is not diplomacy by other means (Karl von Clausewitz, a Prussian military officer, wrote on military strategy). War is the product of failed diplomacy. The legacy of war is international hatred that fuels future wars while the legacy of diplomacy is international harmony that fuels stability. A war over the Taiwan Strait has no winners. All will lose more than they hope to gain.

Henry C K Liu is chairman of the New York-based Liu Investment Group.

THE MEN IN BLACK: HOW TAIWAN SPIES ON CHINA

BY WENDELL MINNICK

TAIPEI—Taiwan has long had an excellent reputation for gathering intelligence on China's activities. President Chen Shui-bian's declaration on November 30 that China had exactly 496 missiles aimed at Taiwan is a good example. He cited the numbers as justification for the forthcoming referendum, asking voters whether China should redirect its missiles. What was staggering to most analysts was not the exact number, but Chen's identification of the number of missiles per base.

Chen identified the number of operational short-range ballistic missiles (SRBM) within 600 kilometers: Leping base (96) and Ganxian (96) in Jiangxi province; Meizhou (96) in Guangdong province; and Yongan (144) and Xianyou (64) in Fujian province. Chen's office later pointed out that intelligence reports indicated that none of China's cruise missiles were available on the bases and that none of the missiles aimed at Taiwan were armed with nuclear warheads. The missiles were identified as four types of SRBM: DF-11 (M-11) DF-11A (Mod 2), DF-15 (M-9) and a DF-15 variant.

So how does Taiwan come up with such precise information?

The answer lies mostly with Taiwan's human intelligence resources, known as HUMINT. The Military Intelligence Bureau (MIB), jokingly referred to as the "Men In Black", recruits from a large pool of over 500,000 Taiwanese businessmen living and working inside China. Because Taiwan businessmen have invested around US$100 billion in China, this makes China hesitant to go around arbitrarily arresting them.

Expatriate Taiwanese are not the only source for the MIB. The mainland Chinese Global Times, published by the People's Daily, reported in 2001 that Taiwanese agents were recruiting local Chinese with sex, money and a third method dubbed the "democratic-justice" technique that appeals to those who feel victimized by Beijing government policies. A lopsided male-to-female ratio, low government salaries and disgruntled government employees in a corrupt bureaucracy, make China a spy recruiter's dream.

It was a disgruntled government employee who played the key role in China's biggest espionage scandal since the end of the civil war in 1949. In 1999 China executed two senior army officers, major general Liu Lian-kun and brigadier general Shao Zheng-zhong, by lethal injection for spying for Taiwan. Liu's motivation for spying was his bitterness at being unfairly impli-

cated in a corruption probe and thus denied a promotion. It is believed that shortly after the scandal, Liu traveled to Hong Kong in the early 1990s and met with the MIB station chief. Liu and Shao were arrested shortly after China's 1996 missile tests over the Taiwan Strait. They had provided information that the three missiles fired near the island were armed with dummy warheads, a fact that was reported in the press after then-president Lee Teng-hui publicly disclosed the information. It angered China into launching a full-scale probe, and Liu and Shao were quickly rounded up. Their deaths served as a reminder of the danger of sharing intelligence with the public.

When Chen Shui-bian made a similar disclosure exposing China's missile capabilities in November, China once again launched a dragnet to find those responsible. It is too early to tell whether they found their prey, but in December reports began to filter out of China of Taiwanese spy networks being swept up. Approximately 24 Taiwanese, and an unknown number of Chinese, were arrested in Guangdong, Hainan, Anhui, and Fujian provinces. One of them, Wang Chang-yung, told reporters from his Fujian jail cell that he had agreed to work for the MIB in 2002 due to financial problems. Another Taiwanese, Chang Keng-huan, also admitted to working for the MIB for money.

Unhappiness in the intelligence community over President Chen's public comment on the 496 missiles has been evident. To make matters worse, it weakened a long-established intelligence-sharing relationship between Taiwan and the US. Information is regularly shared with US intelligence agencies on a number of fronts, including signal intelligence. There is some speculation that Taiwan and the US also share information gathered by Taiwanese businessmen working in North Korea.

One complaint voiced by Taiwanese intelligence officials is that there has been a decrease in intelligence gathered on China since 2000, when Chen was elected as president. As a member—and now head—of the Democratic Progressive Party, Chen was a vocal supporter of an independent Taiwan before the 2000 election. Intelligence officials believe that one method of recruiting agents was the argument that they would be assisting with the eventual reunification of the two countries. With Chen's election came a new perspective that Taiwan was no longer pro-China or pro-unification.

China has also changed the focus of what it dubs a Taiwan spook. During the 1989 Tiananmen Square crackdown—a massacre of peaceful pro-democracy demonstrators—Chinese authorities blamed "Kuomintang agents" for inciting the protests. However, since Chen's election, the label has shifted to "Taiwanese spies". The propaganda move reflects the very real shift, both politically and culturally, of a new Taiwanese identity emerging on the island.

The Military Intelligence Bureau, or "Men in Black", has suffered some setbacks as well—not only in China, but also at home. In November 2003 former MIB officer Tseng Chao-wen and a current MIB officer, Chen Suei-chung, were arrested for spying for China. The two were accused of supplying China with intelligence that Tseng sent by ciphered messages from convenience store fax machines. Police discovered a codebook inside Tseng's home. MIB officials began focusing on Tseng after he made contact with MIB intelligence officers working in Beijing. Then in December another MIB employee, Pan Chin-yang, was arrested while supplying intelligence to China. All were motivated to spy for financial reasons, they said.

The problem with human spies is that they are erratic. Though they can provide details that electronic intelligence gathering devices cannot collect, information is difficult to verify. Taiwan's lack of electronic intelligence devices, compared to their extensive use in the West, only means that it must rely more on the human element. Another problem with espionage is that professional spies are predictable—but the covert world is full of amateurs.

Wendell Minnick *is the Jane's Defense Weekly correspondent for Taiwan and the author of* Spies and Provocateurs: A Worldwide Encyclopedia of Persons Conducting Espionage and Covert Action *(McFarland 1992). He can be contacted at janesroc@yahoo.com.*

Constitutional Diplomacy: Taipei's Pen, Beijing's Sword

ALAN M. WACHMAN

GROUPTHINK WITH CHINESE CHARACTERISTICS

Chen Shui-bian's plan to address deficiencies in the Constitution of the Republic of China (ROC) is, at most, merely a putative source of the escalating tension between the People's Republic of China (PRC) and the ROC. As with so much that emerges from the controversy about Taiwan's appropriate status, the apparent source of friction—in this instance, the constitution—is not really what the disagreement is about. While any change to the ROC Constitution may have symbolic import for both Beijing and Taipei, the dispute across the Taiwan Strait remains fundamentally territorial. The constitutional commotion is just a symptom of what has been an irremediable conflict about whether China encompasses Taiwan or not.

The reason Beijing views Chen's call to amend or rewrite the constitution as provocative is that the PRC leadership has convinced itself that Chen Shui-bian has a "timetable for independence," and that changing the constitution represents one giant step toward that objective.[1] In their minds, Chen is still adhering to the objectives laid out in the 1986 "Political Platform of the Democratic Progressive Party (DPP)" and the DPP's "Resolution Regarding Taiwan's Future" of May 1999. In those documents, the DPP stated a determination to "establish a new constitution drawn up to make the legal system conform to the social reality in Taiwan …" and proclaimed, "Taiwan is a sovereign and independent country. Any change in the independent status quo must be decided by all residents of Taiwan by means of plebiscite."[2]

Beijing fears that Chen Shui-bian plans to use the constitution as a way of foreclosing the option of Taiwan's unification with the mainland. Evidently, it makes little difference to the PRC leadership that Chen Shui-bian has insisted that he has no aim to adjust clauses in the constitution that deal with matters of sovereignty. The PRC also dismisses as disingenuous Chen's assurances that his objective is to rationalize governance by ridding the constitution of anachronistic impediments to democracy. From Beijing's

vantage it is not the constitution, per se, that is the problem, but Chen Shui-bian.

Beijing sees Chen as scheming to shatter the "status quo" in cross–Taiwan Strait relations with incremental steps toward a condition in which any prospect of unification by peaceful means is snuffed out. Beijing fears that Chen's machinations will result in assertions about the definition and governance of Taiwan as a sovereign state—expressed in constitutional terms—that will be put in a referendum before Taiwan's voters who are certain to embrace it as a legal avenue to self-determination.

Even though analysts in Beijing tend to believe that Chen will refrain from a formal declaration of independence, which the PRC has long identified as a casus belli, many PRC observers and officials expects Chen to shepherd his flock toward the endorsement of a constitutional framework that eradicates from the ROC Constitution vestigial institutional and semantic cues that Taiwan is only a part of China.

Using democratic consolidation as a guise, Chen is believed by Beijing to have embarked on a project of constitutional reform that will supplant the ambiguous, but temporarily tolerable, status quo with an unambiguous legal framework. Within such a framework, the PRC leadership is deeply worried that Taiwan—by whatever name—will depict itself as a sovereign state that is happily rid of structural reminders that it was once part of China. That, in Beijing's lexicon, would amount to "one China, one Taiwan," with which the PRC has always said it will not abide.

Beijing is also aware that Chen is likely to play up his devotion to democracy to garner sympathy abroad while casting the PRC as the dictatorial beast, once again seeking to throttle the popular will of a community that wishes to govern itself. In Hong Kong, the populace has embarrassingly exposed the PRC's commitment to the "one country, two systems" framework as premised on Beijing's role as arbiter of what occurs in both systems, vitiating that model for use with Taiwan and highlighting Beijing's insuperable authoritarianism.

Temperamentally disinclined to see sovereignty as flowing from the will of the governed, the leadership in Beijing sees democracy itself as a weapon in Chen Shui-bian's hands that he flouts with Washington and other sympathetic states in mind. The present leadership of the PRC knows, though rarely acknowledges, that the blunders of 1989 cost the PRC dearly and that its missile exercises of 1996 and other displays of distemper toward Taiwan brought few rewards. For that reason, Beijing is genuinely reluctant to be hoist again on its own authoritarian petard. However, the PRC feels Chen Shuibian—abetted by the United States—is rapidly backing it into a corner in which it will have no option other than a military response, if only to preserve its credibility.

Indeed, Beijing now seems to be girding up its loins and mobilizing public support for the eventuality of armed conflict, even if that means a confrontation with the United States and further setbacks to economic and social development in China. Whether this is an elaborate ruse concocted for the purposes of psychological warfare to intimidate Taiwan and jolt Washington into restraining Chen Shui-bian is not easy to discern. What is beyond doubt, though, is that Beijing has done nothing to signal its population that accommodating Taiwan's determination to remain autonomous may serve China's national interest, and enhance the possibility of eventual unification, better than a hostile, armed response will.

Beijing has done nothing to signal its population that accommodating Taiwan's determination to remain autonomous may serve China's national interest, and enhance the possibility of eventual unification.

Negotiating entails compromise. Considering that Beijing has not prepared its citizens to accept anything less from Taiwan than subordination to the "one country, two systems" formula, one may conclude that the leadership is not in an accommodating mood. The message that Beijing broadcasts at home and abroad is that what it defines as "separatism" must be crushed, what it defines as China's sovereignty and territorial integrity must be preserved, and that the PRC is prepared to eat whatever bitterness is thrown in its face in the furtherance of those abstract objectives.

Having been established at the top, these views have not been susceptible to much adjustment from below. The entire policy and propaganda apparatus of the PRC is, on this matter, appears locked into a form of groupthink with Chinese characteristics. While the PRC elite may have a nearly universal and reflexive commitment to unification, or at least an opposition to Taiwan's permanent independence,

one could detect in recent years a wide range of views about how best to handle the issue of Taiwan. Within the past year, though, pluralism has given way to consensus, or at least an enforced uniformity.

Persuasive speculation about the reasons why this has happened focused on the struggle between Hu Jintao, president of the state and head of the Chinese Communist Party (CCP), and Jiang Zemin, who formally retired from those posts in 2002 but retained the chairmanship of the Central Military Commission until he relinquished his hold on it in September 2004.

Now that Hu Jintao has assumed the chairmanship of the Central Military Commission and is characterized as having won the struggle with Jiang Zemin for succession, observers will be eager to see whether Beijing's posture toward Taiwan is adjusted. Even if the untidy transition to the fourth generation of leaders was not the principal cause of Beijing's increasing rigidity and belligerency toward Taiwan, it remains the case that ranks seem to have been closed on a policy of stridency and determination to derail what is imagined to be Chen Shuibian's intentions to steer Taiwan to independence by 2008. Although Hu undoubtedly differs from Jiang in style and may establish a different order of policy priorities, there is no evidence as of this writing that he would treat the cross-Strait relationship differently than did Jiang Zemin and his leadership cohort.

INTENTIONALLY (MIS)READING BETWEEN THE LINES?

The PRC's increasing hostility toward Taiwan generally and Chen Shui-bian particularly is the outgrowth of a process that has been unfolding for years. It began to intensify when former president Lee Teng-hui called on "the people of Taiwan to forge a new constitution and to change the nation's designation from the Republic of China to Taiwan" at the World Taiwanese Congress held on March 15, 2003.[3] Over the summer of 2003, Lee amplified his call for a new constitution and a new national title by saying that the ROC no longer existed. By contrast, Chen Shui-bian seemed eager to make the point that the ROC continues to exist, and that "It is not disputed that I was sworn in as the ROC's 10th president on May 20, 2000, in line with the nation's constitution."[4] Indeed, Chen can often be heard to refer to the Constitution of the ROC and his obligation to serve under it as president of the ROC.

However, on September 28, 2003, Chen Shuibian called for a new constitution when he addressed a convocation of Democratic Progressive Party (DPP) members celebrating the party's 17th anniversary. Chen explained why a new constitution was warranted, by focusing on problems in the constitution that impeded the consolidation of a rational democratic system. On the day after Chen's speech, Chiou Yi-jen, then the secretary-general of the presidential office, stated that "drafting a new constitution does not necessarily mean changing the country's official title" and the presidential spokesman clarified that Chen's call did not invalidate the "five no's" in his first inaugural address.[5] Still, the PRC is convinced that objectives laid out in the DPP platform of 1986 continue to control Taiwan's ruling party and President Chen. Having decided that Chen is entirely untrustworthy, opposing what it sees as efforts by Chen to de-Sinify Taiwan, knowing Chen rejects the concept of "one China," and in the context of ever-more assertive calls by Lee Teng-hui for a Republic of Taiwan, the PRC has blurred the distinction between Chen's call for constitutional reform and Lee's calls to devise a new constitution for a state called Taiwan.

Within days of Chen's remarks about a new constitution, the PRC initiated a campaign to characterize Chen's announcement as having insidious motives. Guo Zhenyuan, a research fellow with the China Institute of International Studies, was quoted in a report carried by the *Xinhua News Agency* as accusing Chen of saying something he had not said. Guo stated that

Chen also showed his "true colors," contradicting his 2000 promise of no declaration of Taiwanese independence, no incorporation of the "two states" remarks into the constitution, no change of the so-called country's name, and no referendum on Taiwan independence during his tenure as leader.[6]

Guo did not explain how he arrived at this conclusion. Similarly, a report in Hong Kong's *Ta Kong Pao* made the connection between Chen's timetable for a new constitution and a "timetable for independence," although it did not specify why it viewed these as the same. The article stated, in part,

This is probably the first time Chen Shui-bian flagrantly set a timetable for promoting "Taiwan's independence." It is also his official call for "Taiwan's independence," provoking all Chinese people, including Taiwanese compatriots. Chen Shui-bian's mentality of "rushing independence" has once again been thoroughly exposed.[7]

From that point forth, the die was cast. Protestations from Chen notwithstanding, the PRC persisted in castigating him for operating from a timetable for independence. As Taiwan's presidential election campaign heated up and as Chen mismanaged the activation of the referendum law, Beijing saw in each step ever more reasons to believe that its analysis of Chen's intentions was accurate. Chen and his administration were, at times, diplomatically injudicious and could have allayed the PRC's anxieties more persuasively by preserving with greater effect the potential for accommodation between Beijing and Taipei. However, Chen has not said anything about a "timetable for independence" and did repeatedly specify many administrative and procedural problems that warranted constitutional change and rationalization. He has also reiterated his commitment to the five no's.

By the fall of 2003, the PRC was operating more on the basis of its anxieties than in response to Chen's own assertions. Beijing had worked itself up into such a fever about

the possibility that Chen would press for a new constitution ratified through a referendum that Hu Jintao evidently came down hard on the Central Leading Group for Taiwan Affairs for its failure to address the Taiwan issue more effectively. According to a report in Hong Kong's *Hsin Pao*, President Hu "pointed out that the group 'did not have a strong will, failed to present clear-cut viewpoints, and did not adopt a firm attitude' in its communication with the U.S. side" about the Taiwan issue. The article makes clear that from that moment, the PRC leadership expected that a referendum, if held, would be the leading edge of a campaign for independence. The formula was clear. The PRC now believes that a new constitution will be drafted in 2006 and implemented, after receiving support through a referendum, in 2008 as Chen's tenure in office ends. On November 22, 2003,

Hu Jintao and Wen Jiabao unprecedentedly convened a Central Committee Politburo meeting to specially discuss the situation in the Taiwan Strait ... The meeting maintained that if no limit was set upon the so-called "plebiscite on legislation" pushed for by Taiwanese authorities, they would have obtained a "legal source" (*fa yuan*) for Taiwan's independence; and if the proposal to "codify the referendum into the constitution" was formally adopted, Taiwan would become a completely independent country.[8]

It is not evident why Hu and Wen regard the amendment of the ROC Constitution as a legal source for Taiwan's independence when they do not regard the ROC Constitution in its present state as a legal source for ROC sovereignty over all China. Indeed, one wonders whether they are bothered more by the potential humiliation that would come from a plebiscite that might expose the lack of commitment by Taiwan's voters to unification or to the establishment of constitutional language reflecting that view. In any case, ever since November the referendum and constitution have been perceived by the PRC leadership as components of a scheme toward independence. The reelection of Chen Shuibian in March 2004, after a most contentious election, was undoubtedly a severe blow to the hopes of the PRC. One response was to pre-empt Chen's inaugural address, punctuated though it was by conciliatory phrases, by issuing a statement on May 17, 2004, three days before the inauguration. In it, the PRC leadership articulated two stark options for Taiwan: acceptance of Beijing's demands or war. The statement read, in part,

The Taiwanese leaders have before them two roads: one is to pull back immediately from their dangerous lurch toward independence, recognizing that both sides of the Taiwan Strait belong to the one and same China and dedicating their efforts to closer cross-Strait relations. The other is to keep following their separatist agenda to cut Taiwan from the rest of China and, in the end, meet their own destruction by playing with fire.[9]

BEIJING'S ARAFATIZATION OF CHEN SHUIBIAN

The paradox, here, is that although the PRC claims it will be mollified if the "Taiwan authorities" mouth the magic "one China" words, the PRC has already decided that it cannot trust Chen Shui-bian and that nothing he says is credible. In that respect, Beijing has painted itself into a corner by marginalizing Chen Shui-bian in a manner that mirrors what Israel under Ariel Sharon seems to have done to Yasir Arafat. One element of this marginalization is to discount what Chen says and to recast it in ways that conform to pre-existing views of what Chen really means. For instance, following Chen's inaugural address of May 20, 2004, *Xinhua News Agency* reported "*Renmin Ribao* Publishes Signed Article, Pointing Out that Chen Shui-bian's May 20 Speech is 'Taiwan-Independence' Rhetoric with a Hidden Agenda." The author reviewed Chen's speech, and commented that although Chen did not use the phrase *yibian, yiguo* [each side, one state], "the whole speech was full of the notion of Taiwan as an 'independent country.' He was actually continuing to advocate 'each side, a (separate) country.' This showed that he had not given up his separatist 'Taiwan-independence' stance." Although Chen's speech did not call for a new constitution to be adopted by referendum, the author asserts that Chen's words "hid connotations of continuous efforts to promote the 'drafting of a new constitution for Taiwan's independence' in 2008." Deconstructing other phrases in Chen's speech, the author concludes, "These few words unmistakably showed what Chen Shui-bian was actually thinking deep down."[10] It is startling that otherwise well-regarded and highly-placed analysts and scholars working at think tanks and research institutions associated with the PRC government express themselves in terms of this rationale: Chen Shui-bian does not mean what he says, but we know what is really in his heart.

The PRC leadership has succumbed to a self-induced panic on the basis of what it imagines Chen Shui-bian really intends by the words it routinely dismisses. Anger and frustration have left Beijing deaf to rhetorical openings from Chen on which it might have built accommodation, and blind to olive branches offered by Taipei to calm roiled waters.

Analysts in the PRC assert that—from Chen's past association with the Taiwan independence movement—it is possible to know what he will do in the future. In its myopic focus on the intentions of the individual leader, rather than on the role of the individual within a broader political system that exercises constraints on ambitions, the PRC ignores the difference between an individual's dreams and the political posture of a person who holds office. Regardless what may be in Chen's heart, he is obliged to function in a political environment in which he has only the power to persuade his constituents and, in any event, must work with a vigorous opposition that will do anything in its power to cut him off at the knees. Moreover, the PRC's view of Chen and the DPP as chugging along on a predetermined path toward independence unmindful of political realities—chief among them

the PRC's belligerence—fails to take note of slight moderation in the DPP's stated ambition. While there are plenty of people in and out of the DPP who probably would like for Taiwan to remain independent if there were no significant costs to doing so, even the DPP has bowed to the need for greater prudence. On its website, the DPP offers the view

> As for the issue of sovereignty, since under present international conditions, it is impossible for either side across the Strait to compromise on this matter, the DPP prefers to avoid discussion with China on this sensitive yet contentious topic while dealing with the more practical and functional matters first.[11]

Beijing's preoccupation with the national leader reflects not just a lack of trust in Chen Shui-bian, but a fundamental misperception of the democratic process. Even in the authoritarian system he commands, Hu Jintao is not at liberty to dictate policy without regard to powerful constituencies. Yet, the PRC leadership seems to regard Chen Shui-bian as acting without concern for popular will or institutional checks that limit his power to simply declare independence—or agree to unification under the "one China" banner—by fiat. Beijing has neglected or discounted the fact that most respondents to public opinion polls in Taiwan express no interest in "independence" and prefer some version of the "status quo." They also seem to minimize the institutional and procedural hurdles that must be overcome to make any change to the constitution.

THE ROC CONSTITUTION:
IF IT AIN'T FIXED, DON'T BREAK IT

The focus on constitutional change as a means of establishing independence is a red herring. Chen Shui-bian and his predecessor, Lee Teng-hui, have repeatedly stated their views that the ROC is already a sovereign state that is independent of the PRC and has been one since 1912.[12] Moreover, the ROC Constitution of 1947 by which Taiwan has been governed contains numerous clauses that underscore the sovereignty of the ROC and might be construed to challenge that of the PRC. While the PRC is enraged by the prospect that Chen will use constitutional reform as a way of asserting Taiwan's sovereignty and rally popular opinion to what it sees as Chen's crusade, there are already ample causes for Beijing's ire in the outdated document that Chen has been seeking to amend.

That Beijing would be so sensitive to the prospect of a new or revised constitution seems odd, considering that it denies that the ROC still exists. The PRC declared that the ROC was "overthrown" in the civil war of the late 1940s, and while it refers to the island as Taiwan, not the ROC, it has tolerated Taiwan existing as the Republic of China.[13] Ironically, even though Beijing refers to Taiwan as "Taiwan," it is unnerved by the prospect of Taiwan referring to itself as "Taiwan." Beijing is evidently more at ease when Taiwan refers to itself as the ROC because that name, at least, implies a connection to China.

That is not to say that the PRC is satisfied with Taiwan remaining autonomous, but it took some comfort from the long-standing KMT position that held open the possibility of unification. This is why the real cause of the present escalation of hostilities is not the constitution, but Beijing's perception of Chen Shui-bian's intentions. If Beijing believed that Chen had a genuine determination to work toward some form of accommodation on the matter of Taiwan's status, everything else he has done or pledged to do would appear much less threatening to the PRC's core interest.

As it stands, the PRC's view of the ROC Constitution seems to be that "if it ain't fixed, don't break it." That is, even though the PRC has long asserted that the ROC no longer exists as a sovereign state and even though the ROC Constitution of 1947 may be offensive in the eyes of the PRC, Beijing would prefer that Chen did not tamper with it, leaving a document that is bad enough alone. Any changes to the constitution draw attention to three discomforting realities:

(1) Beijing has failed to lure the people of Taiwan to see merits in unification;

(2) the people of Taiwan—either by legislative means or by referendum—are capable of affirming their autonomy and intentions to remain apart from China; and

(3) the PRC has failed to deter by threats of force what it deems to be separatist activity.

That leaves Beijing having either to swallow its pride at the cost of its credibility, or to make good on its threats to use force.

An elevation in hostility during the summer of 2004 stemmed in part from the announcement of the Dongshan military exercises that were to be conducted in September by the PRC, with a domination of Taiwan's airspace as the announced objective, as well as in other sharp rhetoric directed both at Taipei and Washington.[14] Although the exercises were reportedly cancelled at the end of August, the antagonism has not abated. The PRC leadership continues to see Chen Shui-bian and his efforts at constitutional "re-engineering" in the most threatening light, to ignore signals to the contrary, and to express resolve to match Chen's efforts, step by step, insuring that as he progresses along his timetable, Beijing will ratchet up its pressure on Taiwan.

With all parties donning their most unyielding masks of indignation and determination, and none now prepared to conciliate, the grave situation threatens to move beyond a point at which last minute solutions can be effective.

In the end, though, Beijing has signaled that if it must, it will put an end to this race and that it will ensure the situation is resolved on its own terms, not Chen's. Ominous warnings from Beijing have prompted Washington to dig in its heels, too, with the usual round of Congressional chest thumping about the sanctity of the Taiwan Relations Act. The PRC has used meetings with visiting U.S. Vice President Dick Cheney and National Security Adviser Condoleezza Rice and a September meeting between Secretary of State Colin Powell and Foreign Minister Li Zhaoxing to underscore the severity of the unfolding situation. With all parties donning their most unyielding masks of indignation and determination, and none now prepared to conciliate, the grave situation threatens to move beyond a point at which last minute solutions can be effective. Dialogue and incentives to accommodate are urgently needed.

Despite the bellicose tone of recent pronouncements from Beijing and Washington and a lively debate on Taiwan that results in what the PRC will see as further provocation, accommodation should still be possible if Taiwan were to act in ways that would persuade Beijing that it has not foreclosed on some types of association with China. Of course, there are vocal political constituencies in Taiwan who oppose any association with China, whether it be the PRC or some other Chinese state. That they cling to a vision of absolute and permanent separation, though, makes them a minority. A far greater number of respondents to public opinion polls on Taiwan have expressed a preference for the elastic concept of the "status quo," and a healthy segment of those people are open to the possibility of association with China under appropriate conditions. Others might be persuaded that an association with China merited consideration if the incentives were structured persuasively, especially as those generations that act now in reaction to the suppression they suffered at the hands of the authoritarian KMT pass power to those generations that have known nothing but democracy in Taiwan.

This leads to a role for the United States. Perhaps it is time for Washington to consider how its own approach to this enduring dispute has impeded accommodation by both Beijing and Taipei. The sympathy Americans may feel for Taiwan's plight in the shadow of PRC threats has prompted policies and postures intended to support and defend Taiwan that have enraged Beijing and, paradoxically, may be contributing to an erosion of Taiwan's security. Beijing may be correct in seeing the robust military assistance that Washington has made available to Taipei as encouraging many in Taiwan to feel complacent about the likelihood of a PRC military assault and dismissive of Beijing's persistent rhetorical belligerence. Political actors in Taiwan who fail to take seriously enough the threat of the PRC may, indeed, have been emboldened to take steps that have contributed to an escalation of cross-Strait tensions. These, in turn, inflame Beijing's sense that it must act even more assertively if only to avoid being seen in Taipei and elsewhere as a "paper tiger."

One approach to disrupt this potentially destructive cycle is to urge Taipei to see that it cannot be independent so long as it is perpetually dependent on Washington to protect it from a military assault by Beijing. In that sense, Taiwan really cannot be independent unless it has the capacity to defend itself or persuade Beijing that what Taiwan experiences as independence may be labeled by Beijing as the endorsement of some version of the "one China" principle. Taiwan now seeks "peace, stability, and development." These goals can only be attained if Beijing's belligerence can be quelled. An enduring condition of peace, stability, and development—to say nothing of independence—cannot be established and sustained if it must depend on a perpetual commitment of security by the United States.

Washington's long-standing good will does not mean Taiwan can expect the American security umbrella to remain extended indefinitely. Selling arms to Taiwan is not a long-term solution, it is a stop-gap measure. However, if Taiwan sees the U.S. commitment as permanent there is little reason for it to apply itself in search of a path to some mutually acceptable accommodation. If it understood American support as coupled to an expectation that Taipei act responsibly in concert with U.S. interests, it might exercise greater restraint and creativity without Washington's prompting. A failure by Taipei to balance its determination to remain autonomous with Beijing's determination to have Taiwan see itself as part of some entity called "China" does not appear to serve well the U.S. interest in avoiding the disruption of peace in the Western Pacific. Similarly, Beijing's unyielding view of Taiwan's status does little to take account of the realities of a burgeoning and distinct political identity on the island and the potential for amity and mutual benefit that would flow from a looser expectation of how the two sides of the Taiwan Strait ought to be bound.

For the solution to this problem to be peaceful, there is only one avenue: compromise. However, not only will Beijing and Taipei each have to compromise on matters of principle, but Washington will, too. Adjusting the posture that the United States has taken toward Taiwan will not be easy. For one thing, selling weapons to Taiwan makes Americans feel good about arming a democratic David to defend itself against an authoritarian Goliath. No right-thinking American political leader wants to take any step that would degrade Taiwan's security or cause Beijing to doubt Washington's inclination to intervene on behalf of Taiwan should the cross-Strait controversy erupt in violence. However, the urge to stay the course must be reconciled with the recognition that absent divine intervention on David's behalf, Goliath was much more likely to prevail.[15] Washington should consider how American weapons sales, military coordination, and other forms of support to Taiwan threaten and infuriate the Chinese Goliath, driving him to battle. Considering how frequently the PRC raises with American officials its objections to U.S. arms sales to Taiwan, it may be worth exploring with interlocutors from Beijing what the PRC is prepared to "give" in order to "get" the United States to modify what it offers to Taipei.

On the matter of the constitution itself, Washington would be correct to ensure that Taipei has no doubt that fiddling with clauses pertaining to sovereignty, territory, or national title—even if they are expressed as "clarifications"—would be seen in Beijing as provocative, would do little to advance the goals of democratic consolidation that Chen Shui-bian has articulated as reasons for constitutional re-engineering, and are therefore both risky and insupportable. This is not a comfortable stance for Washington to adopt, as it suggests that the United States is less committed to allow a democratic process to take its course in Taiwan than it is to avoid the provocation of Beijing. If Taiwan were prepared to bear alone the consequences of triggering a PRC assault, then Washington's view of where prudent self-restraint must trump exuberant national self-expression would be irrelevant. Considering the expectation in Taiwan that the United States be on call to intervene whenever the PRC erupts in hostility, Washington is not amiss in stating its objections to symbolic actions by Taipei that could have devastating costs and Taipei would not be amiss in heeding those warnings.

ENDNOTES

1. "'New Constitution' Means Timetable for Taiwan Independence: Official," *Xinhua News Agency*, April 14, 2004.

2. See "China Policy" on the website of the DPP (reproduced at `http://203.73.100.105/english/pub/LIT_12.asp?ctyp=LITERATURE&catid=1979&pcatid=0`) and "Establishing of a Sovereign and Independent Republic of Taiwan," the first article in the "Party Platform of the Democratic Progressive Party" (reproduced at `http://203.73.100.105/english/pub/LIT_1.asp?ctyp=LITERATURE&pcatid=1975&catid=2121`); "Resolution Regarding Taiwan's Future Passed by the National Party Congress (8th Term, 2nd Meeting) May 8, 1999," (reproduced at `http://www.gio.gov.tw`).

3. "Former President Calls for Formation of New Constitution for Taiwan," *Central News Agency*, March 15, 2003.

4. Sophia Wu, "Chen Says He Assumed the Presidency under the ROC Constitution," *Central News Agency*, August 27, 2003.

5. "KMT Accuses President Chen Shui-bian of Planning Independence Push," *Agence France Presse*, September 29, 2003; "Taiwan Official Says Call For New Constitution Not to Affect 'Five Noes' Pledge," *Central News Agency*, September 30, 2003.

6. "PRC Expert Guo Zhenyuan Calls Chen Shui-bian 'Hopeless Taiwan Independence Element'," *Xinhua News Agency*, September 30, 2003.

7. "TKP Column Blasts Chen Shui-bian's Proposed Timetable for 'Taiwan Independence,'" *Ta Kung Pao*, September 30, 2003.

8. "Wen Jiabao Seeks U.S. Help to Contain Taiwan Independence," *Hsin Pao*, December 11, 2003.

9. "Taiwan Affairs Office of CPC Central Committee, Taiwan Affairs Office of State Council Are Authorized to Issue a Statement on Current Cross-Straits Relation," *Xinhua News Agency*, May 17, 2004.

10. "*Remin Ribao* 11 June Signed Article Refutes Chen Shui-bian Inaugural Speech as Rhetoric," *Xinhua Domestic Service*, June 10, 2004.

11. "China Policy" as presented on the website of the DPP.

12. While Lee Teng-hui seems to have dissociated himself with that view since he left the presidency, Chen Shui-bian has not.

13. People's Republic of China, *White Paper on the Taiwan Question and Reunification of China* (Beijing: The State Council, August 1993).

14. "Three purposes of military maneuver at Dongshan Island," *People's Daily*, July 19, 2004.

15. In the Bible, David rejected King Saul's offer of armaments and went into battle dressed as he was, armed only with a staff, a sling and five smooth stones. David warned Goliath that he came in the name of the Lord and "on this day, the Lord will deliver you into my hand."

Alan M. Wachman is associate professor of international politics at the Fletcher School of Law and Diplomacy, Tufts University. The author is grateful to Bonnie Glaser for a most helpful response to an earlier version of this essay.

They Can't Handle the Truth

Taiwan's media go all out for a story, even if the facts aren't there.
Reformers don't have much clout in a culture that's so freewheeling.

By Mark Magnier
Times Staff Writer

TAIPEI, Taiwan—When Sir Elton John arrived here shortly before midnight in a bright blue track suit and dark glasses, he was greeted at the airport by local reporters who jostled him, slammed cameras in his face and barked questions.

The pop star tried to hide but was soon flushed out and started yelling obscenities.

Not known for taking an insult lying down, the Taiwanese journalists yelled back. Some suggested that he consider going elsewhere.

"We'd love to get out of Taiwan if it's full of people like you. Pig! Pig!" the knighted entertainer screamed last fall.

"The television and the photographers at the airport were the rudest I have ever met, and I've been to 60 countries," John said at his piano bench at a concert a few hours later. "I'm sorry if I offended anyone in Taiwan, I didn't mean to. But to those guys, I meant every word."

Celebrity histrionics aside, Taiwan's media have the reputation of being among the most aggressive in Asia. In a region where print and broadcast reporters are often de facto cheerleaders for governments and billionaires, Taiwan's no-holds-barred journalism is alternately seen as a gutsy check on authority and the embodiment of chaos.

Concerned about the media's excesses and ability to ruin reputations and lives, reformers in and outside the industry are trying to stem the sensationalism, partisanship and corruption that characterize the business. Some argue that the media are merely a reflection of Taiwanese society, which is one of the most freewheeling in Asia.

Foreign luminaries aren't the only ones trying to hide from the island's aspiring Woodwards and Bernsteins, who've been called man-eaters, bloodsuckers and worse. Several years ago, when Taiwan's then-vice president and prime minister, Lien Chan, gave his traveling herd of reporters the slip on a trip to the Dominican Republic and secretly traveled to Ukraine, newspapers summoned all their troops to search for him.

A few months later, then-Foreign Minister John Chang pulled a similar Houdini act during a visit to South Africa. Hounded by angry reporters when he returned to Taipei after a stealth visit to Belgium, Chang defended himself with what is now known here as the "rice cooker" theory of diplomacy. Making policy while one is barraged by reporters, he said, is like trying to boil rice with someone constantly lifting the lid.

Wary of angering those who buy ink by the barrel, however, he quickly apologized and begged the scribes' forgiveness.

The media's willfulness had a deadly outcome, or so some charged, when the daughter of television star Pai Ping-ping was kidnapped a few years ago. The singer criticized the media for following the family in cars, vans and helicopters, even hounding it during the ransom drop.

"Were you helping me or hurting me?" Pai asked at a news conference.

When her daughter was found dead, the accusations grew more pointed. "Reporters are guilty!" screamed placards hoisted by neighbors around Pai's house.

Journalists showed little remorse, citing pressure from their editors.

"If you fail to get this story, jumping from the 14th floor is too good for you," an editor at the United Daily News was quoted—in a well-cited essay on media reform—as saying during a meeting on the newspaper's 14th floor. "You should climb up to at least the 20th floor and jump from there."

In a market of 23 million people, Taiwan has six 24-hour television news channels, 4,185 magazines, 172 radio stations, 135 cable TV channels, 2,524 newspapers and 977 domestic news agencies, the government says. The desperate struggle for ratings results in stories on sex, murder, corruption and kidnappings and not much else, critics charge.

Kuan Chung-hsiang, a journalism professor at Shih Hsin University in Taipei, recounted that one of his top students landed a job at a local TV station but quit a few months later. She'd been told to wear a short skirt and to walk over a hidden camera positioned in a drain for an "investigative" piece about how hidden cameras all over Taiwan were secretly recording lewd scenes. The station couldn't find videos of lewd scenes, so it was staging one.

When the former student strongly objected, Kuan said, her boss asked her, "Do you want conscience or do you want ratings?"

Part of the Taiwanese media's character reflects its evolution, what some refer to as the transition from lapdog to mad dog. Until 1988, major newspapers and TV stations served as government mouthpieces controlled by the ruling Nationalist Party, which had maintained an iron grip for decades.

Less government control has led to privatization, but several important stations are still owned by political parties. In a polarized society where politics is a blood sport—fistfights in the legislature were not uncommon up until a few years ago—media objectivity is spotty at best.

President Chen Shui-bian's ruling Democratic Progressive Party has its own tools to manipulate the media, and, some say, the truth. "The Taiwan media is truly scandalous in its behavior," said Bonnie Glaser of the Center for Strategic and International Studies in Washington. "But the government often joins in. None of them have any scruples."

Journalism watchdogs cite a $250-million budget for "persuading" stations to invite generals and other people the government wants on talk shows, to write dramatic scripts favorable to its policies and to otherwise promote its agenda.

"The Taiwanese government has been doing dollar diplomacy so long overseas, it thinks it's natural to do it at home," said Hu Yu-wei, a journalism professor at National Taiwan Normal University, referring to the government's practice of paying other governments to give it diplomatic recognition.

Small payoffs to journalists for favorable treatment—hardly unusual in many Asian cultures with strong gift-giving traditions—remain a problem, although media experts say the practice is on the wane.

When Lu Shih-hsiang, a professor and head of Taiwan's Foundation for the Advancement of Media Excellence, offered a course on media ethics two years ago, none of his journalism students signed up. Asked why, several said they didn't want to become "schizophrenic," constrained by boring niceties that had no place in the real world.

Double-checking information is a rarity at many news organizations, as are corrections. Reporters acknowledge big rewards in gaining an edge over competitors and little cost for getting it wrong. There's no tradition of libel suits.

"Many reporters don't check their facts," said Chen Chao-jen, a senior reporter with the TVBS network who recounted a story about a bombing in Taipei. Competitors ran a report on their 9 o'clock news saying authorities had arrested a suspect.

"I told my boss it was wrong, but he said write it anyway," Chen said. "Then at 10 o'clock, everyone runs a story saying he's not a suspect."

During last year's presidential election, stations raced to get the results first. Some reported that the Nationalists had garnered 8 million votes. After it was reported that only 6 million people had actually voted, the stations, embarrassed by their error, withheld results and announced that the data had simply stopped coming in.

In a part of the world where national politicians are rarely challenged, however, Taiwanese reporters are as confrontational toward their leaders as they are toward pop stars. Information flows freely, some of it true. Soon after taking office in 2002, Douglas Paal, director of the American Institute in Taiwan, the de facto U.S. embassy, fretted about how quickly sensitive information leaked out.

Taiwanese media enjoy some of the world's strongest press freedoms, according to 2004 surveys by watchdog groups Reporters Without Borders, the International Press Institute and Freedom House.

Some continue to hope, however, that the media will adopt meaningful changes.

"From political shows to news and entertainment, local television programming as a whole is terrible in the extreme, indeed," Lin Chia-lung, an official in the government information office, wrote in an essay urging reform. "The restructuring of the media environment and institutions has become essential."

Calls for change have been growing amid concern that partisanship and excessive commercialism are undermining the media's ability to inform people. A new law requires all political parties to divest their media holdings by the end of this year. The media excellence foundation has encouraged citizens to boycott irresponsible outlets and start filing libel suits. People whose reputations have been besmirched are starting to win verdicts.

Other proposals to improve programming and accountability are also under discussion, including a public network modeled after Britain's BBC or Japan's NHK that would be funded by the government or subscriber fees.

In 2003, the government decided to scrap a program that rated Taiwan's six largest Chinese-language newspapers for accuracy and objectivity after critics charged that it was pursuing its own agenda under the guise of neutrality. In response, policymakers called for better self-policing.

How quickly reforms take hold remains to be seen, but some observers believe that the media are a reflection of broader social forces.

"We have a poor democracy and a poor media," said Chen Hao, senior vice president of CTI Television. "Taiwan is unstable and partisan, and we need to find a middle ground. There's no easy solution."

Until then, celebrities and politicians will have to contend with the media's bulldog tactics. Said Chen, the television reporter: "We give them popularity. That's the price they have to pay."

Special correspondent Tsai Ting-I in Taipei contributed to this report.

Glossary of Terms and Abbreviations

Ancestor Worship Ancient religious practices still followed in Taiwan, Hong Kong, and the People's Republic of China. Ancestor worship is based on the belief that the living can communicate with the dead and that the dead spirits to whom sacrifices are ritually made can bring about a better life for the living.

Brain Drain A migration of professional people (such as scientists, professors, and physicians) from one country to another, usually in search of higher salaries or better living conditions.

Buddhism A religion of East and Central Asia founded on the teachings of Siddhartha Gautama (the Buddha). Its followers believe that suffering is inherent in life and that one can be liberated from it by mental and moral self-purification.

Capitalist A person who has capital invested in business, or someone who favors an economic system characterized by private or corporate ownership of capital goods.

Chinese Communist Party (CCP) Founded in 1921 by a small Marxist study group, its members initially worked with the Kuomintang (KMT) under Chiang Kai-shek to unify China and, later, to fight off Japanese invaders. Despite Chiang's repeated efforts to destroy the CCP, it eventually ousted the GMD and took control of the Chinese mainland in 1949.

Cold War A conflict carried on without overt military action and without breaking off diplomatic relations.

Communism Theoretically, a system in which most goods are collectively owned and are available to all as needed; in reality, a system of government in which a single authoritarian party controls the political, legal, educational, and economic systems, supposedly in order to establish a more egalitarian society.

Confucianism Often referred to as a religion, actually a system of ethics for governing human relationships and for ruling. It was established during the fifth century B.C. by the Chinese philosopher Confucius.

Contract Responsibility System A system of rural production in which the land is contracted by the village to individual peasant households. These households are then responsible for managing the production on their contracted land and, after fulfilling their production contracts with the state, are free to use what they produce or to sell it and pocket the proceeds. Such a system has been in place in China since the late 1970s and has replaced the communes established during the Maoist era.

Cultural Revolution Formally, the Great Proletarian Cultural Revolution. In an attempt to rid China of its repressive bureaucracy and to restore a revolutionary spirit to the Chinese people, Mao Zedong (Tse-tung) called on the youth of China to "challenge authority" and "make revolution" by rooting out the "reactionary" elements in Chinese society. The Cultural Revolution lasted from 1966 until 1976. It seriously undermined the Chinese people's faith in the Chinese Communist Party's ability to rule and led to major setbacks in the economy.

De-Maoification The rooting-out of the philosophies and programs of Mao Zedong in Chinese society.

Democratic Centralism The participation of the people in discussions of policy at lower levels. Their ideas are to be passed up to the central leadership; but once the central leadership makes a decision, it is to be implemented by the people.

ExCo The Executive Council of Hong Kong, consisting of top civil servants and civilian appointees chosen to represent the community. Except in times of emergency, the governor must consult with the ExCo before initiating any program.

Feudal In Chinese Communist parlance, a patriarchal bureaucratic system in which bureaucrats administer policy on the basis of personal relationships.

Four Cardinal Principles The Chinese Communists' term for their commitment to socialism, the leadership of the Chinese Communist Party, the dictatorship of the proletariat, and the ideologies of Karl Marx, Vladimir Lenin, and Mao Zedong.

Four Modernizations A program of reforms begun in 1978 in China that sought to modernize agriculture, industry, science and technology, and defense by the year 2000.

Gang of Four The label applied to the four "radicals" or "leftists" who dominated first the cultural and then the political events during the Cultural Revolution. The four members of the Gang were Jiang Qing, Mao's wife; Zhang Chunqiao, former deputy secretary of the Shanghai municipal committee and head of its propaganda department; Yao Wenyuan, former editor-in-chief of the *Shanghai Liberation Daily*; and Wang Hongwen, a worker in a textile factory in Shanghai.

Great Leap Forward Mao Zedong's alternative to the Soviet model of development, this was a plan calling for the establishment of communes and for an increase in industrial production in both the cities and the communes. The increased production was to come largely from greater human effort rather than from more investment or improved technology. This policy, begun in 1958, was abandoned by 1959.

Great Proletarian Cultural Revolution *See* Cultural Revolution.

Gross Domestic Product (GDP) A measure of the total flow of goods and services produced by the economy of a country over a certain period of time, normally a year. GDP equals gross national product (GNP) minus the income of the country's residents earned on investments abroad.

Guerrilla A member of a small force of "irregular" soldiers. Generally, guerrilla forces are used against numerically and technologically superior enemies in jungles or mountainous terrain.

Han Of "pure" Chinese extraction. Refers to the dominant ethnic group in the P.R.C.

Ideograph A character of Chinese writing. Originally, each ideograph represented a picture and/or a sound of a word.

Islam The religious faith founded by Muhammad in the sixth and seventh centuries A.D. Its followers believe that Allah is the sole deity and that Muhammad is his prophet.

Kuomintang (KMT) The Chinese Nationalist Party, founded by Sun Yat-Sen in 1912. *See also* Nationalists

LegCo Hong Kong's Legislative Council, which reviews policies proposed by the governor and formulates legislation.

Long March The 1934–1935 retreat of the Chinese Communist Party, in which thousands died while journeying to the plains of Yan'an in northern China in order to escape annihilation by the Kuomintang.

Mainlanders Those Chinese in Taiwan who emigrated from the Chinese mainland during the flight of the Nationalist Party in 1949.

Mandarin A northern Chinese dialect chosen by the Chinese Communist Party to be the official language of China. It is also the official language of Taiwan.

Mao Thought In the post-1949 period, originally described as "the thoughts of Mao Zedong." Mao's "thoughts" were considered important because he took the theory of Marxism-Leninism and applied it to the concrete conditions existing in China. But since

Glossary of Terms and Abbreviations

Mao's death in 1976 and the subsequent reevaluation of his policies, Mao Thought is no longer conceived of as the thoughts of Mao alone but as the "collective wisdom" of the party leadership.

May Fourth Period A period of intellectual ferment in China, which officially began on May 4, 1919, and concerned the Versailles Peace Conference. On that day, the Chinese protested what was considered an unfair secret settlement regarding German-held territory in China. The result was what was termed a "new cultural movement," which lasted into the mid-1920s.

Nationalists The Kuomintang (KMT). The ruling party of the Republic of China, now in "exile" on Taiwan.

Newly Industrialized Country (NIC) A term used to refer to those developing countries that have enjoyed rapid economic growth. Most commonly applied to the East Asian economies of South Korea, Taiwan, Hong Kong, and Singapore.

Offshore Islands The small islands in the Formosa Strait that are just a few miles off the Chinese mainland but are controlled by Taiwan, nearly 90 miles away.

Opium A bitter, addictive drug made from the dried juice of the opium poppy.

Opium War The 1839–1842 conflict between Britain and China, sparked by the British import of opium into China. After the British victory, Europeans were allowed into China and trading posts were established on the mainland. The Treaty of Nanking, which ended the Opium War, also gave Britain its first control over part of Hong Kong.

People's Procuracy The investigative branch of China's legal system. It determines whether an accused person is guilty and should be brought to trial.

People's Republic of China (P.R.C.) Established in 1949 by the Chinese Communists under the leadership of Mao Zedong after defeating Chiang Kai-shek and his Nationalist supporters.

Pinyin A newer system of spelling Chinese words and names, using the Latin alphabet of 26 letters, created by the Chinese Communist leadership.

Proletariat The industrial working class, which for Marx was the political force that would overthrow capitalism and lead the way in the building of socialism.

Republic of China (R.O.C.) The government established as a result of the 1911 Revolution. It was ousted by the Chinese Communist Party in 1949, when its leaders fled to Taiwan.

Second Convention of Peking The 1898 agreement leasing the New Territories of Hong Kong to the British until 1997.

Severe Acute Respiratory Syndrome (SARS) A grave respiratory illness that emerged in 2003 as an epidemic in Hong Kong and part of mainland China.

Shanghai Communique A joint statement of the Chinese and American viewpoints on a range of issues in which each has an interest. It was signed during U.S. president Richard Nixon's historic visit to China in 1971.

Socialism A transitional period between the fall of capitalism and the establishment of "true" communism. Socialism is characterized by the public ownership of the major means of production. Some private economic activity and private property are still allowed, but increased attention is given to a more equal distribution of wealth and income.

Special Administrative Region (SAR) A political subdivision of the People's Republic of China that is used to describe Hong Kong's status following Chinese sovereignty in 1997. The SAR has much greater political, economic, and cultural autonomy from the central government in Beijing than do the provinces of the P.R.C.

Special Economic Zone (SEZ) An area within China that has been allowed a great deal of freedom to experiment with different economic policies, especially efforts to attract foreign investment. Shenzhen, near Hong Kong, is the largest of China's Special Economic Zones.

Taiwanese Independence Movement An organization of native Taiwanese who want to establish an independent state of Taiwan.

Taoism A Chinese mystical philosophy founded in the sixth century B.C. Its followers renounce the secular world and lead lives characterized by unassertiveness and simplicity.

United Nations (UN) An international organization established on June 26, 1945, through official approval of the charter by delegates of 50 nations at a conference in San Francisco. The charter went into effect on October 24, 1945.

Yuan Literally, "branch"; the different departments of the government of Taiwan, including the Executive, Legislative, Judicial, Control, and Examination Yuans.

Bibliography

PEOPLE'S REPUBLIC OF CHINA

Periodicals and Newspapers

The following periodicals and newspapers are excellent sources for coverage of Chinese affairs:

Asiaweek
Asian Survey
China Business Review
China Daily
The China Journal
The China Quarterly
The Economist
Far Eastern Economic Review
Journal of Asian Studies
Journal of Contemporary China
Modern China
Pacific Affairs
South China Morning Post

General

Jasper Becker, *The Chinese* (New York: Free Press, 2000).
Compelling portraits of peasants, entrepreneurs, corrupt businessmen and party members, smugglers, and ethnic minorities by a resident journalist. Reveals much about the effect of the government's policies on the lives of ordinary people.

Jung Chang, *Wild Swans: Three Daughters of China* (New York: Simon and Schuster, 1992).
A superb autobiographical/biographical account that illuminates what China was like for one family for three generations.

Kwang-chih Chang, *The Archeology of China*, 4th ed. (New Haven, CT: Yale University Press, 1986).

_____, *Shang Civilization* (New Haven, CT: Yale University Press, 1980).
Two works by an eminent archaeologist on the origins of Chinese civilization.

Peter Hessler, *River Town: Two Years on the Yangtze* (New York: HarperCollins, 2001).
Insights into Chinese culture by a Peace Corps volunteer who lived in a Yangtze River city from 1996 to 1998. The author gains considerable insights into the life of Fuling, a city that will be partly flooded when the Three Gorges Dam is completed.

Anchee Min, *Becoming Madame Mao* (Boston: Houghton Mifflin, 2000).
This novel vividly portrays Mao's wife, Jiang Qing, tracing her life from early childhood through her failed career as an actress, her courtship with Mao Zedong in the caves of Yenan, and her ultimate demise as a member of the notorious Gang of Four. A real page-turner.

History, Language, and Philosophy

Johan Bjorksten, *Learn to Write Chinese Characters* (New Haven, CT: Yale University Press, 1994).
A delightful introductory book about writing Chinese characters, with many anecdotes about calligraphy.

William Theodore De Bary, ed., *Sources of Chinese Tradition*, Vols. I and II (New York: Columbia University Press, 1960).
A compilation of the major writings (translated) of key Chinese figures from Confucius through Mao Zedong. Gives readers an excellent understanding of intellectual roots of development of Chinese history.

William Theodore De Bary and Weiming Tu, eds. *Confucianism and Human Rights* (New York: Columbia University Press, 1998).
Articles debate whether the writings of Confucius and Mencius (a Confucian scholar) are relevant to today's human rights doctrine (as defined by the United Nations).

John DeFrancis, *Visible Speech: The Diverse Oneness of Writing Systems* (Honolulu, HI: University of Hawaii Press, 1989).
Discusses the evolution of the Chinese written language and compares it with other languages that use "visible" speech.

Patricia Buckley Ebrey, *The Cambridge Illustrated History of China* (New York: Cambridge University Press, 1996).
A beautifully illustrated book on Chinese history from the Neolithic Period to the People's Republic of China. Includes photos of artifacts (such as bronze vessels) and art (from Buddhist art to modern Chinese paintings), which enrich the historical presentation.

John King Fairbank and Merle Goldman, *China: A New History,* Enlarged Edition (Cambridge, MA: Harvard University Press, 1998).
Examines forces in China's history that define it as a coherent culture from its earliest recorded history to the present. Examines why the ancient and sophisticated China had fallen behind other areas by the nineteenth century. The Chinese Communist Revolution and its aftermath are reviewed.

William Hinton, *Fanshen: A Documentary of Revolution in a Chinese Village* (New York: Random House, 1968).
A gripping story based on the author's eyewitness account of the process of land reform carried out by the CCP in the north China village of Long Bow, 1947 to 1949.

Edgar Snow, *Red Star Over China* (New York: Grove Press, 1973).
This classic, which first appeared in 1938, is a journalist's account of the months he spent with the Communists' Red Army in Yan'an in 1936, in the midst of the Chinese Civil War. It is a thrilling story about the Chinese Revolution in action and includes Mao's own story (as told to Snow) of his early life and his decision to become a Communist.

Jonathan D. Spence, *The Search for Modern China* (New York: W. W. Norton & Co., 1990).
A lively and comprehensive history of China from the seventeenth century to 1989. Looks at the cyclical patterns of collapse and regeneration, revolution and consolidation, and growth and decay.

235

Song Mei Lee-Wong, *Politeness and Face in Chinese Culture* (New York: Peter Lang, 2000).

Part of a series on cross-cultural communication, this book discusses how politeness is portrayed in speech and how it relates to a central concept in Chinese culture: "face" and "losing face."

Politics, Economics, Society, and Culture

Julia F. Andrews, *Painters and Politics in the People's Republic of China, 1949–1979* (Berkeley, CA: University of California Press, 1994).

A fascinating presentation of the relationship between politics and art from the beginning of the Communist period until the eve of major liberalization in 1979.

Susan D. Blum and Lionel M. Jensen, eds., *China Off Center: Mapping the Margins of the Central Kingdom* (Honolulu, HI: University of Hawaii Press, 2002).

Arguing that there are many "Chinas," these articles offer new insights into the complexity and diversity of China. Interpretative essays on topics such as linguistic diversity, regionalism, homosexuality, gender and work, popular music, magic and science. Ethnographic reports on minorities.

Ma Bo, *Blood Red Sunset* (New York: Viking, 1995).

Perhaps the most compelling autobiographical account by a Red Guard during the Cultural Revolution. Responding to Mao Zedong's call to youth to "make revolution," the author captures the intense emotions of exhilaration, fear, despair, and loneliness. Takes place in the wilds of Inner Mongolia.

Susan Brownell and Jeffrey Wasserstrom, eds., *Chinese Femininities and Chinese Masculinities: A Reader* (Berkeley, CA: University of California Press, 2002).

A reader that investigates various issues through the lens of feminist and gender theory.

Thomas Buoye, Kirk Denton, Bruce Dickson, Barry Naughton, and Martin K. Whyte, *China: Adapting the Past, Confronting the Future* (Ann Arbor, MI: The University of Michigan Center for Chinese Studies, 2002).

Articles on China's geography and pre-1949 history, including environmental history, Confucianism, and the Boxer Uprising. It also examines the last few decades, including homosexuality, the Internet, and culture; several short stories.

Nien Cheng, *Life and Death in Shanghai* (New York: Grove Press, 1987).

A gripping autobiographical account of a woman persecuted during the Cultural Revolution because of her earlier connections with a Western company, her elitist attitudes, and her luxurious lifestyle in a period when the Chinese people thought the rich had been dispossessed.

Deirdre Chetham, *Before the Deluge: The Vanishing World of the Yangtze's Three Gorges* (New York: Palgrave Macmillan, 2002).

A portrait of life along the Yangtze River just before it is flooded to fill up the nearly completed Three Gorges Dam. Examines the policies that led to the dam, the criticisms of it, and the hopes and fears of what this dam might generate other than electricity.

Deborah S. Davis, ed., *The Consumer Revolution in Urban China* (Berkeley, CA: University of California, 2000).

Articles cover the impact of China's consumer revolution on urban housing, purchases of toys, clothes, and leisure activities for children, and bridal consumerism. Also covers aspects of urban consumerism.

Michael S. Duke, ed., *World of Modern Chinese Fiction: Short Stories & Novellas From the People's Republic, Taiwan & Hong Kong* (Armonk, NY: M. E. Sharpe, Inc., 1991).

A collection of short stories written by Chinese authors from China, Taiwan, and Hong Kong during the 1980s. The 25 stories are grouped by subject matter and narrative style.

Elizabeth Economy, *The River Runs Black: The Environmental Challenge to China's Future* (Ithaca: Cornell University Press, 2004).

The central government's inability to cope with the growing environmental crisis has led to serious social, economic, and health issues, as well as a steadily rising involvement of citizens in non-governmental organizations. Such civic participation may lead to greater democratization and the development of civil society.

Barbara Entwisle and Gail E. Henderson, eds., *Re-drawing Boundaries: Work, Households, and Gender in China* (Berkeley, CA: University of California Press, 2000).

Looks at how gender inequality affects types of work, wages, and economic success. Examines issues of work and gender in China's cities and countryside and among the "floating" population.

B. Michael Frolic, *Mao's People: Sixteen Portraits of Life in Revolutionary China* (Cambridge, MA: Harvard University Press, 1980).

A must read. Through composite biographies of 16 different types of people in China, the author offers a humorous but penetrating view of "unofficial" Chinese society and politics. Biographical sketches reflect political life during the Maoist era, but the book has enduring value for understanding China.

Merle Goldman and Elizabeth J. Perry, eds., *Changing Meanings of Citizenship in Modern China* (Cambridge, MA: Harvard University Press, 2002).

Studies of citizenship in China over the last century. Focuses on the debate over the relationship of the individual to the state, the nation, the community, and culture.

Liang Heng and Judith Shapiro, *Son of the Revolution* (New York: Vintage, 1984).

A gripping first-person account of the Cultural Revolution by a Red Guard. Offers insights into the madness that gripped China during the period from 1966 to 1976.

Ellen Hertz, *The Trading Crowd: An Ethnography of the Shanghai Stock Market* (Cambridge, England: University of Cambridge Press, 1998).

An anthropologist examines the explosion of "stock fever" since the stock market opened in Shanghai in 1992. Looks at the dominant role of the state in controlling the market. The result is a stock market quite different from those in the West.

Alan Hunter and Kim-kwong Chan, *Protestantism in Contemporary China* (New York: Cambridge University Press, 1993).

Examines historical and political conditions that have affected the development of Protestantism in China.

Linda Jakobson, *A Million Truths: A Decade in China* (New York: M. Evans & Co., 1998).

Reveals the many contradictions and complexities of Chinese society from 1987 to 1997.

William R. Jankowiak, *Sex, Death, and Hierarchy in a Chinese City* (New York: Columbia University Press, 1993).

Written by an anthropologist with a discerning eye, this is one of the most fascinating accounts of daily life in China. Particularly strong on rituals of death, romantic life, and the on-site mediation of disputes by strangers (e.g., with bicycle accidents).

Maria Jaschok and Suzanne Miers, eds., *Women and Chinese Patriarchy: Submission, Servitude and Escape* (New York: Zen Books, 1994).

Examines Chinese women's roles, the sale of children, prostitution, Chinese patriarchy, Christianity, and feminism, as well as social remedies and avenues of escape for women. Based on interviews with Chinese women who grew up in China, Hong Kong, Singapore, and San Francisco.

Yarong Jiang and David Ashley, *Mao's Children and the New China* (New York: Routledge, 2000).

More than 20 ex-Red Guards who participated in the Cultural Revolution were interviewed in Shanghai in the mid-1990s. They reminisce about their lives then, revealing much about life in Shanghai during the critical period in China's political history.

Ian Johnson, *Wild Grass: Three Stories of Change in Modern China* (New York: Pantheon Books, 2004).

The author portrays three ordinary citizens who, by testing the limits of reform, may cause China to become a more open country.

Lane Kelley and Yadong Luo, *China 2000: Emerging Business Issues* (Thousand Oaks, CA: Sage Publications, 1998).

Looks to the emerging business issues for Chinese domestic firms and foreign firms.

Conghua Li, *China: The Consumer Revolution* (New York: Wiley, 1998).

An impressive account of China's rapidly growing consumer society. Looks at the forces that are shaping consumption, China's cultural attitudes toward consumerism, consumer preferences of various age groups, and the rapid polarization of consumer purchasing power.

Zhisui Li, *The Private Life of Chairman Mao* (New York: Random House, 1994).

A credible biography of the Chinese Communist Party's leader Mao Zedong, written by his physician, from the mid-1950s to his death in 1976. Wonderful details about Mao's daily life and his relationship to those around him.

Jianhong Liu, Lening Zhang, and Steven F. Messner, eds., *Crime and Social Control in a Changing China* (Westport, CT: Greenwood Press, 2001).

Focuses on crime in the context of a rapidly modernizing China. Shows the deeply rooted cultural context for Chinese attitudes toward crime, criminals, and penology that might well interfere with reform.

Stanley B. Lubman, *Bird in a Cage: Legal Reform After Mao* (Stanford, CA: Stanford University Press, 1999).

Traces the victories and frustrations of legal reform since 1979, but is based on a thorough examination of the pre-reform judicial system.

Michael B. McElroy, Christopher P. Nielsen, and Peter Lydon, eds., *Energizing China: Reconciling Environmental Protection and Economic Growth* (Cambridge, MA: Harvard University Press, 1998).

Research reports address the dilemmas, successes, and problems in China's efforts to reconcile environmental protection with economic development. Addresses issues such as energy and emissions, the environment and public health, the domestic context for making policy on energy, and the international dimensions of China's environmental policy.

Ramon H. Myers, Michel C. Oksenberg, and David Shambaugh, eds., *Making China Policy: Lessons From the Bush and Clinton Administrations* (New York: Rowman and Littlefield, 2002).

Examines the policy of the United States toward China during the George Bush and Bill Clinton administrations (1989–2000). Includes an account of China's perception and response to America's China policies.

Andrew J. Nathan and Perry Link, eds., and Liang Zhang, compiler, *The Tiananmen Papers: The Chinese Leadership's Decision to Use Force Against Their Own People—In Their Own Words* (New York: Public Affairs, 2001).

Widely believed to be authentic documents that reveal what was said among China's top leaders behind closed doors during the Tiananmen crisis in 1989. These leaked documents lay out the thinking of China's leaders about the students and workers occupying Tiananmen Square for about six weeks—and how they eventually decided to use force.

Suzanne Ogden, *Inklings of Democracy in China* (Cambridge, MA: Harvard University Asia Center and Harvard University Press, 2002).

Asks whether liberal democracy is possible or even appropriate in China, given its history, culture, and institutions. Looks at a broad array of indicators. Argues for fair and consistent standards for evaluating freedom and democracy in China and for comparing it with other states.

Suzanne Ogden, Kathleen Hartford, Lawrence Sullivan, and David Zweig, eds., *China's Search for Democracy: The Student and Mass Movement of 1989* (Armonk, NY: M. E. Sharpe, 1992).

A collection of wall posters, handbills, and speeches of the prodemocracy movement of 1989. These documents capture the passionate feelings of the student, intellectual, and worker participants.

Elizabeth J. Perry and Mark Selden, *Chinese Society: Change, Conflict and Resistance* (New York: Routledge, 2000).

A collection of articles on the resistance generated by economic reforms since 1979. Topics include suicide as resistance, resistance to the one-child campaign, and religious and ethnic resistance.

James Seymour and Richard Anderson, *New Ghosts, Old Ghosts: Prisons and Labor Reform Camps in China* (Armonk, NY: M. E. Sharpe, 1998).

A look inside labor camps in China's northwestern provinces, including details about prison conditions and management, the nature of the prison population, excesses perpetrated in prisons, and the fate of released prisoners.

David Shambaugh and Richard H. Yang, *China's Military in Transition* (Oxford: Clarendon Press, 1997).

Collection of articles on China's military covers such topics as party–military relations, troop reduction, the financing of defense, military doctrine, training, and nuclear force modernization.

Stockholm Environment Institute and United Nations Development Program (UNDP) China, *China Human Development Report 2002: Making Green Development a Choice* (New York: Oxford University Press, 2002).

Examines the key issues for sustainable development in China. Also looks at the government's response and the creation of environmental associations to address the issues.

James and Ann Tyson, *Chinese Awakenings: Life Stories From the Unofficial China* (Boulder, CO: Westview Press, 1995).

Lively verbal portraits of the lives of Chinese people from diverse backgrounds (for example, "Muddy Legs: The Peasant Migrant"; "Turning Iron to Gold: The Entrepreneur"; "Bad Element: The Shanghai Cosmopolite").

United Nations Development Program, *China: Human Development Report* (New York: UNDP China Country Office, annual).

Provides measurements of the effect of China's economic development on human capabilities to lead a decent life. Areas examined include health care, education, housing, treatment and status of women, and the environment.

Chihua Wen, *The Red Mirror: Children of China's Cultural Revolution* (Boulder, CO: Westview Press, 1995).

A former editor and reporter presents the heartrending stories of a dozen individuals who were children when the Cultural Revolution started. It shows how rapidly changing policies of the period shattered the lives of its participants and left them cynical adults 20 years later.

Jianying Zha, *China Pop: How Soap Operas, Tabloids, and Bestsellers Are Transforming a Culture* (New York: W. W. Norton, 1995).

A Chinese Mainlander examines the impact of television, film, weekend tabloids, and best-selling novels on today's culture. Some of the material is based on remarkably revealing interviews with some of China's leading film directors, singers, novelists, artists, and cultural moguls.

Yuezhi Zhao, *Media, Market, and Democracy in China: Between the Party Line and the Bottom Line* (Urbana, IL: University of Illinois Press, 1998).

Raises the basic question of whether the expected value of a "free press" will be realized in China if the party-controlled press is replaced by private entrepreneurs and a state-managed press is required to make a profit.

Tibet and Minority Policies

Robert Barnett, ed., *Resistance and Reform in Tibet* (Bloomington, IN: Indiana University Press, 1994).

An informative and quite balanced collection of articles on the highly emotional and politicized topic of Tibet.

Melvyn C. Goldstein, *The Snow Lion and the Dragon: China, Tibet, and the Dalai Lama* (Berkeley, CA: University of California Press, 1997).

The best book on issues surrounding a "free" Tibet and the role of the Dalai Lama. Objective presentation of both Tibetan and Chinese viewpoints.

Melvyn C. Goldstein and Matthew T. Kapstein, eds., *Buddhism in Contemporary Tibet: Religious Revival and Cultural Identity* (Hong Kong: Hong Kong University Press, 1997).

An excellent, nonpolemical collection of articles by cultural anthropologists on Buddhism in Tibet today. Studies of revival of monastic life and new Buddhist practices in the last 20 years are included.

Hette Halskov Hansen, *Lessons in Being Chinese: Minority Education and Ethnic Identity in Southwest China* (Seattle, WA: University of Washington Press, 1999).

Examines Chinese efforts to achieve cultural and political integration through education of a minority population in Chinese cultural values and communist ideology.

Donald S. Lopez, *Prisoners of Shangri-la: Tibetan Buddhism and the West* (Chicago, IL: University of Chicago Press, 1998).

Explodes myths about Tibetan Buddhism created by the West. Shows how these myths have led to distortions that do not serve well the cause of greater autonomy for Tibet.

Orville Schell, *Virtual Tibet: Search for Shangri-la from the Himalayas to Hollywood* (New York: Henry Holt and Co., 2000).

Examines the journals of those hoping to find a spiritual kingdom in Tibet. Notes the perilous journeys undertaken for the last 200 years in pursuit of this quest.

Foreign Policy: China

Elizabeth Economy and Michel Oksenberg, *China Joins the World: Progress and Prospects* (New York: Council on Foreign Relations, 1999).

An outstanding collection of essays on where China fits into the international institutional structure in such matters as arms control, human rights, trade and investment, energy, environmental protection, and the information revolution.

Alastair Iain Johnston and Robert S. Ross, eds., *Engaging China: The Management of an Emerging Power* (New York: Routledge, 1999).

A collection of articles on how various governments, including Korea, Singapore, Indonesia, Japan, Taiwan, the United States, and Malaysia, have tried to "engage" an increasingly powerful China.

Samuel S. Kim, ed., *China and the World*, 4th ed. (Boulder, CO: Westview Press, 1998).

Examines theory and practice of Chinese foreign policy with the United States, Russia, Japan, Europe, and the developing world. Looks at such issues as the use of force, China's growing interdependence with other countries, human rights, the environment, and China's relationship with multilateral economic institutions.

Richard Madsen, *China and the American Dream: A Moral Inquiry* (Berkeley, CA: University of California Press, 1995).

Looks at the emotional and unpredictable relationship that the United States has had with China from the nineteenth century to the present.

James Mann, *About Face: A History of America's Curious Relationship With China, From Nixon to Clinton* (New York: Alfred A. Knopf, 1999).
A journalist's account of the history of U.S.–China relations since Nixon. Through examination of newly uncovered government documents and interviews, gives account of development of the relationship, with all its problems and promises.

Robert S. Ross, ed., *After the Cold War: Domestic Factors and U.S.–China Relations* (Armonk, NY: M. E. Sharpe, 1998).
Examines how domestic factors, such as public opinion and interest groups, affect the development of U.S.–China policy.

Robert S. Ross, *Negotiating Cooperation: The United States and China, 1969–1989* (Stanford, CA: Stanford University Press, 1995).
The difficulties of engineering a cooperative relationship between the United States and China are presented from a "realist" framework that assumes that the primary concern of both sides is with national security and the "strategic balance." The Soviet threat is considered critical to Sino–American efforts to cooperate.

David Zweig, *Internationalizing China: Domestic Interests and Global Linkages* (Ithaca, NY: Cornell University Press, 2002).
Case studies on issues that connect domestic interests to China's foreign policy and international linkages, and the decreasing role of bureaucrats in regulating the internationalization of China.

HONG KONG

Periodicals and Newspapers

Hong Kong Commercial Daily
Hong Kong News Online
Hong Kong Standard
South China Morning Post

Politics, Economics, Society, and Culture

Robert Ash, Peter Ferdinand, Brian Hook, and Robin Porter, *Hong Kong in Transition: One Country, Two Systems* (New York: Routledge Curzon, 2003).
Investigates changes since the 1997 handover in Hong Kong's business environment, including the role of public opinion and government intervention, and the evolving political culture.

Ming K. Chan and Alvin Y. So, eds., *Crisis and Transformation in China's Hong Kong* (Armonk, NY: M. E. Sharpe, 2002).
Examines political and social changes in Hong Kong since it was returned to China's sovereignty in 1997.

Joseph Y. S. Cheng, ed., *The Other Hong Kong Report, 1997* (Hong Kong: Hong Kong University Press, 1997).
An annual publication that takes a different perspective from the official governmental annual report on Hong Kong. Timely articles on Hong Kong's legal system, human rights, the new middle class, the environment, housing policy, and so on.

Robert Cottrell, *The End of Hong Kong: The Secret Diplomacy of Imperial Retreat* (London: John Murray, 1993).
Exposes the secret diplomacy that led to the signing of the "Joint Declaration on Question of Hong Kong" in 1984, the agreement that ended 150 years of British colonial rule over Hong Kong. Thesis is that Britain was reluctant to introduce democracy into Hong Kong before this point because it thought it would ruin Hong Kong's economy and lead to social and political instability.

Michael J. Enright, Edith E. Scott, and David Dodwell, *The Hong Kong Advantage* (Oxford: Oxford University Press, 1997).
Examines the special relationship between the growth of Hong Kong's and mainland China's economies, such topics as the role of the overseas Chinese community in Hong Kong and the competition Hong Kong faces from Taipei, Singapore, Seoul, and Sydney as well as from such up-and-coming Chinese cities as Shanghai.

Wai-man Lam, *Understanding the Political Culture of Hong Kong: The Paradox of Activism and Depoliticization* (Armonk, N.Y.: M.E. Sharpe, 2004).
Through case studies of protest, Lam challenges the view of a politically apathetic Hong Kong populace. Looks at role of ideology, nationalism, gender, civil rights, and economic justice as motivating political participation in Hong Kong.

C. K. Lau, *Hong Kong's Colonial Legacy: A Hong Kong Chinese's View of the British Heritage* (Hong Kong: Chinese University Press, 1997).
Engaging overview of the British roots of today's Hong Kong. Special attention to such problems as the "identity" of Hong Kong people as British or Chinese, the problems in speaking English, English common law in a Chinese setting, and the "strictly controlled" but rowdy Hong Kong "free press."

Jan Morris, *Hong Kong: Epilogue to an Empire* (New York: Vintage, 1997).
Witty and detailed first-hand portrait of Hong Kong by one of its long-term residents. Gives the reader the sense of actually being on the scene in a vibrant Hong Kong.

Christopher Patten, *East and West: China, Power, and the Future of Asia* (New York: Random House), 1998.
The controversial last governor of Hong Kong gives a lively insider's view of the British colony in the last 5 years before it was returned to China's sovereignty. Focuses on China's refusal to radically change Hong Kong's political processes on the eve of the British exit. Argues against the idea that "Asian values" are opposed to democratic governance, and suggests that "Western values" have already been realized in Hong Kong.

Mark Roberti, *The Fall of Hong Kong: China's Triumph and Britain's Betrayal* (New York: John Wiley & Sons, Inc., 1994).
A fast-paced, drama-filled account of the decisions Britain and China made about Hong Kong's fate beginning in the early 1980s. Based on interviews with 150 key players in the secret negotiations between China and Great Britain.

Ming Sing, *Hong Kong's Tortuous Democratization: A Comparative Analysis* (New York: Routledge Curzon, 2004).
An examination of the governance in Hong Kong since the 1940s, and the constraints to democratization. Looks beyond the limits imposed by Beijing to other forces, including lack of public support and weak pro-democracy forces, to explain why democracy has not yet emerged.

Alvin Y. So, *Hong Kong's Embattled Democracy: A Societal Analysis* (Baltimore: Johns Hopkins University Press, 1999).
Traces Hong Kong's development of democracy.

Steven Tsang, *A Modern History of Hong Kong* (New York: I.B. Tauris, 2004).

History of British colonial rule from before the Opium Wars. Examines problems in creating the rule of law and an independent judiciary in Hong Kong, and the impact of trade with China on Hong Kong's society and economy.

Frank Welsh, *A Borrowed Place: The History of Hong Kong* (New York: Kodansha International, 1996).

Best book on Hong Kong's history from the time of the British East India Company in the eighteenth century through the Opium Wars of the nineteenth century to the present.

TAIWAN

Periodicals and Newspapers

Taipei Journal

Taipei Review

Politics, Economics, Society, and Culture

Bonnie Adrian, *Framing the Bride: Globalizing Beauty and Romance in Taiwan's Bridal Industry* (Berkeley, CA: University of California Press, 2003).

A fascinating ethnographic study of Taipei's bridal photography as a narrative on contemporary marriages, inter-generational tensions, how the local culture industry and brides use global images of romance and beauty, and the enduring importance of family and gender.

Muthiah Alagappa, ed., *Taiwan's Presidential Politics: Democratization and Cross-Strait Relations in the Twenty-first Century* (Armonk, NY: M. E. Sharpe, 2001).

Focuses on Taiwan's presidential elections in March 2000 and the impact of those elections one year later on the democratic transition from a one-party-dominant system to a multiparty system. Also examines the degree to which Taiwan under the leadership of Chen Shui-bian was able to consolidate democracy.

Christian Aspalter, *Understanding Modern Taiwan: Essays in Economics, Politics, and Social Policy* (Burlington, VT: Ashgate, 2001).

A collection of articles on Taiwan's "economic miracle" and such topics as Taiwan's "identity," democratization, policies on building nuclear-power plants and the growing antinuclear movement, labor and social-welfare policies, and the role of political parties in developing a welfare state.

Melissa J. Brown, *Is Taiwan Chinese? The Impact of Culture, Power, and Migration on Changing Identities* (Berkeley: University of California Press, 2004).

Author explores the meaning of idenity in Taiwan. From 1945–1991, Taiwan's government claimed that Taiwanese were ethnically and nationally Chinese (and even claimed legal authority over the mainland). Since 1991, the government has, in a political effort to claim national and cultural distance from the mainland, moved to a position asserting that their identity has been shaped by a mix of aboriginal ancestry and culture, Japanese cultural influence, and Han Chinese cultural influence and ancestry. Examines cultural markers of identity, such as folk religion, footbinding and ancestor worship as well as how identities change.

Richard C. Bush, *At Cross Purposes: U.S.-Taiwan Relations* (Armonk, N.Y.: M. E. Sharpe, 2004).

The former head of the American Institute in Taiwan (1997–2002) examines why President Roosevelt decided that Taiwan ought to be returned to China after World War II, the U.S. position on the Kuomintang's repressive government rule, the nature of the U.S. "2-China" policy from 1950 to 1972, and the resent basis for U.S. military and political relations with Taiwan.

Fen-ling Chen, *Working Women and State Policies in Taiwan: A Study in Political Economy* (New York: Palgrave, 2000).

A study of the impact of social welfare and state policies on the relationships between men and women since 1960. "Gender ideology" has changed and, with it, women's views of the workplace and their role in society. Examines related issues of childcare, wages, the women's movement, and women in policy-making system.

Ko-lin Chin, *Heijin: Organized Crime, Business, and Politics in Taiwan* (Armonk, N.Y.: M. E. Sharpe, 2003).

An examination of the connection between Taiwan's underworld (*hei*) and business/money (*jin*) to politics that has accompanied Taiwan's efforts to democratize since emerging from martial law after 1987. Looks at ways in which *heijin* politics have undercut democratization through vote buying, political violence, bid rigging, insider trading, and violence.

Bruce J. Dickson and Chien-min Chao, eds., *Assessing the Lee Teng-hui Legacy in Taiwan's Politics: Democratic Consolidation and External Relations* (Armonk, NY: M. E. Sharpe, 2002).

Focuses on the impact of Lee Teng-hui presidency (1996–2000) on democratic consolidation, the role (and demise) of the Nationalist Party and the rise of the Democratic Progressive Party, and the economy. Also examines President Lee's impact on security issues.

A-Chin Hsiau, *Contemporary Taiwanese Cultural Nationalism* (New York: Routledge, 2000).

Traces the development of Taiwanese cultural nationalism. Includes the impact of Japanese colonialism, post–World War II literary development, and the spawning of a national literature and national culture.

Chen Jie, *Foreign Policy of the New Taiwan: Pragmatic Diplomacy in Southeast Asia* (Northampton, MA: Edward Elgar, 2002).

Outstanding book on Taiwan's foreign policy (1949–2000). Shows patterns in Taiwan's diplomacy and provides basis for theories and insights about Taiwan's policies, frustrations, sensitivities, and motivations in international affairs. Also covers Taiwan's policy toward the millions of "overseas Chinese."

David K. Jordan, *Gods, Ghosts, and Ancestors: The Folk Religion of a Taiwanese Village* (Berkeley, CA: University of California Press, 1972).

A fascinating analysis of folk religion in Taiwan by an anthropologist, based on field study. Essential work for understanding how folk religion affects the everyday life of people in Taiwan.

Robert M. Marsh, *The Great Transformation: Social Change in Taipei, Taiwan, Since the 1960s* (Armonk, NY: M. E. Sharpe, 1996).

An investigation of how Taiwan's society has changed since the 1960s when its economic transformation began.

David Shambaug, ed., *Contemporary Taiwan* (Oxford: Clarendon Press, 1998).

Broad coverage of society, the economy, and politics in Taiwan today, including the impact of globalization and regionalization on Taiwan's technology, the impact of economic development on the environment, and Taiwan's policy toward reunification with mainland China.

Scott Simon, *Sweet and Sour: Life-Worlds of Taipei Women Entrepreneurs* (Lanham, Maryland: Rowman & Littlefield Publishers, 2003).

Examines the contradictions and tensions that characterize the lives of Taiwan's female entrepreneurs, who spear-headed Taiwan's economic 'miracle.' Presents portraits of these women, including street vendors, a hair dresser, a café owner, a fashion designer, and more. Sheds light on urban life and on impact of patriarchal culture on male-female relations.

Index

Index